2nd International Conference on Renewable Energy and Environment Engineering (REEE 2019)

E3S Web of Conferences Volume 122 (2019)

Munich, Germany
19 – 22 August 2019

Editor:

Nidia Caetano

ISBN: 978-1-5108-9602-4

Printed from e-media with permission by:

Curran Associates, Inc.
57 Morehouse Lane
Red Hook, NY 12571

Some format issues inherent in the e-media version may also appear in this print version.

For additional information, please contact EDP Sciences – Web of Conferences at the address below.

EDP Sciences – Web of Conferences
17, Avenue du Hoggar
Parc d'Activité de Courtabœuf
BP 112
F-91944 Les Ulis Cedex A
France

Phone: +33 (0) 1 69 18 75 75
Fax: +33 (0) 1 69 28 84 91

contact-edps@webofconferences.org

Additional copies of this publication are available from:

Curran Associates, Inc.
57 Morehouse Lane
Red Hook, NY 12571 USA
Phone: 845-758-0400
Fax: 845-758-2633
Email: curran@proceedings.com
Web: www.proceedings.com

TABLE OF CONTENTS

Preface

The 2nd International Conference on Renewable Energy and Environment Engineering (REEE 2019) has been successfully held in Munich, Germany on August 19-22, 2019.

The proceedings are the result of REEE 2019, which presents 30 accepted articles selected from more than 70 papers submitted to the conference from 20 countries and districts including Thailand, China, Morocco, Romania, Macau, Spain, Iraq, South Africa, UK, USA, Colombia, South Korea, Germany, Turkey, Poland, United Arab Emirates, Indonesia, Mexico, New Zealand, Korea, Belgium, Kazakhstan, Czech Republic, India, etc. All of the papers were subjected to peer-review by the program technical committee members and international reviewers. All the papers are presented orally or by poster in parallel sessions.

We also honored to have one very interesting keynote speech and three invited speeches by Prof. Elmar Bollin, Offenburg University of Applied Sciences, Germany; Prof. Dr. Jin Su Jeong, Universidad de Extremadura, Spain; Prof. Dr. Jorge Loredo, University of Oviedo, Spain and Prof. Dr. Aurel Lunguleasa, Transilvania University of Brasov, Romania

On behalf of the organizing committee, I would like to express my sincere gratitude to guest editors, program technical committee members and international reviewers, who provided constructive criticism and stimulated comments and suggestions to the authors. It's a tough work to select papers from the submissions, and without their hard work the proceedings will not see the light. Besides, we also want to thank the chairs, who are dedicated to make sure to run the conference smoothly, the keynote speakers and invited speakers who shared with us their precious research findings. The proceeding is a credit to a large group of people, and everyone should be proud of the outcome.

We truly believe the participants will find the discussion fruitful, and will enjoy the opportunity for setting up future collaboration. We also hope REEE will one day become the leading conference in this specific academic area.

Yours Sincerely,

Conference Chair
Prof. Elmar Bollin
Offenburg University of Applied Sciences, Germany

Statement of Peer review

In submitting conference proceedings to E3S Web of Conferences, I certify to the Publisher that I adhere to the **Policy on Publishing Integrity** of the journal in order to safeguard good scientific practice in publishing.

1. All articles have been subjected to peer review administered by the proceedings editors.
2. Reviews have been conducted by expert referees, who have been requested to provide unbiased and constructive comments aimed, whenever possible, at improving the work.
3. Proceedings editors have taken all reasonable steps to ensure the quality of the materials they publish and their decision to accept or reject a paper for publication has been based only on the merits of the work and the relevance to the journal.

Title, date and place of the conference

2019 The 2nd International Conference on Renewable Energy and Environment Engineering

Munich, Germany
August 19-22, 2019

Proceedings editor(s): Prof. Nidia Caetano

Date and editor's signature 18 Sept. 2019

17, avenue du Hoggar - PA de Courtabœuf – BP 112 - 91944 Les Ulis Cedex A (France)
Tél. : 33 (0)1 69 18 75 75 - Fax : 33(0)1 69 07 45 17 – www.edpsciences.org

E3S Web of Conferences **122**, 01001 (2019) https://doi.org/10.1051/e3sconf/201912201001
REEE 2019

Biomass - alternative renewable energy source and its conversion for hydrogen rich gas production

Elena David[1,*], *Janez* Kopac[2], *Adrian* Armeanu[1], *Violeta* Niculescu[1], *Claudia* Sandru[1], *Viorel* Badescu[3]

[1]National Research Institute for Cryogenic & Isotopic Technologies, Uzinei No.4; P.O.Raureni, P.O.Box 7; 240050 Rm.Valcea, Romania
[2]Faculty of Mechanical Engineering, University of Ljubljana, Askerceva 6, 1000 Ljubljana, Slovenia
[3] Polytechnic University of Bucharest, Splaiul Independenţei 313, Bucharest, 060042, Romania

Abstract. This paper presnts biomass as a renewable energy source and defines the resources as well as the ways through biomass energy is converted into fuels, the technologies used for extracting the energy from biomass as well as the advantages and disadvantages that appear by using of biomass as a energy source. In addition,it is known hydrogen is an important alternative energy vector and a bridge to a sustainable way fot the energy future. Hydrogen is an energy carrier and can be obtained by different production technologies from a large variety of primary energy sources. At present, many researches are focused on getting energy from biomass, a sustainable and non-polluting way to replace fossil fuels, because the biomass can be considered as the best option with high potential, which meets energy requirements and could insure fuel supply in the future. Biomass and residual biomass can be used to produce hydrogen rich gas sustainably. Biomass pyrolysis and gasification offers an efficiency and economical route for the renewable hydrogen production and this is also discussed in the paper.

1. Introduction

Biomass is a source of raw material and this term describes all organic matter resulted by photosynthesis. Biomass includes all land and water-based vegetation and trees, and also all waste biomass such as forestry and agricultural residues, municipal bio-solids (sewage), municipal solid waste (MSW), animal wastes (manures), and certain types of industrial wastes. The energy markets are heavily dependent on fossil fuels which are limited reserves, so new sources must be found to replace fossil fuels. In this context, biomass is other naturally source containing carbon resource that is enough in quantity to be used as a substitute for fossil fuels [1].Through the photosynthesis process, chlorophyll into plants storages the energy of the sun by converting CO_2 from the environment into carbohydrates (i.e., complex compounds composed of C, H , and O). When these complex compounds are burned, they turn back into CO_2 and H_2O and release the energy they contain. By the comparison, in this way, the biomass could function as a sort of natural battery for the storing of solar energy [2] and a potential source of H_2 as a future solution [3]. The use of the biomass energy form has played a key role during the mankind evolution and until relatively recently it constituted the only form of energy which was exploited by people and is still the main energy source for more than half the population in the world for the domestic energy needs [4]. One of the simple forms of biomass use is a basic open fire to provide heat for cooking, warming water or home warming. In the present there are technologies for the extracting energy and converting it into useful power or heat in an efficient way with positive environment impact and there is expected to be the driving forces that stimulate the transformation of biomass into one of the dominant energy resources [5]. Unlike fossil fuels, the biomass is renewable and a short time is needed to replace what is used as an energy resource. Biomass also is the renewable energy source that releases CO_2 in use, but however the CO_2 release is compensated by the fact that the biomass growing uses the CO_2 from environmenr to capture energy during photosynthesis process. When the biomass resource is being used sustainably, there are not carbon emissions over the time frame of a cycle of biomass production [6]. Figure 1 presents two environmentally friendly schemes for a biomass energy cycle and the biomass way is used for energy generation [2]. In the present, many researches have been focused on sustainable and environmental friendly energy production from the biomass to replace fossil fuels [7]. On the other hand, hydrogen production is one of the most promising alternative energy technologies. Hydrogen is a secondary form of energy that must be manufactured like electricit, it is an energy carrier [6,7]. Hydrogen is produced from a variety of primary energy sources and by different production technologies. The most H_2 is currently obtained from non-renewable sources such as natural gas, oil, and coal [8] and only 4% and 1% is generated from water using electricity and biomass, respectively [9]. H_2 produced of renewable primary energy sources such as biomass is

* Corresponding author: elenadavid2004@yahoo.com

https://doi.org/10.1051/e3sconf/201912201001

ideal for gradually replacing fossil fuels [10] because biomass and biomass derived fuels can be used to produce hydrogen sustainably. Using biomass instead of fossil fuels to produce H_2 reduces the CO_2 amount released into the atmosphere, since the CO_2 released when the biomass is gasified was previously absorbed from the atmosphere and fixed by photosynthesis process in the plant growing [10].

2. Processes used for energy extracting from biomass

Biomass can be converted to solid, liquid, gaseous fuels, thermal energy or other valuable chemical products through different conversion processes. Generally, the manly bio-power technologies are comprised of combustion and co-firing, gasification, pyrolysis, fermentation and anaerobic digestion.

The direct combustion is the simplest method of energy extracting from biomass and industrial biomass combustion systems burn many kinds of biomass fuel, including wood, agricultural wastes, municipal solid waste (MSW), wood pulping liquor, refuse-derived fuel etc. Biomass is burned to produce steam, then the steam turns a turbine and the turbine drives a generator that produces electricity. Because of potential ash build-up, which fouls boilers, the efficiency is reduced and the costs increase, and due to this only certain kinds of biomass materials are used for direct combustion.

The gasification is a process which transforms a solid fuel at high temperatures and a limited oxygen amount, into a gaseous fuel. The gas produced by this process (Figure 1) is a mixture of gases such as CO, CO_2, N_2, H_2, and CH_4.

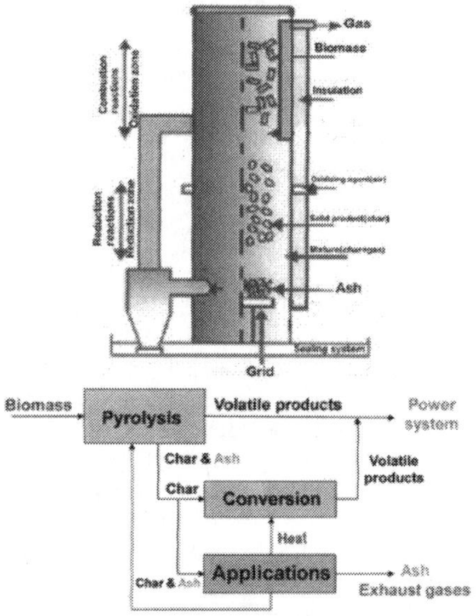

Figure 1. Schemes of a gasification process

The gas produced is then used to drive a high efficiency and combined cycle gas turbine. The gasification process has several advantages over combustion process, one of the resultant gases, CH_4, can be considered in a similar way as natural gas, and used for the same applications. Another advantage of gasification process is that it produces a fuel with fewer impurities because they have been removed and therefore causes fewer pollution problems when it is burnt. Under suitable parameters, gasification process can also produce syngas, a mixture of CO and H_2 that can be used to obtain hydrocarbon (e.g., CH_4, CH_3OH, etc.) for replacement fossil fuels. H_2 itself is a potential fuel, without pollution, that can substitute petroleum and oil [7].

Pyrolysis process in simplest form, represents heating of the biomass to release the volatile products and leaving behind bio-oil and charcoal (Figure 2).

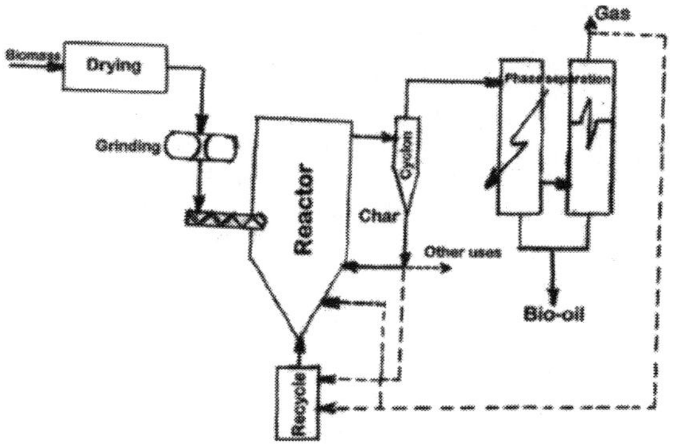

Figure 2. A scheme of representation of the pyrolysis process.

This process has doubled the energy density of the raw material because bio-oil and charcoal, which represents more than half the weight of the initial biomass, contains almost the same amount of energy and makes the fuel easier transportable. Also, the charcoal burns at a higher temperature than the raw biomass, making it much more useful for different applications. More new pyrolysis methods are developed recently to storage volatile products that are otherwise lost to the system. From the collected volatiles can be produced a gas which is rich in hydrogen (as a potential fuel).

Fermentation process uses yeasts and other microorganisms to ferment the sugar of biomass into ethanol. The ethanol obtained through fermentation process is then mixed with diesel to produce diesehol, a product especially used by trucks and buses [7].

Biomass digestion process works by utilizing anaerobic bacteria and these microorganisms usually live at the bottom of sludges or in other places where the air is not present, consuming dead organic matter to produce CH_4 and H_2. By feeding organic matter such as sewage sludge into a tank, called digester, and adding bacteria, it can be collected the produced gas to use as an energy source.

2.1 Advantages of biomass energy

Some of the advantages of using biomass as a source of energy are: (i) biomass energy is an abundant, environmental friendly, secure and renewable source of energy; (ii) biomass does not increase CO_2 to the

 https://doi.org/10.1051/e3sconf/201912201001

atmosphere because it absorbs the same amount of carbon in growing as it releases when is used as a fuel; (iii) a major advantage of biomass is that it can be used to generate electricity using the same power plants that are used for the burning of fossil fuels; (iv) biomass fuels are sustainable, the green plants from which biomass fuels are obtained fix CO_2 as they grow, and consequently does not increase the level of carbon in the atmosphere. In addition, using refuse as a fuel avoids polluting landfill disposal; (v) alcohols and other fuels produced by biomass are viable, efficient, and relatively clean burning; (vi) biomass is easily available and can be grown with relative ease in all the world.

2.2. Impediments to biomass energy use

The use of biomass energy encounters several impediments, such as:
(i) Biomass is still an expensive source of energy, both in terms of producing biomass and convertion it into product fuels, because a large quantity of biomass is needed;
(ii) On a small scale there is probably a loss of energy as a lot of energy must be used for the growing the plant mass and also the biomass is difficult to store in its raw form;
(iii) An other disadvantage of biomass use is that direct biomass combustion could be harmful to the environment as burning biomass releases CO_2 , which contributes to the warming of the atmosphere and possible climate change. Also biomass burning creates soot and other air pollutants;
(iv).Excessive wood consumption can destroy forests, soils are easily eroded and do not hold rainfall.Also increased runoff could cause flooding downstream; (v) when plant and animal wastes are used to produce fuel, these cannot be added to the soil as fertilizer and a soil without fertilizer is depleted of nutrients and produce fewer crops; (iv) biomass contains less energy than a similar volume of fossil fuels.

3. Hydrogen production routes from biomass

The chemical structure and main organic components in biomass are very important in the development of processes for producing fuels and valuable chemicals[11]. The chemical components of the lignocellulose are divided into four major components: cellulose, hemicellulose, lignin and extractives [12]. Generally, the first three components have high molecular weights and contribute much mass, while the last component is of small molecular size, and available in little quantity [13]. Hydrogen can be produced from biomass, but this technology needs further development. The production of H_2 from biomass has many advantages [14]. such as the independence from oil imports, stable pricing level, the CO_2 balance improved by around 30%.

The processes available for H_2 production from biomass are divided into two categories: thermochemical and biological routes. Hydrogen can be produced from bio-renewable feedstocks by thermochemical conversion processes such as gasification, steam gasification, pyrolysis, steam reforming of bio-oils, and also supercritical water gasification. Biological hydrogen production can be classified into the following groups: (i) biophotolysis of water using green algae and blue-green algae (cyanobacteria), (ii) photo-fermentation, (iii) dark-fermentation, and (iv) hybrid reactor system. The advantage of the thermochemical process is that its overall efficiency (thermal to hydrogen) is higher ($\eta \approx 50\%$) and production cost is lower [15]. The H_2 yield that could be obtained from biomass is low, 15-18% , based on dry biomass weight [15]

3.1. Hydrogen production from biomass by gasification process

At temperatures of approximately 600-900°C, the biomass undergoes thermal decomposition and forms gas-phase products that usually contains CO, H_2, CO_2, CH_4, H_2O, and other gaseous hydrocarbons (C_2-C_7) [16]. Composition of gas product resulted from the biomass gasification depends on the gasification process, agent of gasifying, and the composition of the feedstock [17]. Biomass gasification is generally given by the reaction:

$$\text{Biomass} + O_2 \text{ (or } H_2O) \rightarrow CO, CO_2, H_2O, H_2, CH_4 + C_{2\text{-}7} + \text{char} + \text{tar} + \text{ash} \quad (1)$$

In a gasification process that uses biomass as a feedstock, the first step is a thermos-chemical decomposition of the cellulose, hemicellulose and lignin compounds with the production of char, tar and volatiles [18]. Further the gasification of char and tar determines some other equilibrium reactions to be possible to occur. The possible products resulted from a gasification process are shown in Fig. 3 [19].

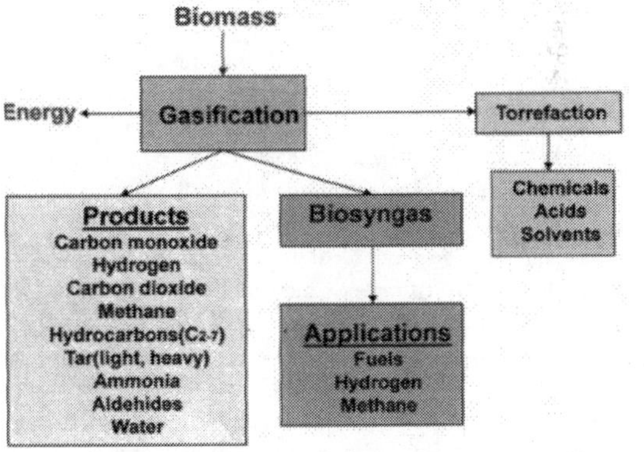

Figure 3. Possible products obtained from a gasification process.

When the biomass is gasified, the tar that is formed together with the gas phase is difficult to separate and remove with a physical dust removal method [20]. The distribution of the products and gas composition depend on many parameters including the temperature of gasification and the reactor type. The most common gasifier types are fixed bed (updraft, downdraft or horizontally fixed beds), fluidized bed, and also entrained flow gasifiers. All these gasifiers need to include significant gas conditioning along with the removal of tar and inorganic impurities and the subsequent conversion of CO to H_2 by water gas shift reaction. Table 1 presents the typical gas composition data as resulted from wood and charcoal downdraft gasifier that operates on low to medium moisture content biomass [21]

Table 1. Gas composition data as obtained from wood and charcoal downdraft gasifiers operated on low to medium moisture content fuels (wood 20%, charcoal 7%) [21].

Comp	H_2(%)	CO_2(%)	CH_4(%)	CO(%)	N_2(%)	HV (MJ/m^3)
W gas	12-20	9-15	2-3	17-22	50-54	5-5.9
C gas	4-10	1-3	0-2	28-32	55-65	4.5-5.6

Source: Ref.[21]

Gasification processes provide the possibility to transform renewable biomass feedstocks into clean gas fuel or synthesis gas. The synthesis gas includes mainly H_2 and CO, also called as bio-syngas [22] and this is a gas rich in H_2 and CO produced by gasification process of biomass. Bio-syngas can be obtained from biomass by non-catalytic,catalytic, and also steam gasification processes. The steam gasification of biomass is a promising process used for thermos-chemical H_2 production. The steam reforming and so-called dry or CO_2 reforming are usually promoted by the use of catalysts and occur according to the following reactions:

$$CnHm + nH_2O \leftrightarrow CO + (n + m/2)\, H_2 \quad (2)$$

$$C_nH_m + nCO_2 \leftrightarrow (2n)\, CO + (m/2)\, H_2 \quad (3)$$

Conversion of biomass to synthesis gas by steam gasification is a challenge because of the diversity of the raw material (structure, composition, physical properties, reactivity, etc.) and because of the severe parameters (temperature, heating rate, residence time, biomass/catalyst ratio, etc.) that are required [23]. The H_2 yield obtained from steam gasification process increases with increasing water-to-biomass ratio (W/S). The yields of H_2 resulted from the pyrolysis and the steam gasification processes increase with increasing of temperature. Usually, the gasification temperature is higher than that of pyrolysis[24]. Also, the effect of catalyst on gasification products is very important, the use of the catalyst do not affect the gas yields,but the gas composition is strongly influenced, in the sense that the H_2 and CO_2 content increases, while that of CO decreases.

3.2. Hydrogen production from biomass by pyrolysis process

In the recent years , hydrogen production by biomass pyrolysis and/or gasification became a subject large researched in the field of biomass utilization[25]. Previous studies [26, 27] have reported that the steam introduction during biomass gasification process has resulted in an increasing of H_2 yield due to the tar and higher hydrocarbon reforming, as well as the water gas shift reaction. Using the steam as an oxidizer, the following chemical reaction can be written:

$$CH_{1.31}O_{0.69} + 1.31\, H_2O \rightarrow 1.97\, H_2 + CO_2 \quad (4)$$

with the biomass chemical formula calculated from the elemental analysis, Table 2.

Table 2. Proximate and ultimate analyses of walnut shells

Element	UA (wt.%)	PA(wt.%)	
C	49.26	fixed carbon	10.76
H	5.38	moisture	7.85
O	45.08	volatiles	80.21
N	0.28	ash	1.18

According to reaction (4), a minimum of steam/biomass (S/B) ratio of 0.97 is determined for the theoretical maximum H_2 yield. In the present the researches are mostly focused on varying operating parameters, using different types of reactor, using various types of catalysts, etc.[25,26,27]. Unfortunately, a little attention is considered to reducing the cost through improving the steam supply mode [27] even if it consumes a large quantity of energy of the whole process. Normally, the raw biomass contains a big amount of moisture, which even exceeds 50 wt % (wet basis). In the conventional biomass steam gasification (CBSG) such moisture is removed through a pre-drying process before gasification. Then, the pre-dried biomass is gasified with steam,which is usually obtained from an auxiliary system (steam generator or boiler). Both the pre-drying process and the steam generation involve high energy consumption and make CBSG process more expensive and complex. From all of these considerations, in this part of paper a method of hydrogen-rich gas production by biomass pyrolysis in an autogenerated steam atmosphere is proposed. In this method, the biomass with high moisture content is directly used as feedstock of pyrolysis reactor. The method utilizes steam autogenerated from biomass moisture as a reactant to combine with the intermediate products resulted from the pyrolysis process to produce additional gas (hydrogen). In the present work, to obtain some basic informations, the effect of several parameters such as pyrolysis temperature, heating rate, moisture content, and stripping gas flow on the pyrolysis process was investigated. The residual biomass, including walnut shell (WS) was employed as raw material in this study. The raw walnut shells in hemispherical shape with about 25 mm diameter and 3 mm thickness were supplied by the Research Station for Fruit Growing Valcea. These were grinded at a particle size: <0.15 mm.The proximate analyses of raw walnut shells was done according to ASTM D3174-04 for ash analysis and ASTMD3175-89a for volatile matter. The ultimate analysis to determine the quantity of C, H, N, (O calculated from difference) was made according to ASTM D5291-96 using a FLASH-2000 Elemental Analyzer. All chemicals used in this study were of analytical grade. To study the effect of moisture content, the walnut shells were dried naturally,by sun drying to different moisture contents: (a) wet walnut shell sample,WSW, the as-received walnut shell, with a moisture content of ≈ 47.37 wt % (wet basis); (b) a partially pre-dried fraction of the as received wet walnut shell sample, WSPD, with a moisture content of ≈ 33.68 wt % (wet basis); and (c) totally pre-dried walnut shell sample, WSTD, with a moisture content of ≈7.85 wt % (wet basis).

4. Conclusions

Biomass represents a vaste renewable energy source and there are various ways by which biomass energy can be converted into fuels.The technologies used for extracting the energy from biomass as well as the advantages and disadvantages that appear by using of biomass as a energy source were presented in this study. In addition, hydrogen-rich gas production from biomass pyrolysis in an autogenerated steam atmosphere was studied. The tests were carried out to establish the influence of some process parameters (such as the reactor temperature, moisture content, flow rate of the stripping gas, and also the heating rate) on the pyrolysis process and hydrogen content from gas product.The obtained results lead to the following conclusions: (i) The direct pyrolysis of wet biomass without step of predried treatment promotes hydrogen-rich gas production. Under the condition of fast-heating rate (fast pyrolysis) and without the using of N2 stripping gas, the pyrolysis of WSW sample releases higher H2 yield than those released from WSTD sample;(ii) In order that the steam resulting from its own moisture content of the biomass to be used to produce additional hydrogen, a high heating rate is required (fast pyrolysis) because under fast-heating conditions, both drying and pyrolysis steps can occur in a shorter time and thus are improved the interactions between the autogenerated steam and the intermediate products (volatile and char) and more hydrogen is produced; (iii) The stripping gas flow rate greatly influence the hydrogen yield. With the decreasing of the stripping gas flow rate the hydrogen amount increases, this being much more evidente in the case of fast pyrolysis process, because higher reactor temperature favors hydrogen production; (iv) This study demonstrates that a new steam supply way is possible by combination of the pre-drying step with the autogenerated steam step and thus makes this method much more economic.

The financial Support for conducting of this research was provided by the Ministry of Research and Innovation from Romania, National Plan of Research & Development in Energy and Environmental field. Project no. PN 19 11 03 01//2019

References

1. D. Klass, An Introduction to Biomass Energy a Renewable Resource (www.bera1.org)
2. T.N. Veziroglu, S.Sahin.,. Energy Convers Manage, **49** (2008)
3. M. Balat, Int J Hydrogen Energ **33**(2008);
4. S.V. Mohan, Y.V.Bhaskar,P.N. Sarma, Water Res ., **41**(2007)
5. D. Das, N. Khanna, T.N. Veziroglu, Chem Ind Chem Eng Q, **14**(2008)
6. S.Yolcular, Energy Sources Part A, **31**(2009)
7. Biomass-The Growing Energy Source, (www.science. org.au/ nova)
8. M. Balat, Energy Sources Part B, **2**(2007)
9. S.Yaman, Energy Convers Manage, **45**(2004)
10. A. Demirbas, Energy Sources Part A, **31**(2009)
11. A. Demirbas, Energy Sources Part A, **30**(2008)
12. S.M.Swami, V. Chaudhari,D.S. Kim, S.J.Sim,M.A. Abraham, Ind Eng Chem Res, **47**(2008)
13. H. Balat, Energy Educ Sci Technol Part A, **24**(2010)
14. M. Balat, Energy Sources Part A, **30**(2008)
15. M.F. Demirbas, Appl Energy, **86**(2009)
16. S.J.Yoon, Y.C.Choi, J.G.Lee, Energy Convers Manage, **51**(2010)
17. H.E.M. Stassen, H.A.M.Knoef, Biomass Technology Group, University of Twente; 1993.
18. A.H. Demirbas, Energy Educ Sci Technol Part A, **26**(2010)
19. C. Dupont, G.Boissonnet, J.M.Seiler, P.Gauthier, D.Schweich, Fuel, **86**(2007)
20. M. Ni, Y.C.L. Dennis, K.H.L. Michael, K. Sumathy, Fuel Process. Technol. **87**(2006)
21. P.M. Lv, J. Chang, Z.H. Xiong, H.T. Huang, C.Z. Wu, Y. Chen, J.X. Zhu, Energy Fuels 17(2003)
22. S.T.Chaudhari, A.K. Dalai, N.N. Bakhshi, Energy Fuels, **17**(2003)
23. M.R.Mahishi, D.Y. Goswami, J. Int. J. Hydrogen Energy, **32**(2007)
24. A.E. Pűtűn, E. Apaydın, E. Pűtűn , Energy, **27**(2002)
25. E.Apaydi-Varol, E.Pűtűn, A.E. Pűtűn, Fuel, **86**(2007)
26. E.Pűtűn , B.B. Uzun, A.E. Pűtűn , Bioresour. Technol. , **97**(2006)
27. A.Domınguez, J.A. Menendez, J.J. Pis, J. Anal. Appl. Pyrolysis, **77**(2006)

Iron Oxide-Coated Gravel Fixed Bed Column Study Performance to Remove Mixed Metals from Landfill Leachate

Ibrahim Yildiz[1,*] and *Banu* Sizirici[2]

[1]Khalifa University of Science and Technology, Chemistry Department, Abu Dhabi, P.O Box: 127788, United Arab Emirates
[2] Khalifa University of Science and Technology, Civil Infrastructure and Environmental Engineering Department, Abu Dhabi, P.O Box: 127788, United Arab Emirates

Abstract. Iron oxide-coated gravel as an adsorbent was employed in continuous fixed bed column study to remove Fe(II), Ni(II), and Zn(II) simultaneously in synthetic leachate samples. Experimental and modeled adsorption capacities derived from the breakthrough curves showed the adsorption capacity order of Zn(II)>Fe(II)> Ni(II). Iron oxide-coated gravel column removed 58.24% of Zn(II), 47.71% of Fe(II), and 39.45% of Ni(II). Desorption process was studied in order to test the regeneration capability of iron oxide-coated gravel. It was seen that 99.64 % of Ni(II), 99.54% of Fe(II) and 6.75% of Zn (II) were recovered through the first cycle of adsorption/desorption. In the second cycle, the recovery rates dropped to 81.4% for Ni(II), 80% for Fe(II) and 4% for Zn(II). Based on these results, iron oxide coated gravel has potential to remove mixed metal ions simultaneously in aqueous solutions.

1 Introduction

High metal concentrations in landfill leachates stem from increasing amounts of electronic wastes, batteries, paints and treated woods in mainstream solid wastes[1, 2]. It was reported that improper management of leachate can pollute environment to a significant extent [3-5]. For that reason, reduction of these metals with effective treatment methods bears critical importance.

Removal of inorganic and organic materials from wastewater samples has been widely carried out through low cost adsorbents such as limestone, zeolite, gravel, sand, peat and industrial waste [6]. Furthermore, external coating of these materials with iron oxide causes changes in the morphology of the surfaces, and this produces novel adsorption sites having high levels of surface defects and reactive surface sites with greater intrinsic reactivity[7-9]. Iron oxide-coated adsorbents removed metals, oxyanionic metals and bacteria [9-20]. Fixed bed studies conducted with iron oxide-coated zeolite showed the removal order of Pb(II) > Cu(II) > Cd(II) > Cr(II) > Zn(II) [7]. In addition, a fixed bed column composed of iron oxy hydroxide-coated brick showed the removal order of Pb(II) > Cd(II) > Zn(II) > Fe(II) ≈ Ni(II) ≈ Cu(II)[21]. Iron oxide-coated sand in a column showed the adsorption order of Cr(VI)>As(V) [22] A column study utilizing zeolite-supported microscale zero-valent iron as adsorbent showed more adsorption for Pb(II) than Cd(II) [23]. Another study showed that almost all Cu(II) and 80% of Zn(II) could be removed in a column reactor containing iron oxide-coated gravel [19]. In a similar way, iron oxide-coated sand in a fixed bed column system removed 80% for Cu, 90% for Pb, and 98% for Cd[24]. Morever, iron oxide-coated sand and limestone in a column removed 98.5% of iron and 95% of arsenic[25]. Iron oxide-coated peat in up-flow columns simultaneously removed more than 99% of Cu and 90% of Zn[26]. Nano-sized iron oxide-coated sand in a continuous column study removed 90.8% of turbidity, 73.3% of Pb, 75.8% of Zn, 85.6% of Cd and 67.5% of PO_4 in a synthetic urban runoff [27]. A previously reported study showed that iron oxide-coated gravel (IOCG) in a column study showed initial removal of 96.5% of Cu(II), 94.8% of Pb(II), 90% of Cd(II), 84% of Fe(III) and 67% of Al(III) at pH 7 [28].

A limited number of studies reported the removal efficiency of IOCG in aqueous solutions for individual ions and mixed metals. Since multiple metal ions coexist together in waste water, studies aiming at simultaneous removal of mixed metals will carry more importance towards the implementation of the results in large scale systems. In this study, IOCG as an adsorbent was employed in a fixed bed column study to remove Fe(II), Ni(II), and Zn(II) simultaneously from high strength synthetic leachate samples. Adsorption/desorption studies were conducted to evaluate the reusability of IOCG. It is expected that the results of this study will provide a framework in the prediction and evaluation of simultaneous metal ions uptake with IOGC in industrial applications.

2 Materials and Methods

2.1 Filter media preparation

* Corresponding author: ibrahim.yildiz@ku.ac.ae

https://doi.org/10.1051/e3sconf/201912201002

The gravel used in this study was purchased from Marinara Transport, Abu Dhabi/UAE. It was made up of local rock fragments. The gravel with an average diameter of 2 mm-2.38 mm was chosen in order prevents clogs and high head losses in the filters. Surface coating of the gravel was performed following our one of previous studies [5]. The characterization of gravel and IOCG was performed using scanning electron microscope (SEM), energy-dispersive X-ray (EDX), x-Ray diffraction (XRD), and Brunauer-Emmett-Teller (BET) surface area analysis, and the details were reported and explained in our previous studies [5, 28, 29].

2.2 Synthetic leachate preparation

The synthetic grade leachate was prepared following an established reported protocol[5]. All chemicals were purchased from Fisher Scientific and were of analytical reagent grade. An Atomic Absorption Spectrometer (AAS), Varian FS220, was used to measure the concentrations of the metal ions. The pH of the leachate samples was measured with a Hach multimeter, HQD4D.

2.3 Column adsorption studies

In this study, acrylic flow cell columns having end plate assembles and screens made of two mesh stainless steel were used. The measurements were run in duplicates and the average values were used in the data analysis. The empty columns were cleaned with 3% of nitric acid solution to remove any residual organic contaminants. Samples of synthetic leachate were mixed with a magnetic stirrer and pumped through the columns upward using a peristaltic pump (Welch Model 3200) to be able to prevent channeling from gravity as shown in Figure 1.

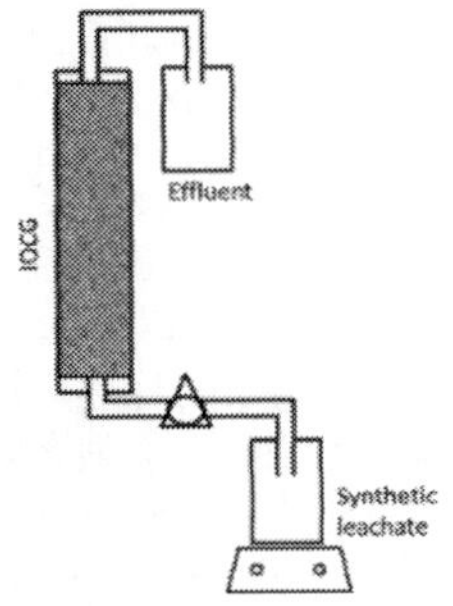

Figure 1. Column set up

The operational parameters with the column specifications are listed in Table 1. Leachate flow was stopped at time t when the concentration of a particular metal ion in the effluent reached to 98% of its initial concentration according to the formula $C_t = 0.98xC_o$ where C_t is the metal concentration at time t and C_o is the initial metal concentration in the leachate sample. The pH of the samples were adjusted to 7 for the maximum efficiency of metal removal based on the result of a previous study using IOCG[5]. Effluent samples were collected at different time periods, filtered using 0.45 µm syringe filter, and then analyzed with AAS to measure the concentrations of metal ions.

Table 1. Operational parameters and column specifications for column setup

Operational Parameter	Value
Detention time (packed)	5 min
Total porosity*	0.25
Feed flow rate	15 mL/min
Operating temperature	23±1 °C
IOCG weight	107.06 gr
Active bed volume	76.96 cm³
Total height of column	35 cm
Active height of column	15 cm
Inner diameter of the column	2.54 cm

*p_t: $(1-V_s/V_t)$, p_t: total porosity, V_s: soil particle volume, V_t: Total volume

The total adsorbed metal ion, q_{total} (mg), is equal to the area under the plot of the adsorbed ion concentration C_{ad} ($C_{ad}=C_0-C_t$) (mg/L) versus time (min) and can be calculated from Eq. (1):

$$q_{total\,(mg)} = \frac{Q.A}{100} = \frac{Q}{1000}\int_0^{t_{total}} C_{ad}dt \quad (1)$$

The dynamic adsorption capacity, q_e (mg/g), was calculated from Eq (2):

$$q_e = \frac{q_{total}}{m} \quad (2)$$

The removal efficiency of ions was calculated from the ratio of total adsorbed metal ions in the column to the total amount of metal ions sent to the column based on Eq. (3):

$$Efficiency = \frac{q_{total}}{C_0\frac{Q}{1000}t_{total}}x100 \quad (3)$$

The modeled column adsorption capacities were calculated based on the Thomas model given below,

$$Ln\left(\frac{C_0}{C_t} - 1\right) = k_{Th}q_0\frac{m}{Q} - k_{Th}C_0t \quad (4)$$

The linear plots of ln $[(C_0/C_t)-1]$ versus t based on the results in column experiments were generated to derive the k_{th} and q_0 values which correspond to the slope and intercepts in the graphs.

2.4 Column desorption studies

The ability to regenerate IOCG was tested through two cycles of adsorption/desorption process. Exhausted IOCG containing adsorbed metal ions were passed through 0.1 M of HCl solution with a 15 mL/min flow rate. Effluent samples at different time periods were collected in order to determine the desorption ratio[5]. Desorption ratio was calculated based on the following formula:

$$\text{Desorption ratio(\%)}: \frac{\text{Amount of metal ions desorbed}}{\text{Amount of metal ions adsorbed}} x100 \quad (5)$$

3 Materials and Methods

3.1 Fixed-bed column experiments

Adsorption of heavy metals by IOCG in a fixed bed column study was presented as C_t/C_o per day as shown in Figure 2. For a constant flow rate, 96.5% removal for Zn (II) and 77% removal for Fe(II) in 5 days and 78% removal for Ni(II) in 3 days were observed during the study. Removal rates dropped afterwards. Percent removal became 0 after 21 days of operation for Fe(II), 15 days for Ni(II) and 17 days for Zn(II).

Dynamic adsorption capacity along with exhaustion times and total percent removal for metal ion on IOCG were given in Table 2. A previous study with batch experiments showed that 98.8% of Fe(II) , 88% of Ni(II), and 94% of Zn(II) can be removed[5]. Furthermore, when compared with a previous batch study, it was found that the removal efficiency of IOCG in fixed-bed columns decreased as a result of decrease in contact time and decrease in contact between adsorbate and adsorbent. This situation resulted in lower retention times for the adsorption process to reach equilibrium which was also argued in another study[30].

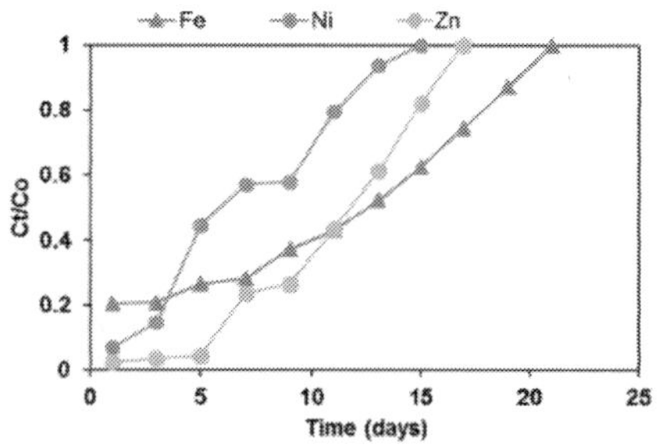

Figure 2. Breakthrough curves corresponding to Fe(II), Ni(II), and Zn(II) ions in leachate samples

Table 2. Exhaustion times, percent removals and dynamic adsorption capacity

Ions	q_e (mg/g)	t_e (days)	%Removal
Ni(II)	2.5	15	39.45
Fe(II)	11.29	21	47.71
Zn(II)	13.24	17	58.24

The dynamic adsorption capacity order for metals was Zn(II)>Fe(II)>Ni(II) which correlated with the solubility of hydroxides (pK_{sp}: Ni:15.2, Zn: 16.5, Fe^{2+}:16.31) as well as the hydrated radius (Zn: 4.30A°, Ni:4.04A°, Fe^{+2}:4.28A°) [31, 32] . It is apparent that the solubility products (K_{sp}) and the hydrated radius of metals are inherently related to their adsorption capabilities. Ions having smaller hydrated ionic radius are expected to diffuse easier and faster onto the surface of the adsorbent, and as a result they can accumulate in the cracks and channels more as compared to metal ions with larger hydrated radius. Ions with smaller hydrated ionic radius are able to move closer to the adsorbent surface and easily enter the channels in the adsorbent for preferential adsorption[31]. It could be argued that Zn(II) has a strong affinity towards IOCG on the basis of its higher percent removal rate and its dynamic adsorption capacity. Lower removal efficiency of Ni(II) could be explained with the concept of charge repulsion. In aqueous solutions with pH<8, nickel ions prefers Ni^{2+} form, and they compete with H^+ ions for the negatively charged surfaces on the IOCG. As a result, reduced Ni(II) removal rate could be observed[33]. In parallel, another study reported that Ni(II) adsorption was considerably lower in the presence of Cu(II) which competed with Ni(II) for available adsorption sites. Another reason for the lower Ni (II) removal rate could be lower retention time for Ni(II) to interact with IOCG.

It was noted that the shape of the breakthrough curve is not classical S shape. Another study reported that mixed metal solutions give different breakthrough curves[5]. Furthermore, different breakthrough curves could be obtained for a particular metal ion when other metal ions are present due to the differences in their adsorption abilities [2, 34].

The breakthrough behavior of metal ions in the leachate samples were analyzed and described in a satisfactory manner using the Thomas model. Using this model, q_0 and k_{th} corresponding to the three metal ions were calculated. It was shown that IOCG displayed high loading capacity toward these ions in the column studies in which the uptake rate ranged between 15.34 mg/g to 2.95 mg/g as shown in Table 3. A previous study showed that IOCG achieved metal ion uptakes of 66.82 mg/g for Cu(II), 26.5 mg/g for Pb(II), 23.33 mg/g for Cd(II), 1.87 mg/g for Al(III), and 0.56 mg/g for Fe(III)[28]. Oher studies showed similar results; iron oxide-coated Australian zeolite performed metal ion uptakes of 0.89 mg/g for Cu(II), 0.93 mg/g for Cd(II), and 0.83 mg/g Zn (II); iron-impregnated activated carbon displayed 2.746 mg/g for Cu(II) uptake[7, 8, 14]. The correlation coefficients of Thomas model (R^2 ranging between 0.95 to 0.97) for these ions were very high, and it indicated that this model was a suitable tool to study IOCG adsorption capacities.

Table 3. Constants corresponding to the removal of metal ions in leachate using Thomas model

Ions	q_o (mg/g)	k_{th} (L/mg.h)	R^2
Ni(II)	2.95	0.009805	0.95
Fe(II)	11.104	0.0012	0.97
Zn(II)	15.34	0.0023	0.96

A high correlation was observed between experimental and modeled adsorption capacities indicating the applicability of the Thomas model to predict IOCG sorption on metals shown in Table 4.

Table 4. Comparison of experimental and modeled adsorption capacities

Ions	q_o (mg/g)	q_e (mg/g)
Ni(II)	2.95	2.5

Fe(II)	11.104	11.29
Zn(II)	15.34	13.24

3.2. Desorption experiments

The efficiency of metal ion recovery was tested through two cycles of adsorption/desorption process, and in the first cycle the recovery rates of 99.64 % for Ni(II), 99.54% for Fe(II) and 6.75% for Zn (II) were obtained as shown in Figure 6. In the second cycle, the recovery rates dropped to 81.4% for Ni(II), 80% for Fe(II) and 4% for Zn(II). The apparent decline in the recovery of metal ions in the second cycle could be attributed to the dissolution of iron oxide coating upon exposure to 0.1 M HCl solution for the regeneration process [7, 35]. Nevertheless, the adsorption capacity of regenerated IOCG is still high, and this finding verifies its potential for multiple cycles of metal uptake.

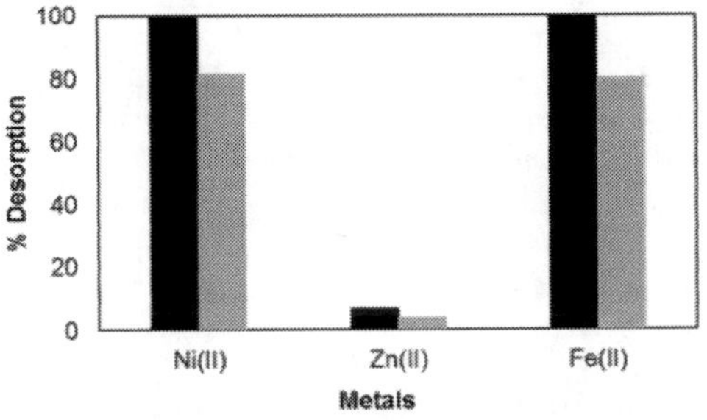

Figure 6. Desorption of metals (black color refers to first cycle and grey color refers to second cycle of adsorption/desorption process

Nomenclature

q_e = dynamic adsorption capacity, mg/g
C_0 = initial concentration of heavy metal (mg/L)
V = volume of the solution (L)
m = mass of adsorbent (g).
q_e = max. amount adsorbed, mg/g IOCG
k_{Th} = Thomas rate constant (L/mg.h),
q_0 = Thomas modeled column adsorption capacity (mg/g),
C_t = outlet concentration at time t (mg/L),
Q = flowrate (mL/min) and
t = filtration time (min)

4. Conclusion

In this study IOCG was incorporated into column experiments in order to test its efficiency to remove multiple metal ions simultaneously in high strength leachate samples. Zn(II) ion showed the strongest affinity toward IOCG based on its high breakthrough time. Charge repulsion could be invoked to elaborate the lower removal rate of Ni(II). Loading capacities of metal ions under dynamic conditions were derived using breakthrough data, and the results showed that IOCG could effectively remove these ions. The experimental and modeled adsorption capacity order was Zn(II)> Fe(II) > Ni(II) with the average removal rate of 58.24% for Zn(II), 47.71% for Fe(II), and 39.45% for Ni(II). The two cycles of adsorption/desorption study demonstrated that IOCG could be regenerated and reused in metal uptake studies albeit with an apparent decline in the adsorption capacity in the second cycle due to the dissolution of its iron oxide coating. Notwithstanding this observation, it seems obvious that IOCG is a potential material in the removal of metal ions simultaneously from landfill leachate. It could be incorporated into various column designs in fixed bed reactors, and it could be adapted to pilot or large-scale systems in industrial applications. In summary, our work could provide and inspire an initial framework for further studies.

References

1. Mojiri, A., et al., *Metals removal from municipal landfill leachate and wastewater using adsorbents combined with biological method.* Desalination and Water Treatment, 2016. **57**(6): p. 2819-2833.
2. Daso, A.P., et al., *Polybrominated diphenyl ethers (PBDEs) and 2,2 ',4,4 ',5,5 '-hexabromobiphenyl (BB-153) in landfill leachate in Cape Town, South Africa.* Environmental Monitoring and Assessment, (2013). **185**(1): p. 431-439.
3. Sizirici, B. and B. Tansel, *Parametric fate and transport profiling for selective groundwater monitoring at closed landfills: A case study.* Waste Management, (2015). **38**: p. 263-270.
4. Lins, C., et al., *Removal of ammonia nitrogen from leachate of Muribeca municipal solid waste landfill, Pernambuco, Brazil, using natural zeolite as part of a biochemical system.* Journal of Environmental Science and Health Part a-Toxic/Hazardous Substances & Environmental Engineering, (2015). **50**(9): p. 980-988.
5. Sizirici, B. and I. Yildiz, *Adsorption capacity of iron oxide-coated gravel for landfill leachate: simultaneous study.* International Journal of Environmental Science and Technology, (2017). **14**(5): p. 1027-1036.
6. Ali, I., *New generation adsorbents for water treatment.* Chemical reviews, (2012). **112**(10): p. 5073-5091.
7. Thuy Chung, N., et al., *Simultaneous adsorption of Cd, Cr, Cu, Pb, and Zn by an iron-coated Australian zeolite in batch and fixed-bed column studies.* Chemical Engineering Journal, (2015). **270**: p. 393-404.
8. Tiwari, D., Lalhmunsiama, and S.-M. Lee, *Iron-impregnated activated carbons precursor to rice hulls and areca nut waste in the remediation of Cu(II) and Pb(II) contaminated waters: a physico-chemical studies.* Desalination and Water Treatment, (2015). **53**(6): p. 1591-1605.
9. Kan, C.-C., et al., *Adsorption of Mn2+ from aqueous solution using Fe and Mn oxide-coated*

sand. Journal of Environmental Sciences-China, (2013). **25**(7): p. 1483-1491.

10. Kundu, S. and A.K. Gupta, *Adsorption characteristics of As(III) from aqueous solution on iron oxide coated cement (IOCC).* Journal of Hazardous Materials, (2007). **142**(1-2): p. 97-104.
11. Nguyen, T.V., et al., *Arsenic removal by iron oxide coated sponge: Experimental performance and mathematical models.* Journal of Hazardous Materials, (2010). **182**(1-3): p. 723-729.
12. Buamah, R., B. Petrusevski, and J.C. Schippers, *Adsorptive removal of manganese(II) from the aqueous phase using iron oxide coated sand.* Journal of Water Supply Research and Technology-Aqua, (2008). **57**(1): p. 1-11.
13. Park, S.J., et al., *Bimetallic oxide-coated sand filter for simultaneous removal of bacteria, Fe(II), and Mn(II) in small- and pilot-scale column experiments.* Desalination and Water Treatment, (2015). **54**(12): p. 3380-3391.
14. Lee, S.-M., C. Laldawngliana, and D. Tiwari, *Iron oxide nano-particles-immobilized-sand material in the treatment of Cu(II), Cd(II) and Pb(II) contaminated waste waters.* Chemical Engineering Journal, (2012). **195**: p. 103-111.
15. Unob, F., et al., *Reuse of waste silica as adsorbent for metal removal by iron oxide modification.* Journal of Hazardous Materials, (2007). **142**(1-2): p. 455-462.
16. Sud, D., G. Mahajan, and M.P. Kaur, *Agricultural waste material as potential adsorbent for sequestering heavy metal ions from aqueous solutions - A review.* Bioresource Technology, (2008). **99**(14): p. 6017-6027.
17. Zamil, S.S., et al., *Correlating metal ionic characteristics with biosorption capacity of Staphylococcus saprophyticus BMSZ711 using QICAR model.* Bioresource Technology, (2009). **100**(6): p. 1895-1902.
18. Tiwari, D., et al., *Manganese-modified natural sand in the remediation of aquatic environment contaminated with heavy metal toxic ions.* Chemical Engineering Journal, 2011. **171**(3): p. 958-966.
19. Norris, M.J., et al., *Treatment of heavy metals by iron oxide coated and natural gravel media in Sustainable urban Drainage Systems.* Water Science and Technology, (2013). **68**(3): p. 674-680.
20. Kumar, A., et al., *Iron oxide - coated fibrous sorbents for arsenic removal.* Journal - American Water Works Association, (2008). **100**(4): p. 151-A4.
21. Allandin, O., et al., *Adsorption capacity of iron oxyhydroxide-coated brick for cationic metals and nature of ion-surface interactions.* Applied Clay Science, (2014). **90**: p. 141-149.
22. Mak, M.S., I.M. Lo, and T. Liu, *Synergistic effect of coupling zero-valent iron with iron oxide-coated sand in columns for chromate and arsenate removal from groundwater: influences of humic acid and the reactive media configuration.* Water Research, (2011). **45**(19): p. 6575-6584.
23. Kong, X., et al., *Evaluation of zeolite-supported microscale zero-valent iron as a potential adsorbent for Cd 2+ and Pb 2+ removal in permeable reactive barriers.* Environmental Science and Pollution Research, (2017). **24**(15): p. 13837-13844.
24. Benjamin, M.M., et al., *Sorption and filtration of metals using iron-oxide-coated sand.* Water Research, (1996). **30**(11): p. 2609-2620.
25. Devi, R.R., et al., *Removal of iron and arsenic (III) from drinking water using iron oxide-coated sand and limestone.* Applied Water Science, (2014). **4**(2): p. 175-182.
26. Kasiuliene, A., et al., *Treatment of metal (loid) contaminated solutions using iron-peat as sorbent: is landfilling a suitable management option for the spent sorbent?* Environmental Science and Pollution Research, (2019).
27. Khiadani, M., M. Zarrabi, and M. Foroughi, *Urban runoff treatment using nano-sized iron oxide coated sand with and without magnetic field applying.* Journal of Environmental Health Science and Engineering, (2013). **11**(1): p. 43.
28. Sizirici, B. and I. Yildiz, *Simultaneous adsorption of divalent and trivalent metal cations by iron oxide-coated gravel.* International Journal of Environmental Science and Technology, (2018): p. 1-10.
29. Sizirici, B., et al., *Adsorptive removal capacity of gravel for metal cations in the absence/presence of competitive adsorption.* Environmental Science and Pollution Research, (2018). **25**(8): p. 7530-7540.
30. Li, C. and P. Champagne, *Fixed-bed column study for the removal of cadmium (II) and nickel (II) ions from aqueous solutions using peat and mollusk shells.* Journal of Hazardous Materials, (2009). **171**(1-3): p. 872-878.
31. Sounthararajah, D.P., et al., *Adsorptive removal of heavy metals from water using sodium titanate nanofibres loaded onto GAC in fixed-bed columns.* Journal of Hazardous Materials, (2015). **287**: p. 306-316.
32. Barnum, D.W., *Hydrolysis of cations. Formation constants and standard free energies of formation of hydroxy complexes.* Inorganic Chemistry, (1983). **22**(16): p. 2297-2305.
33. Mahmoud, M.R. and N.K. Lazaridis, *Simultaneous Removal of Nickel(II) and Chromium(VI) from Aqueous Solutions and Simulated Wastewaters by Foam Separation.* Separation Science and Technology, (2015). **50**(9): p. 1421-1432.
34. Zulfadhly, Z., M.D. Mashitah, and S. Bhatia, *Heavy metals removal in fixed-bed column by the macro fungus Pycnoporus sanguineus.* Environmental Pollution, (2001). **112**(3): p. 463-470.

https://doi.org/10.1051/e3sconf/201912201002

35. Han, R., et al., *Characterization and properties of iron oxide-coated zeolite as adsorbent for removal of copper(II) from solution in fixed bed column.* Chemical Engineering Journal, (2009). **149**(1-3): p. 123-131.

Assessment of the Romanian onshore and offshore wind energy potential

Florin Onea* and *Liliana* Rusu

Department of Mechanical Engineering, ''Dunarea de Jos'' University of Galati, Domneasca Street, 47, Galati 800008, Romania

Abstract. A general assessment of the wind energy potential from the eastern part of Romania was carried out in this work by taking into account onshore and offshore wind conditions. First, a perspective of the importance of the renewable resources into the Romanian electricity system was provided, from which was noticed that the wind production cover almost 15% from the demand (reported to 2017). From the analysis of the wind data significant differences were noticed between the onshore and offshore regions, the last one presenting more important wind resources, that significantly increase during the winter time (with almost 22%). According to the performance reported by an offshore wind turbine (rated at 3 MW), such generator will operate at full capacity in a maximum 9% from the time, if we consider a hub height of 119 m.

1 Introduction

The onshore wind energy is one of the most dynamical sectors, being estimated that this is one of the cheapest renewable solutions that can compete at this moment with the conventional generation solutions in terms of the costs [1]. Definitely, the evolution of this industry will finally reach a saturation, taking into account that the suitable sites for such projects will be limited. Therefore, some new environments need to be taken into account, this being the case of the coastal and marine regions. The vast water areas allow the development of large capacity projects, being also expected more important wind resources than onshore [2–5]. In some cases, the offshore regions seem to be defined by similar wind resources as onshore, the only difference being related to the decision of the developers to assembly larger farms in order to cover the higher costs associated to such projects [6].

Europe is a leader in terms of wind electricity production, exceeding some important sectors, such as fuel oil (in 2007), nuclear sector (in 2013), hydropower (in 2015) or coal sector (in 2016). It was estimated that at the end of 2017 a total capacity of 937 GW was installed in this region, from which approximately 3.1 GW were offshore, this being related to an increase of 101% compared to 2016. Romania is defined by a wind power capacity of 3.03 GW, which increased only with 5 MW in 2017. Reported to electricity demand, was estimated that almost 12.2% is covered at this moment from the operating wind farms [7].

Regarding the Black Sea wind conditions, the previous research indicated that the western part of this basin present in general more important wind resources [8,9]. As for the Romanian coastal environment, it is estimated that the sites located on the northern areas may present interest for the development of a wind project [10,11].

In this context, the aim of the present study is to identify the differences between the Romanian onshore and offshore regions (if any) and also to estimate the performances of some offshore wind turbines that may operate in this coastal sector.

2 Methods and materials

The wind conditions used in the present work are related to the ERA-Interim database that is assembly by the European Centre for Medium-range Weather Forecast. These values are provided at 10 m height above the sea level (U10 parameter), being processed for the 20-year time interval (1999-2018). Four data per day were considered (00-06-12-18 UTC), and correspond to a spatial resolution of 0.125° x 0.125°, which is the finest grid available on the ECMWF server (https://www.ecmwf.int/en/forecasts/datasets/reanalysis-datasets/era-interim).

In order to assess the energy potential of a particular site, in general the wind conditions are evaluated at a hub height that starts from 80 m. A logarithmic law can be used in this case, as follows [12,13]:

$$U80 = U10\frac{\ln(z_{80}) - \ln(z_{10})}{\ln(z_{10}) - \ln(z_0)} \tag{1}$$

where, U80 - wind speed at 80 m, U10 - initial wind speed (at 10 m), z_0 - roughness of the sea surface (0.0002 m), z_{10} and z_{80} - reference heights.

* Corresponding author: florin.onea@ugal.ro

https://doi.org/10.1051/e3sconf/201912201003

3 Results

Figure 1 presents the share of renewable electricity in the European Union (EU) and Romania, considering the time interval from 1990 to 2017 [14]. At this point, it is important to mention that, Romania entered in the EU at the beginning of 2007. Reported on a European level, in general Romania reveals a much higher percentage share, being reported only in 2012 similar results. Definitely, there is an upward trend, more renewable sources being added to the electricity network, as we can observe in the case of the year 2016 (close to 45% for Romania). Wind and solar come together in this report, but it is important to mention that the onshore wind capacity is more relevant for this statistic. The interval from 2009 to 2017 seems to define these two sectors, being registered significant progress that is similar or exceeds the EU level (the year 2014 and year 2015).

A first perspective of the Romanian wind conditions is provided in Figure 2. From the spatial distribution of the U80 parameter, we can see that there is a clear separation between the onshore and offshore areas, the second one presenting resources that are more important. The onshore conditions gradually increase as we approach to the shoreline, being encountered areas with 5.5 m/s. As we go offshore it is possible to reach a maximum of 7.5 m/s between the 29° and 30° longitude lines. In Figure 1b, the wind conditions associated to some reference sites (onshore and offshore) is presented. In general, the onshore sites reveal values located close to 4 m/s, compared to the offshore sites (O1, O2 and O3) where an average of 6.8 m/s may be noticed. From the distribution of the wind roses reported for the site O2 (Figure 1c) we can observe that the northeast sector is the dominant one (close to 6%), where the presence of the wind conditions that exceed 12 m/s is more visible.

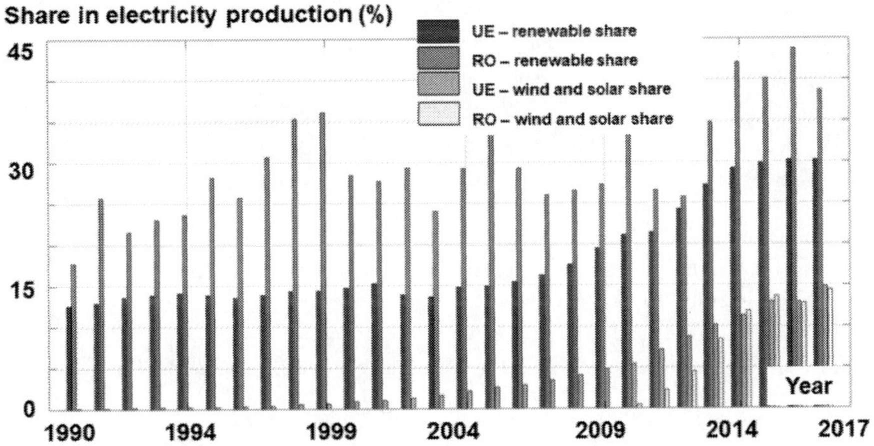

Figure 1. Share of renewable contribution to the EU and Romanian electricity production.

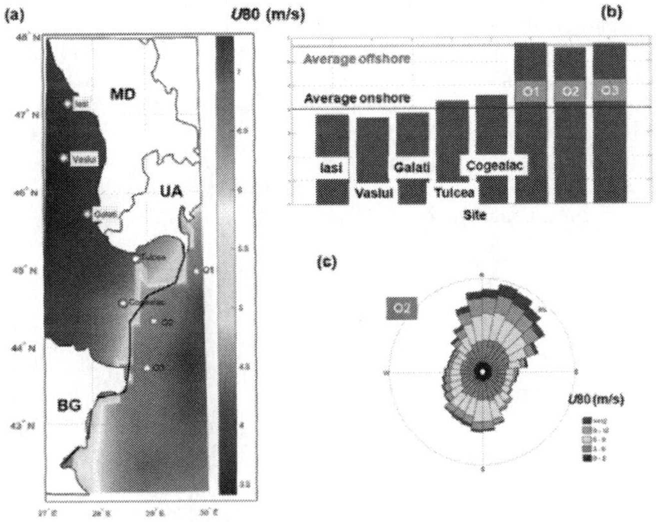

Figure 2. Distribution of the U80 parameter from the ERA-Interim data for the Romanian environment (from 1999 to 2018), where: (a) spatial distribution of the average values; (b) average values for some onshore and offshore sites; (c) wind roses for the site O2.

Figure 3 illustrates the differences between the full-time distribution (Figure 1a) and the main seasons, namely spring (March–April–May), summer (June–July–August), autumn (September–October–November) and winter (December–January–February). During the spring, we may expect an increase of the conditions with a maximum 7% onshore compared to a decrease of 8% near the Romanian coast. In the summer, all the regions show a decrease (between 8% and 19%), compared to the autumn where only the offshore regions may report an increase of 4%. A maximum increase of 22% is expected offshore during the wintertime, while only 4% is expected for the onshore regions.

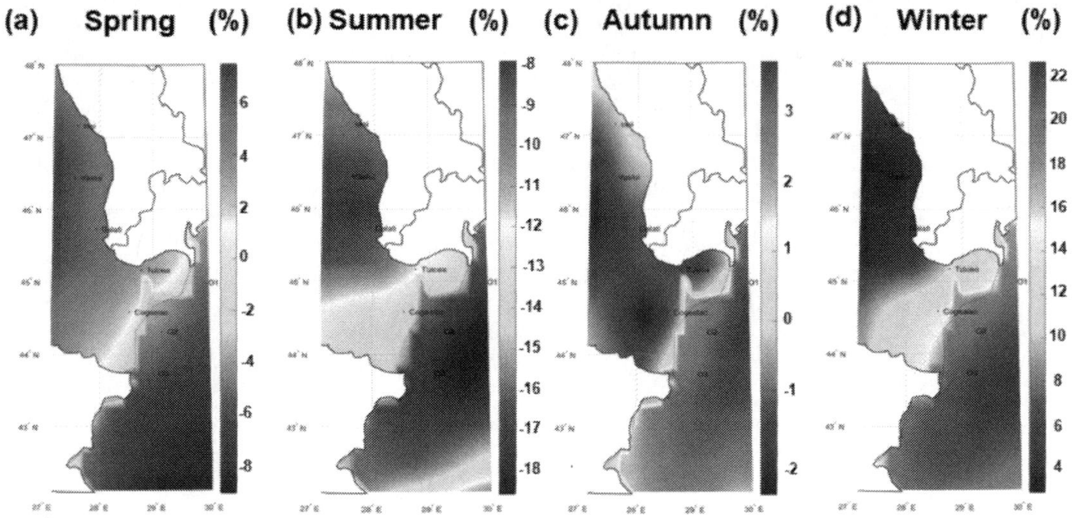

Figure 3. *U*80 differences (in %) between the average values related to the total time distribution (all the values) and the four main seasons: (a) spring; (b) summer; (c) autumn; (d) winter.

At this moment, there is no offshore wind project in the Black Sea, and therefore the next step will be to identify the performances of such system that may operate in the Romanian sector. By looking on the wind speed reported at the offshore sites O1, O2 and O3 we can see that the average wind speed (at 80 m) is located between 6.5 m/s and 5 m/s. According to the IEC recommendations these sites are related more to the C3 and C4 classes [15], and by looking on the operating turbines it seems that the system V112-3.0 [16] will be more suitable for this environment. The power curve of this turbine is presented in Figure 4, being defined by a cut-in of 3 m/s and by a rated speed of 12 m/s at which a maximum 3 MW is generated. For the hub heights, the manufacturer indicates three versions, namely: 84 m, 94 m and 119 m.

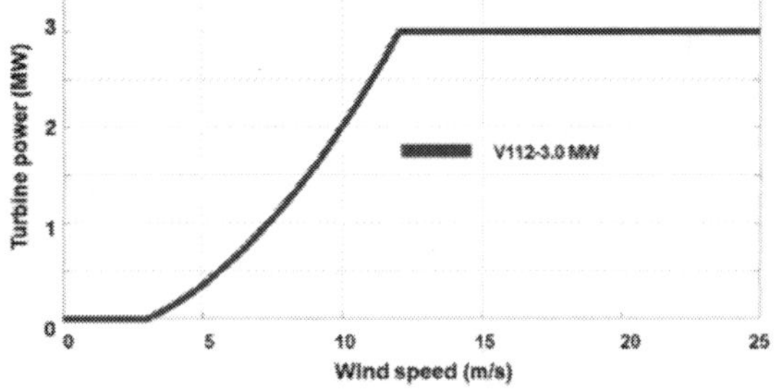

Figure 4. Representation of the V112-3.0 power curve.

The inactivity period of the selected turbine (wind speeds < cut-in) is presented in Figure 5, by taking into account various hub heights. The considered sites are representative for the Romanian coastal areas (north-centre-south) being defined at approximately 30 km offshore. As expected the downtime values decrease as the hub height increase, much higher values being accounted for the site O2 located in the center of the target area. The sites O1 and O3 reveal similar results, being reported a 12% for the hub height of 84 m, compared to a minimum of 11.3% in the case of the site O3 (hub height 119 m).

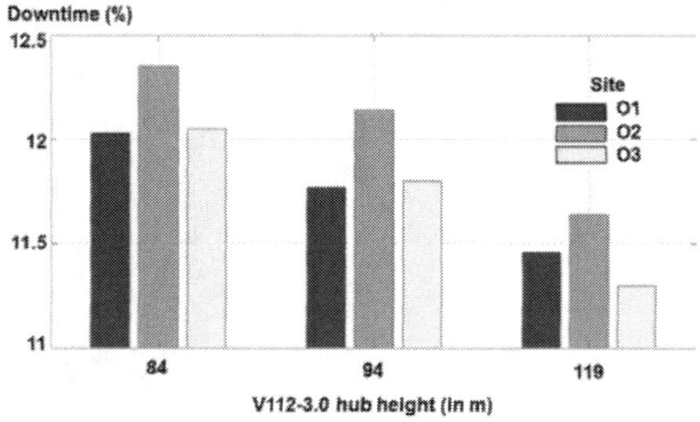

Figure 5. Downtime interval (%) reported by the V112-3.0 system for the sites O1, O2 and O3.

Figure 6 present the percentage of the time, during which the selected turbine will operate on a full capacity (3 MW). As we can notice, the values oscillate between 6% and 9%, better results being noticed for the hub height at 119 m. The differences between the sites are more visible, the site O3 reporting much better results.

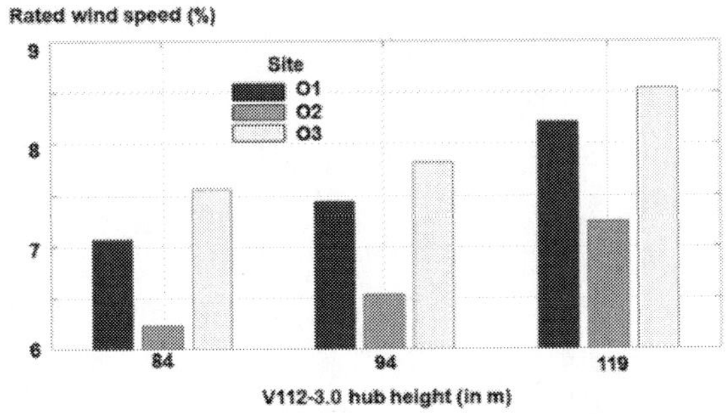

Figure 6. The efficiency of the V112-3.0 system, considering only the wind data higher than the rated speed (12 m/s).

4 Conclusions

A general assessment of the Romanian wind energy potential was provided in this work, in order to highlight the benefits coming from this energy sector. At this moment the wind projects are operational only on the Romanian onshore areas, being expected that in the near future this sector to expand to the offshore regions. By looking at the statistical data was highlighted the fact that there is interest for this type of energy, being reported an increase in the installed number of turbines.

From the analysis of the ERA-Interim was possible to highlight the spatial differences between the onshore and offshore areas, the second one presenting significantly much higher values. The accuracy of this model can be discussed in the vicinity of the coastline, since there is a sharp transition between onshore and offshore. Considering the total time values, various seasonal patterns can be noticed, better performances being expected during the winter time, when an increase with almost 22% of the wind speed is reported close to the offshore regions.

In the case of a wind turbine rated at 3 MW, we may expect downtimes periods that vary from 11% to 12.5%, while the turbine will work at full capacity at values that do not exceed 9%.

Finally, we can conclude that the Romanian coastal areas are an attractive environment for the development of a wind project (onshore or offshore). An example is the successful development of the already operational Fantanele-Cogealac project, which is one of the largest in the world (16th place – April 2019).

Acknowledgments

This work was supported by a grant of Ministry of Research and Innovation, CNCS – UEFISCDI, project number PN–III–P1–1.1–PD–2016–0235, within PNCDI III. The ERA-Interim data used in this study have been obtained from the ECMWF data server.

References

[1] H. Klinge Jacobsen, P. Hevia-Koch, C. Wolter, Nearshore and offshore wind development: Costs and competitive advantage exemplified by nearshore wind in Denmark, Energy for Sustainable Development. 50 (2019) 91–100. doi:10.1016/j.esd.2019.03.006.

[2] J. Li, X. (Bill) Yu, Onshore and offshore wind energy potential assessment near Lake Erie shoreline: A spatial and temporal analysis, Energy. 147 (2018) 1092–1107. doi:10.1016/j.energy.2018.01.118.

[3] E. Rusu, F. Onea, Joint Evaluation of the Wave and Offshore Wind Energy Resources in the Developing Countries, Energies. 10 (2017) 1866. doi:10.3390/en10111866.

[4] F. Onea, E. Rusu, Sustainability of the Reanalysis Databases in Predicting the Wind and Wave Power along the European Coasts, Sustainability. 10 (2018) 193. doi:10.3390/su10010193.

[5] F. Onea, L. Deleanu, L. Rusu, C. Georgescu, Evaluation of the wind energy potential along the Mediterranean Sea coasts, Energy Explor. Exploit. 34 (2016) 766–792. doi:10.1177/0144598716659592.

[6] P. Enevoldsen, S.V. Valentine, Do onshore and offshore wind farm development patterns differ?, Energy for Sustainable Development. 35 (2016) 41–51. doi:10.1016/j.esd.2016.10.002.

[7] Wind in Power: 2017 European Statistics | WindEurope, (n.d.). https://windeurope.org/about-wind/statistics/european/wind-in-power-2017/ (accessed May 11, 2019).

[8] D. Ganea, E. Mereuta, L. Rusu, Estimation of the Near Future Wind Power Potential in the Black Sea, Energies. 11 (2018) 3198. doi:10.3390/en11113198.

[9] L. Rusu, D. Ganea, E. Mereuta, A joint evaluation of wave and wind energy resources in the Black Sea based on 20-year hindcast information, Energy Explor. Exploit. 36 (2018) 335–351. doi:10.1177/0144598717736389.

[10] D.M. Niculescu, E.V.C. Rusu, An overview of the wind power potential in the Romanian coastal environment - moving from onshore to offshore, in: W. Cui, E. Rusu (Eds.), 2018 3rd International Conference on Advances on Clean Energy Research (Icacer 2018), E D P Sciences, Cedex A, 2018: p. UNSP 01007.

[11] F. Onea, L. Rusu, A Study on the Wind Energy Potential in the Romanian Coastal Environment, Journal of Marine Science and Engineering. 7 (2019) 142. doi:10.3390/jmse7050142.

[12] A. Raileanu, F. Onea, E. Rusu, Assessment of the Wind Energy Potential in the Coastal Environment of Two Enclosed Seas, Ieee, New York, 2015.

[13] F. Onea, E. Rusu, Efficiency assessments for some state of the art wind turbines in the coastal

environments of the Black and the Caspian seas, Energy Explor. Exploit. 34 (2016) 217–234. doi:10.1177/0144598716629872.

[14] World Energy Statistics | Enerdata, (n.d.). https://yearbook.enerdata.net/ (accessed June 30, 2018).

[15] E. Hau, H. von Renouard, Wind Turbines, Springer Berlin Heidelberg, Berlin, Heidelberg, 2006. doi:10.1007/3-540-29284-5.

[16] Vestas V112-3.0 MW - Vestas - PDF Catalogs | Technical Documentation | Brochure, (n.d.). http://pdf.directindustry.com/pdf/vestas/vestas-v112-30-mw/20680-244373.html (accessed May 14, 2019).

https://doi.org/10.1051/e3sconf/201912201004

Advanced algorithmic model for poly-optimization of biomass fuel production from separate combustible fractions of municipal wastes as a progress in improving energy efficiency of waste utilization

Krzysztof Gaska[1,*], *Agnieszka* Generowicz[2], *Mykhailo* Lobur[3], *Nazariy* Jaworski[3], *Józef* Ciuła[4], and *Myroslava* Vovk[5]

[1]Faculty of Energy and Environmental Engineering, Silesian University of Technology, Gliwice, Poland
[2]Faculty of Environmental Engineering, Cracow University of Technology, Cracow, Poland
[3]CAD Systems Department, Lviv Polytechnic National University, Lviv, Ukraine
[4]Institute of Technology, State University of Applied Sciences in Nowy Sącz, Nowy Sącz, Poland
[5]Department of Mathematics, Lviv Polytechnic National University, Lviv, Ukraine

Abstract. Usage of generated fuel components from segregated municipal waste groups in combined combustion processes with traditional fuels results in significant savings by partial replacement of fossil fuels and the reduction of environmental inconvenience that arises directly from the migration of waste to the environment. This article develops technological process of fuel components production which is controlled by advanced embedded SCADA control systems, providing monitoring of process parameters, process visualization and implementation, with the help of logic controllers of freely programmable PLC and complex control algorithms (including adaptive, predictive and inferential control algorithms) by implementation of the polymorphic model of fuel components obtaining process multi-stream optimization.

1 Introduction

Growing demand for heat and electricity used in many industries and utilities leads to a sharp decline in the natural resources of traditional fuels (coal, lignite, oil and natural gas). Excessive exploitation of natural energy sources results, in deterioration of the environment and irreversible climate change. The consequence of energy carriers, products or semi-finished products extraction used in various technological processes is the formation of secondary post-technological wastes of different groups. The source of waste (outside the industry) is other entity that produces unsuccessful streams of hazardous waste (medical waste, chemical reagents, including pesticides, etc.). The urban enthalpy of devaluation allows for their reuse after appropriate treatment [1-3].

Due to the use of molded fuel components from separated waste groups in combined combustion processes with conventional fuels one can achieve significant saving thanks to the partial replacement of fossil fuels and the reduction of environmental disadvantages that arise directly from the migration of waste to the environment (soil, atmospheric air, surface water and underground water) [8]. Particular attention should be paid to the possibility of using certain groups of hazardous waste in these processes (pesticides, medical waste). Conducting technological processes by using preformed fuel components in thermally coupled systems should be conditioned by technical criteria, which means that they should not interfere with the main processes, and in the global plan, with the factors of sustainable development. The physical process of preparing a fuel mixture requires careful analysis of the physical and chemical properties of the components, molded fuel components and technological conditions. The technological process of fuel components production is controlled by advanced SCADA-type control systems which provide process parameters monitoring, visualization and implementation with the help of freely programmable PLC logic controllers and complex control algorithms (including adaptive, predictive and inferential control algorithms).

2 Fuel components production process multi-stream optimization model

Optimization task of the complex technological process of manufacturing fuel components, formed from separated fractions from combustible municipal waste includes two main threads:

A) optimization of light fraction (calorific value) production consisting of isolation of heavy, non-metal components from the waste flow (in the optical separator). They have a high calorific value of Wd exceeding 40 MJ/kg in the case of PE (mainly PE, PET, PP plastics). It can be done by using an improved algorithm for forecasting MPC [4]

* Corresponding author: krzysztof.gaska@polsl.pl

control (based on the modified MC SVM (Multi-Class Vector Support Machine) algorithm) implemented in the control layer of the SCADA optimization hierarchical control system;

B) optimization of the hybrid fuel components obtaining process which consists in mixing of light fraction (obtained as a result of optimization of the optical separation process – A thread) with fossil fuels that corresponds to the target function, i.e. maximization under restrictions that are imposed on alternatives of optimization problem solution and parameters of implemented technological processes (rotary kilns, boilers, fluidized-bed boilers, etc.). The process optimization is carried out with the help of a modified simplex algorithm.

The choice of mechanical processes and their parameters, as well as the relationship between processes, is based on the knowledge of the morphological composition, the frequency and amount of waste delivered, as well as the concept of further processing of the recovered fractions. Physical properties such as humidity (mechanical sorting is the best for dry waste) and the density of waste are irrelevant. Depending on these parameters, the process line will take another form. These settings may differ not only by the devices used in them but also from the point of view of their number and location in the technological process. Only proper selection of the dividing line separate elements guarantees achievement of the predicted recovery rates of fraction from mixed municipal waste. In this regard, knowledge of the processes used to purchase individual waste fractions is crucial for the development and exploitation of waste segregation lines [5-7].

All elements of the system are connected with a control system based on a complex algorithm defined for a specific fraction of waste. In the context of ensuring the proper conduct of the process, the molded fuel production plant should be considered as a complex, non-linear, multidimensional control object.

The integrated multithreaded process control algorithm, embedded in a programmable logical PLC controller (controlled by the SCADA system), performs the task of solving the optimization problem defined in the form of a support vector machine model (SVM) of non-metallic fraction components (preferably ballistic and optical division), which reproduces the structures of predefined classes – subsets generated by the greedy algorithm of the Greedy Set Cover approximation containing fraction components described by a set of attributes (including physical and chemical properties, fuel properties, emissions, texture, structure, spectroscopic spectra).

Each subgroup is the fuel class component of the fraction composition. It is formed by the optimization effect using Greedy Set Cover approximation algorithm, which maximizes the target function (Wd-value of the heating) when performing the technological constraints imposed on the linear programming tasks (humidity, fraction of chlorine, heavy metals). Reference classes are described by a set of attributes (texture, structure, spectrum) which allow the classification algorithm of the lower level of classification management to be classified using the SVM network model (MC SVM) based on a set of functions assigned to predefined classes represented by disjoint subsets of coverage by the Greedy Set Cover algorithm, and finally verify that the identified component/fraction belongs to the template class on the incident matrix.

The optimization of lightweight fraction production process is implemented in the control layer of the SCADA hierarchical control system (fig. 1,2).

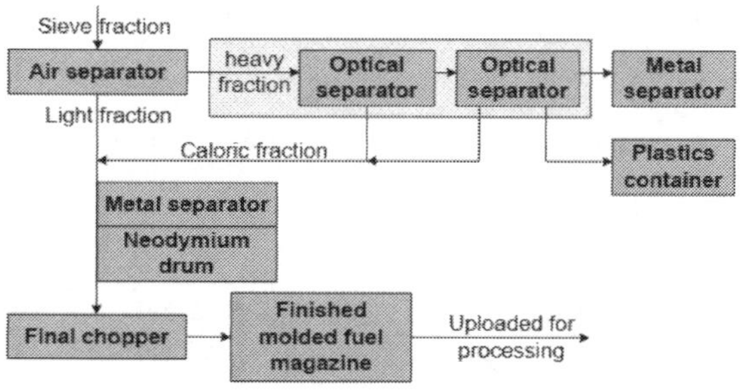

Fig. 1. Block diagram of ballistic and optical separation processes separated from the technological line.

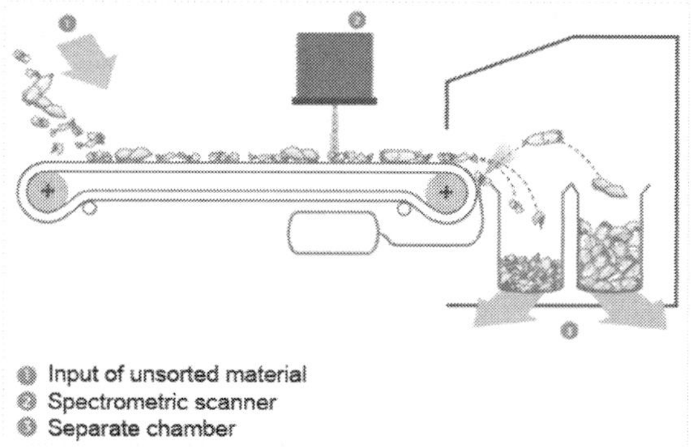

Fig. 2. Optical separator.

3 Mathematical formulations

Output signals of the preprocessor in the form of the following components of the Fourier descriptors, after the conversion, providing the invariant to scale, rotation and displacement, are input signals for the multilayer neural network, functioning as a pattern recognition system and simultaneously classification, that is, assigning a template to the corresponding class. The number of network input nodes is equal to the number of Fourier descriptors that are considered in the classification. Given that, each source neuron is a single class, their number is also constant and equals many classes. The classifier learns on a set of training data representing sequential classes of recognizable images (detecting on the basis of spectra, texture, contour, structure). In recovery mode, the classified template after passing all the phases of the preprocessor enters the network input and stimulates the output neuron that corresponds to this class.

At the stage of patterns recognition, through their noises, the source signals of the network neurons can accept continuous values from the interval [0;1] instead of the expected binary values, and the first corresponds to the recognized class.

In the classification question, classificatory MC SVM is based on a one-way neural network that implements various types of activation functions, including linear, polynomial, radial and sigmoid. The purpose of the classification is to maximize the separation limit between the two different classes described in the set of pairs (x_i, d_i), where x_i is the input vector and d_i is a given point (for two classes it reaches the value of 1 for class 1 or –1 for class 2). Assuming the linear separation of both classes, the equation dividing the two classes can be preserved in the form:

$$\begin{gathered} g(x) = w^T x + b = 0, \\ w = [w_1, w_2, \ldots, w_N]^T, \\ x = [x_1, x_2, \ldots, x_N]^T. \end{gathered} \tag{1}$$

In this equation, taking N inputs, the weight vector w is N-dimensional; weight b is polarization. Solutions of the equations that determine the class affiliation is:

$$\begin{gathered} w^T x_i + b > 0 ?\!\to d_i = \ , \\ w^T x_i + b < 0 ?\!\to d_i = - \ , \end{gathered} \tag{2}$$

or after the transformation:

$$\begin{gathered} w^T x_i + b \geq 1 ?\!\to d_i = \ , \\ w^T x_i + b \leq -1 ?\!\to d_i = - \ . \end{gathered} \tag{3}$$

If a pair of points (x_i, d_i) corresponds to the above equation with the equals sign, then the vector $x_i = x_{sv}$ creates the so-called support vector. They are those data points that are closest to the optimal hyperplane and are the most difficult to classify.

The problem of studying SVM linear networks, i.e. the choice of weights of synaptic connections for linearly separated training data, is to maximize the separation limit (fig. 3). This is a quadratic programming problem with linear weight constraints, which is solved by means of the Lagrange multiplier method by minimizing Lagrange function.

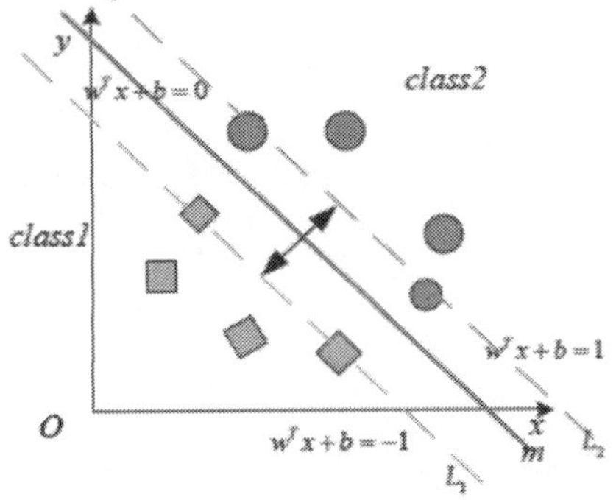

Fig. 3. Interpretation of the margin of separation between two data classes.

Taking into account that the task of training is to maximize the Lagrange function against Lagrange multipliers, the primary problem is converted into a double, which is formulated as follows:

$$\begin{gathered} max\left\{\sum_{i=1}^{p}\alpha_i - \frac{1}{2}\sum_{i=1}^{p}\sum_{j=1}^{p}\alpha_i\alpha_j d_i d_j x_i^T x_j\right\}, \\ \alpha_i \geq 0, \quad \sum_{i=1}^{p}\alpha_i d_i. \end{gathered} \tag{4}$$

The solution of the aforementioned optimization problem to Lagrange multipliers allows one to determine the equation of an optimal hyperplane, which is determined by the x-weight vector and the polarization parameter b.

When one is solving the problem of classification of indivisible samples, the problem is to determine the optimal hyperplane, which minimizes the probability of classification error on the training set with the maximum possible differentiation. As in the case of linearly separable regularities, the primary problem is reduced to a dual problem, which is formulated in the following way:

$$\begin{gathered} max\left\{\sum_{i=1}^{p}\alpha_i - \frac{1}{2}\sum_{i=1}^{p}\sum_{j=1}^{p}\alpha_i\alpha_j d_i d_j x_i^T x_j\right\}, \\ 0 \leq \alpha_i \leq C, \quad \sum_{i=1}^{p}\alpha_i d_i = 0, \end{gathered} \tag{5}$$

for i = 1,2, ..., p and the value of the constant C is accepted by user. Therefore, from the solution of the dual problem one can obtain the expression for weight vector of the optimal hyperplane in the form:

$$w_0 \sum_{i=1}^{N_{sv}} \alpha_i d_i x_i. \tag{6}$$

Summation is applying only to learning components for which the Lagrange multipliers are different from zero. They support vectors, the number of which is N_{sv}. Equation of the optimal hyperplane depends only on supporting vectors. Other vectors from the training set do not affect on the outcome of the decision. The equation defining the output *y(x)* of the SVM linear network with optimal weights is expressed by the following relation:

$$y(x) = ?\!w_0 x + b = \sum_{i=1}^{N_{sv}} \alpha_i d_i x_i^T x + b \tag{7}$$

This is a linear equation for the input variables described by the vector x and scales that depend on the Lagrangian non-zero multipliers, their corresponding N_{sv}, support vectors xi and given values d_i.

Linear non-vaporization of patterns does not mean that they are not separated at all. As a rule, if x is an input vector that describes a picture, then after resetting it into K-dimensional space, it is represented by a set of signs $\varphi_j(x)$ for j=1,2, ..., K. As a result of this transformation the hyperplane equation in a linear space is determined by the formula:

$$g(x) = w^T \varphi(x) + b = \sum_{j=1}^{N} w_j \varphi_j(x) + b = 0, \tag{8}$$

where w_j means the weights leading from $\varphi_j(x)$ to the output neuron, the weight vector w is K-dimensional and b is the weight of the polarization. Properties of the process described by the functions $\varphi_j(x)$ assumed the role of the individual variables x_j. The value of the output signal is described by the equation:

$$y(x) = w^T \varphi(x) + b. \tag{9}$$

The solution of primary problem is accomplished by transforming into a dual task, as well as in networks with linear separation of samples by minimizing the Lagrange function:

$$\begin{gathered} J(x, b, \varepsilon, \alpha, \mu) = \frac{1}{2} w^T w + \\ + C \sum_{i=1}^{p} \left[d_i (w^T \varphi(x_i) + b - 1 + \epsilon_i \right] \alpha_i = \sum_{i=1}^{p} \mu_i \epsilon_i . \end{gathered} \tag{10}$$

Solution of this optimization problem in the first step equals to assuming partial derivatives of the Lagrange function relative to w, b and ε to zero. The primary problem turns into a dual problem defined by the Lagrange multipliers α_i with the following form:

$$\begin{gathered} max \left\{ \sum_{i=1}^{p} \alpha_i - \frac{1}{2} \sum_{i=1}^{p} \sum_{j=1}^{p} \alpha_i \alpha_j d_i d_j K(x_i^T x_j) \right\}, \\ 0 \le \alpha_i \le C, \quad \sum_{i=1}^{p} \alpha_i d_i = 0. \end{gathered} \tag{11}$$

The function $K(x_i, x_j)$ in the formulated double problem is the scalar product of the vector function $\varphi(x) = [\varphi_1(x), \varphi_2(x), ..., \varphi_K(x)]$T:

$$K(x_i^T x_j) = \varphi^T(x_i)\varphi(x_j). \tag{12}$$

This product defines the so-called kernel function.

Finally, an expression which specifies the output of the nonlinear SVM network can be written as:

$$y(x) = w^T \varphi(x) + b = \sum_{i=1}^{N_{sv}} \alpha_i d_i K(x, x_i) + b, \tag{13}$$

which depends on the kernel function $K(x_i, x_j)$ and not from the activation function $\varphi(x)$.

Due to the classification problem, which requires a division of data into more classes, it is necessary to perform several classifications by using the methods "one against all" and "against each other". In the "one vs all" method in M classes are defined M SVM networks that determine exactly one class. The method requires preparation of M SVM, each of which learns on another set of data. After training all M networks there is a stage of reproduction, in which the same x vector is fed into each SVM network and the output signals (functions of M decisions) of all trained SVM are determined. The SVM network was chosen to assign model classification through generalizing abilities. The SVM network is somewhat sensitive to selected hyperparameters that determine the number of neurons in the hidden layer.

The presented algorithm for classification of waste in the optical separator, based on the SVM network in the classical version, allows one to split fractional components on the basis of certain classes of standards described by the vector of parameters characterizing spectral bands, texture, structure and form. Due to the need to increase the quality control indicator of technological process carried out in an optical separator in the context of obtaining products with precisely defined physico-chemical, combustible and emission properties, the SVM algorithm was modified by extending the library of templates of primary classes (determined on the basis of spectra, textures and geometric characteristics) library of certain class standards, which are described by physico-chemical and inflammable properties (heat, humidity, content of volatile parts).

The parent algorithm, that is greedy approximation algorithm which solves the problem of covering a plurality, is responsible for the formation of class templates corresponding to optimized non-interlocked subsets of a set whose components are components of the waste fractions directed to the optical separator.

Reference classes are described by a set of attributes (texture, structure, spectrum) that allows lower level classification algorithm to be classified using the SVM network model (MC SVM) based on a set of functions intended for predefined classes represented by disjoint subsets of coverage by using Greedy Set Cover algorithm.

The modification of algorithm for solving the problem of optimization consists in realization of a method which allows one to check that a certain component (on the basis of image analysis) belongs to a class standard determined on the basis of fuel and physico-chemical properties and meets the criterion of the formulated target function, that is:

$$\max_{u \in S} W_d^T u, \quad S = \left\{ u \in R^n : Au \le b_{max}, Au \ge b_{min}, u \ge 0 \right\}, \tag{14}$$

where W is the vector of the dimension $|j|$ (number of fractions components involved in the process of fuel component forming) with the components w_j, indicating the value of resulting fuel components heating. The process of optimization is reduced to the problem of linear programming (linear target function and linear constraints) with constraints imposed on the decision variables defined by the vector U, with the components u_j, which determines the set of admissible solutions:

$$U = (u_j) = \left[u_1, u_2, \ldots u_j \right]^T . \tag{15}$$

with dimension $|j|$ such u_j means the share of j-th fuel fraction.

The target function specified in the general form can be written as follows:

$$W_d(u) = w_{d_1}u_1 + w_{d_2}u_2 + \ldots + w_{d_j}u_j = \sum_{j=1}^{n} w_{d_j}u_j \Rightarrow \max, \quad (16)$$

while the system of boundary equations in the form:

$$B_{\min} \le AU \le B_{\max}, \quad B_{\min} = (b_k^{\min}) = \left[b_1^{\min}, b_2^{\min}, \ldots b_k^{\min}\right]^T, \quad (17)$$

means a one-dimensional matrix (vector of left boundary equations) of dimension $|k|$ such that components mean the minimal participation of b in fomation:

$$\left(b_k^{\min}\right) = \begin{cases} b_1^{\min} = \text{chlorine (Cl)}, \\ b_2^{\min} = \text{sulfur (S)}, \\ b_3^{\min} = \text{PCB}, \\ b_4^{\min} = \text{metal}, \\ b_5^{\min} = \text{mercury (Hg)}, \\ b_6^{\min} = \text{cadmium (Cd)}, \end{cases} \quad (18)$$

while:

$$B_{\max} = (b_k^{\max}) = \left[b_1^{\max}, b_2^{\max}, \ldots b_k^{\max}\right]^T, \quad (19)$$

means a one-dimensional matrix (vector of right sides of boundary equations) dimension $|k|$ such that components mean the maximum content of harmful substances. The values of the individual components of the vector are similar to $B_{\min} = (b_k^{\min})$.

4 Greedy Set Cover collection optimization problem solving method

The method parameter is a pair (X, F) consisting of a finite set X (a set of waste fraction components) and a family of F subsets X (corresponding to predefined class standards determined on the basis of fuel, physico-chemical properties and correspond to the criterion of the formulated target function, i.e. $\max_{u \in S} W_d^T u$, such that each element of the set X belongs to at least one member of the family F: $X = \bigcup_{S \in F} S$. In this case, the subset $S \in F$ covers its elements. The subfamily $\xi \subseteq F$, whose elements cover the whole set X: $X = \bigcup_{S \in \xi} S$, is the solution of the method.

The list below is a realization (in the pseudocode) of the Greedy Set Cover algorithm:

GreedySetCover(X, F):

1. $U \leftarrow X$;
2. $\xi \leftarrow \varnothing$;
3. **while** $U \neq \varnothing$:
4. **choose** $S \in F$ **which maximizes** $|S \cap U|$;
5. $U \leftarrow U - S$;
6. $\xi \leftarrow \xi \cup \{S\}$.

In each phase, U means a set of elements that are not known. The ξ contains built lid. Line 4 is the stage at which a greedy solution is made, i.e. subset S is selected. It covers as many elements as possible that are without covering (components of fractions). After selecting S, its elements are removed from U and S is added to ξ. When algorithm ends, the ξ is a subframe of F covering X.

5 Data model

A mathematical model for optimization of process of producing a light fraction (calorific value) consisting of separation (in an optical separator) of heavy non-metal components with high heating values Wd, from a fraction of waste flow is represented by a matrix (column vector). $Z=(z_j)$ corresponds to components that store indexes of the corresponding class templates generated by the greedy optimization algorithm. $P^k=(p_{i,j})$ with the components $p_{i,j}$ containing the i-th component (fraction) contained in the waste stream is sent to the optical separator into the reference class. The number of predefined classes is determined by the number found by the Greedy Set Cover algorithm, i.e. by greedy approximation of the given coverage with the optimal subsets P^k which correspond to the criterion (14).

The algorithm that exceeds SVM checks in each iteration whether a specific component (based on the spectroscopic spectrum analysis)/fraction belongs to a class standard determined on the basis of fuel and physico-chemical properties and meets the criterion of the formulated target function.

Summing up, the algorithm implemented in the monitoring layer of SCADA control layer must reproduce the structures of predefined class templates – Greedy Set Cover subsets containing the fraction components described by a set of attributes, including physical and chemical properties, fuel properties, Wd, texture, spectroscopic spectrum, structure. I.e., each subgroup representing the class of fractional components that generates fuel is formed by the effect of optimization of greedy set covering algorithm in accordance with criteria for Wd and constraints on optimization problem solution variables [12-14].

6 Component-oriented model

In connection with the necessity of the managerial decisions sequence making problem solving (generating optimal trajectories of control decisions for the actuators of the optical separator), a polymorphic component model of a class was presented. It represents the algorithm of MS SVD, taking into account changes in its definition, with a logical representation of knowledge about an object for which the learning process is based on consistent verification, updating of knowledge and use of the results of this update to verify and reconstruct data structures.

The class type *TMultiClassSupportVectorMachine* contains the definition of a field with a logical representation of knowledge about management of technological process and methods of implementing algorithms that implement the tasks of specialization and adaptation of class component, in response to changing the state of equipment and technological parameters. Specialization follows the learning process in space of states and events by validating, reconstructing object data structures and reconfiguring control algae.

The component-oriented model of heavy screens separation in the optical separator is computer implementation of mathematical models (sets of data and algebraic expressions) describing individual components (objects, processes, relationships) [9-11]. The essence of the model definition is the use of special types of classes, structure of which allows one to display the specific properties of object components and to perform operations on data structures in the form of fields and methods. Object fields have another important advantage that allows one (via appropriate assignments) to refer to components created in other programs, such as Matlab, which run in different address spaces and provide a number of component categories stored in executable libraries (DLLs or EXE). They can be used to develop their own programs, directly using predefined interfaces or by defining child objects. The purpose of these links is to use the properties and methods (components of objects of predefined libraries).

The MC SVM algorithm "one vs all" can be written as follows:

Input: category N, $D=\{D_1,D_2, ..., D_N\}$ for training samples; testing sample T;
Output: Categories T;
Algorithm:

1. // *educational section*
2. **for** $n = 1$–N:
   a. positive Sample ← D_N,
      negative Sample ← other samples other than D_N;
   b. save the data SVM_n classifier;
3. //*test section*
4. **for** $n = 1$–N:
   a. use the M SVM_n classifier to calculate the value of $f(x)_n$;
5. compare all $f(x)_n$, output a maximum of $f(x)_n$.

7 Results and discussion

In order to verify numerical model and to validate its computer code, sample problems were simulated and the results were compared with the analytical solutions.
Validation of the model wass carried out with the following assumptions:

1. For the fuel formation process two components are used: dark coal P1 [kg i/kg] and combustible municipal waste P2 [kg i/kg] of the following chemical composition (c, s, h, o, n, w, p, cl):
P1: (0,610; 0,003; 0,041; 0,0897; 0,0001; 0,0714; 0,1842; 0,0074),
P2: (0,5001; 0,0092; 0,0676; 0,2186; 0,1358; 0,02; 0,0471; 0,0016)

2. Optimisation constraints imposed on fuel formation process are used as given following:
$s<0,01$ and $cl<0,003$

As a result of computer simulations and laboratory experiments the model gives the results that are very close to the obtained from the real process. Calculated Root Mean Square Error (RMSE) was as follows:
a. for mass fraction of chlorine in formed fuel : *RMSE=0,00019 (kg Cl/kg)*
b. for mass fraction of sulphur in formed fuel : *RMSE=0,00038 (kg S/kg)*
c. for *(Wd)* of formed fuel : *RMSE=47,81 (kJ/kg)*

Application of the advanced algorithm for fuel formation process allowed the identification of the optimal mass fractions of components P1 and P2 in formed fuel (within the range of 0,01 - 0,24 kg i/kg) and (within the range of: 0,76 - 0,99 kg i/kg) with the constraints on the decision variables. Optimum value of the objective function (*Wd*) was found within the range of 22,86 to 25,12 MJ/kg. The acceptable agreement between our simulation results and the measured data demonstrates that the model can reasonably simulate the real process of fuel production from separate combustible fractions of municipal wastes. The algorithm has been implemented in the SCADA system, in one of the companies producing fuels from waste. Energy savings of 15% per year have been achieved.

8 Conclusions

The integrated multithreaded process control algorithm, embedded in a programmable logical PLC controller (controlled by the SCADA system), performs the task of solving the optimization problem defined in the form of neural network model (Support Vector Machine) of non-metallic fraction components, which reproduces the structures of predefined classes, i.e. subsets generated by the greedy algorithm of the Greedy Set Cover approximation containing fraction components described by a set of attributes (including physical and chemical properties, fuel properties, emissions, texture, structure and spectra).

Each subset, representing the model class of component fuel fractions, is generated by using the optimization effect of greedy algorithm of the Greedy Set Cover approximation, maximizing the purpose of the target (Wd-calorific value) when performing the technological constraints imposed on the solution of variable tasks linear programming. Reference classes are described by a set of attributes (texture, structure and spectrum) that allow the classification algorithm of lower level classification to be classified using the SVM network model (MC SVM) based on a set of functions intended for predefined classes represented by disjoint subsets of the Greedy Set Cover algorithm coverage and finally verify that the identified component/fraction belongs to the matrix of the class on the incident matrix.

References

1. S. Amponsah, S. Salhi, The investigation of a class of capacitated arc routing problems: the collection of garbage in developing countries, *Waste Management*, 24:711-721 (2004)
2. O. Apaydin, M. Gonullu, Route optimization for solid waste collection: *Trabzon (Turkey) case study. Global NEST Journal*, 9(1):6-11 (2007)
3. K. Barlishen, B. Baetz, Development of a Decision Support System for Municipal Solid Waste Management Systems Planning, *Waste Management and Research*, 14:71-86 (1996)
4. K. Gaska, A. Generowicz, I. Zimoch, J. Ciuła, Z. Iwanicka, A high-performance computing (HPC) based integrated multithreaded model predictive control (MPC) for water supply networks. *Architecture Civil Engineering Environment*, 10 (4):141-151 (2017)
5. K. Gaska, A. Generowicz, I. Zimoch, J. Ciuła, D. Siedlarz, A GIS based graph oriented algorithmic model for poly-optimization of waste management system. *Architecture Civil Engineering Environment*, 4:152-159 (2018)
6. K. Gaska, A. Generowicz, J. Ciuła, A HPC (High Performance Computing) based model for optimisation of the waste management systems in metropolitan regions. *Technical Transactions* 3:153-163 (2018)
7. K. Gaska, K. Pikon, Methodology of biomass fuels creation processes; *Twenty-sixth Annual International Conference on Incineration and Thermal Treatment Technologies - IT3'07 Conference*, May 14-18, 2007, Phoenix, USA (ISBN: 9780923204822)
8. D. Kowalski, B. Kowalska, T. Bławucki, P. Suchorab, K. Gaska, Impact Assessment of Distribution Network Layout on the Reliability of Water Delivery. *Water*, 11:480 (2019), doi:10.3390/w11030480
9. A. Generowicz, K. Gaska, G. Hajduga, Multi-criteria Analysis of the Waste Management System in a Metropolitan Area, *E3S Web of Conferences, 10th Conference on Interdisciplinary Problems in Environmental Protection and Engineering*. 2018, doi: 10.1051/e3sconf/20184400043
10. J. Ciuła, K. Gaska, A. Generowicz, G. Hajduga, Energy from landfill gas as an example of circular economy, *E3S Web of Conferences, Vol. 30, The First Conference of the International Water Association IWA for Young Scientist in Poland "Water, Wastewater and Energy in Smart Cities"*, Cracow, Poland, 2018, doi: 10.1051/e3sconf/20183003002
11. K. Gaska, A. Generowicz, Advanced computational methods in component-oriented modelling of municipal solid waste incineration processes, *Architecture Civil Engineering Environment*, 10(1):117-130 (2017)
12. S. Werle, Impact of feedstock properties and operating conditions on sewage sludge gasification in a fixed bed gasifier, *WASTE MANAGEMENT & RESEARCH*, 32(10):954-960 (2014), doi: 10.1177/0734242X14535654
13. M. Smol, J. Kulczycka, A. Henclik, K. Gorazda, Z. Wzorek, The possible use of sewage sludge ash (SSA) in the construction industry as a way towards a circular economy. *Journal of Cleaner Production*, 95:45-54 (2015)
14. M. Smol, J. Kulczycka, Z. Kowlaski, Sewage sludge ash (SSA) from small and large incineration plants as a potential source of phosphorus – Polish case of study, *Journal of Environmental Management*, 184(3):617–628 (2016)

Pyrolysis oil production from polypropylene plastic waste using molybdenum modified alumina-silica catalysts

Sasiradee Jantasee[1,*], *Natacha* Phetyim[1], *Komm* Petchinthorn[2], *Tunyahpat* Thanupongmanee[1], and *Nuntiporn* Sripirom[1]

[1]Chemical and Materials Engineering Department, Faculty of Engineering, Rajamangala University of Technology Thanyaburi, 12110 Pathum Thani, Thailand
[2]International College, Rajamangala University of Technology Krungthep, 10120 Bangkok, Thailand

Abstract. The production of pyrolysis oil from polypropylene plastic waste was examined over molybdenum modified alumina-silica catalysts (Mo/Al-Si). The reactions were carried out with 1 L of batch reactor under atmospheric pressure at 430 °C. The pyrolysis oil yield was in the order, 10% Mo/Al-Si > 5% Mo/Al-Si > the absence of catalyst. The 10% Mo/Al-Si was highest activity due to the stronger acidity facilitating the pyrolysis reaction. It accelerated the reaction to produce the pyrolysis oil at lower temperature. Comparison of the pyrolysis oil properties to the standards of the diesel fuel from Thai Department of Energy Business shows that the color and the distillation were within standards. Moreover, the results reveal that the kind of raw material affected the product yield of pyrolysis process.

1 Introduction

Recently, the economic and industrial advance in Thailand have substantially progressed and increased in existence of continuous improvement. With the larger amount of populations, this eventually results in the higher demands of using more resources.

In addition, Thailand has also found the problems of more garbage. Statistics from the Pollution Control Department (PCD) found that in 2017, Thailand has total garbage of 26.6 million tons per year with the average 73,560 tons per day, but only able to eliminate 31.1 % or 8.3 million tons per year. 4.9 million tons were able to recycle and 7.2 million tons were eliminated incorrectly leaving 6.2 million tons of waste remain in the area each year [1]. By moving to landfill and burning them in the hygienic furnace, these two methods can cause both soil and air pollution as a broad environmental problem [2]. Hence, it becomes a government policy to reduce the amount of accumulated waste in local authorities.

The waste to energy technology is investigated to process the potential materials in waste form such as plastic, biomass and rubber tire in an attempt to create the diesel oil for certain purposes [3-5]. Pyrolysis process is one of the options of waste to energy technology to deliver pyrolysis oil to replace fossil fuel [6,7]. Its ability to handle unsort and dirty plastic is the main advantage of the pyrolysis process. Furthermore, the process itself does not create toxic or environmental harmful emission unlike incineration.

The products from pyrolysis process have 3 states namely liquid (pyrolysis oil), gas and solid waste [8]. This method will help to decrease the amount of accumulated plastic waste. In fact, it will help to increase the value of the unwanted waste.

This research interested in developing the alumina-silica catalyst that was modified with molybdenum to increase the effectiveness in pyrolysis process of polypropylene plastic waste, in order to produce diesel fuel with equivalent standards.

2 Experimental

2.1 Synthesis of alumina-silica catalyst

Desired amount of ammonia was added into the deionization water. This solution was mixed with cetyltrimethyl ammonia bromide, and stirred for 30 minutes. Then, 6.3 g of tetraethyl orthosilicate was added into the mixture and continuously stirred for 1 h. After that the solid was filtered, dried at 120 °C for 12 h, and calcined at 700 °C for 6 h [9].

2.2 Preparation of molybdenum modified alumina-silica catalyst

Molybdenum modified alumina-silica catalysts were prepared by an incipient wetness impregnation method. First, ammonium heptamolybdate tetrahydrate at concentration of 5% and 10% by weight of molybdenum was dissolved in deionization water. The solution was slowly dropped on the alumina-silica and mixed together. Then, the resulting catalysts were dried at 110 °C for 12 h, and calcined at 700 °C for 6 h.

2.3 Pyrolysis test and product analysis

* Corresponding author: sasiradee.j@en.rmutt.ac.th

 https://doi.org/10.1051/e3sconf/201912201005

The polypropylene plastic waste was collected from the strapping band. 300 g of raw material was cut into a particle size of 5-7 mm, and added to 1 L of glass reactor with or without 0.5 g of the catalysts. The pyrolysis was carried out under atmosphere pressure. The temperature was raised from room temperature to 430 °C. The reaction time was 3 h after occurring the first drop of pyrolysis oil. The pyrolysis oil and solid waste were weighted for yield percentage calculation. For comparison the pyrolysis oil properties to the diesel fuel standard properties announced by Thai Department of Energy Business, the properties of pyrolysis oil including flash point, viscosity, specific gravity, color value and distillation were tested according to the standards as shown in Table 1.

Table 1. American Society for Testing and Materials (ASTM) standards for testing of fuel oil properties [10].

Qualification test	Standard
Flash point	ASTM D93
Viscosity at 40 °C	ASTM D445
Color	ASTM D1500-96
Specific gravity	ASTM D1298
Distillation temperature at 90% recovery	ASTM D86
Cetane index	ASTM D976-06

3 Results and discussion

3.1 Effects of molybdenum content on pyrolysis of polypropylene plastic waste

From Table 2, the products from pyrolysis test have 3 states, liquid, solid and gas. The pyrolysis process of polypropylene plastic waste that did not use catalyst and used catalyst at 5% and 10% molybdenum modified alumina-silica catalysts were investigated. The results show percentage of pyrolysis oil were 81.67%, 83.65% and 86.63% respectively. For the solid waste had percentage yield between 1.38 to 6.31% by weight, the system that did not use catalyst had the most solid waste of polypropylene plastic as seen in Figure 1.

Fig. 1. Solid from the pyrolysis of polypropylene plastic waste without catalyst.

When adding the molybdenum modified alumina-silica catalysts, it resulted in a decreasing in the amount of solid waste. The filling of molybdenum modified alumina-silica catalyst helps to increase a quantity of pyrolysis oil. The percentage of molybdenum modified alumina-silica catalyst by 10% presented higher percentage of pyrolysis oil than the 5% molybdenum on alumina-silica catalyst because the more molybdenum added possibly can help to increase the acidity of the catalyst [9]. The pyrolysis process requires a surface catalyst with high acidity properties. Therefore, it is suitable to use this catalyst in the experiment. From the experiment of difference in the use of molybdenum modified alumina-silica catalysts and pyrolysis without catalyst found that the durations of the first drop of oil were different. By increasing the percentage of molybdenum modified alumina-silica catalyst, the result shows the first drop of oil occurring faster than not using catalyst.

Table 2. The pyrolysis of polypropylene plastic waste results.

Catalysts	Yield (%)			Time for the first oil drop (min)
	Gas	Pyrolysis oil	Solid	
No catalyst	12.02	81.67	6.31	42
5% Mo/Al-Si	13.79	83.65	2.56	31
10% Mo/Al-Si	11.99	86.63	1.38	28

The results of properties test of pyrolysis oil from polypropylene plastic waste are on the Table 3. It shows that all pyrolysis oil had the color value that met the standard of diesel fuel, less than 7.5. The color of all pyrolysis oils are presented in Figures 2 to 4.

Fig. 2. Liquid from the pyrolysis of polypropylene plastic waste without catalyst.

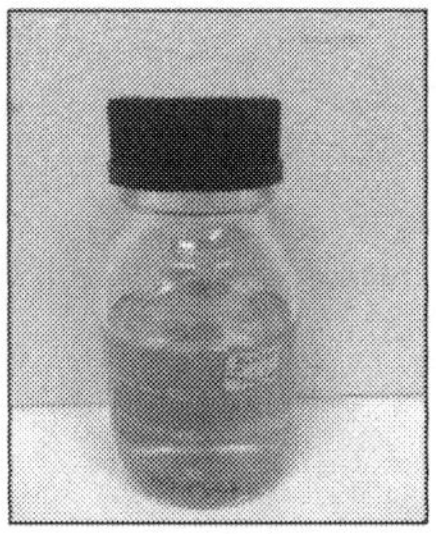

Fig. 3. Liquid from the pyrolysis of polypropylene plastic waste with 5% molybdenum modified alumina-silica catalyst.

Fig. 4. Liquid from the pyrolysis of polypropylene plastic waste with 10% molybdenum modified alumina-silica catalyst.

As for specific gravity that nearly reached the standard criteria of diesel oil which had specific gravity of 0.78 - 0.79. As for their flash points did not pass the standard criteria of diesel fuel, it depends on the composition of plastic type. Low flash point implied that the pyrolysis oils contained highly volatile materials [11]. Flash point can be improved by removing lighter components from the pyrolysis oils. Simultaneously, the viscosity values of all pyrolysis oils were still lower than the standard, and the distillation value at 90% recovery passed all of the diesel standard. The cetane index was calculated from the distillation temperature at 50% recovery and density. The use of catalysts resulted in the decease of cetane index.

Table 3. The properties of pyrolysis oil from polypropylene plastic waste.

	No catalyst	5% Mo/Al-Si	10% Mo/Al-Si	Diesel standard
Color	0.5	0.5	0.5	<7.5
Specific gravity	0.78	0.79	0.79	0.81-0.87
Flash point (°C)	27	28	29	>52
Viscosity (cSt)	1.49	1.43	1.47	1.80-4.10
Distillation (°C)	285.8	251.1	279.1	<357
Cetane index	60.0	48.0	42.5	>50

For the carbon composition analysis of the pyrolysis oil by Gas Chromatography-Mass Spectrometry (GC-MS) can separate the chemical structures in the oil into 6 groups such as straight chain paraffin, branched chain paraffin, double bonding compounds, triple bonding compounds, alicyclic compounds and aromatic compounds. All pyrolysis oils mainly consisted of double bonding compounds and alicyclic compounds, these were 40% and 50% respectively. The pyrolysis of polypropylene plastic waste without catalyst produced the pyrolysis oils containing a complex mixture of C_6 – C_{25} organic compounds. For the pyrolysis oils produced from polypropylene plastic waste with the 5% and 10% molybdenum modified alumina-silica catalysts consisted of a complex mixture of C_5 – C_{15} organic compounds.

3.2 Effects of types of raw materials on pyrolysis with 10% molybdenum modified alumina-silica catalyst

From the pyrolysis of polypropylene plastic waste, the alumina-silica with 10% molybdenum catalyst had the greatest pyrolysis oil yield. Therefore, in this part the pyrolysis of polyethylene, polystyrene plastic waste, and used engine oil with 10% molybdenum modified alumina-silica catalyst were conducted at the same conditions with the previous part in order to investigate the effects of different raw materials on pyrolysis. A drinking water bottle cap was used as a high density polyethylene plastic waste. The polystyrene was collected from a foam food container. The used engine oil (API SN 0W20) was supported by the Thanyaburi Honda Cars Center. The pyrolysis oil created from the different raw materials had the color as illustrated in Figure 5 to 8.

Fig. 5. Liquid from the pyrolysis of polyethylene plastic waste.

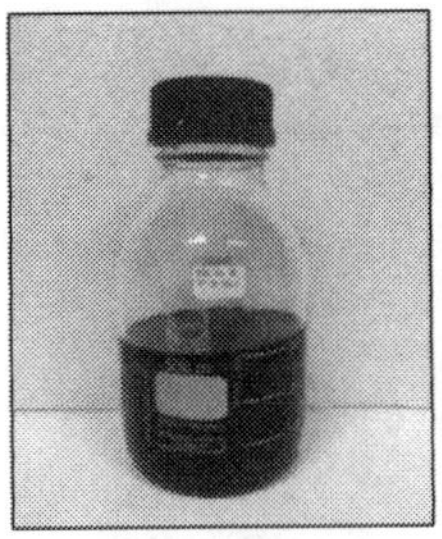

Fig. 6. Liquid from the pyrolysis of polystyrene plastic waste.

Fig. 7. Liquid from the pyrolysis of used engine oil plastic waste.

Fig. 8. Liquid from the pyrolysis of mixed polypropylene, polyethylene, polystyrene, and used engine oil.

The products yields are presented in Table 4. The pyrolysis oil obtained from polyethylene, polystyrene and used engine oil were 75.71%, 93.13%, and 65.72% respectively. The result reveals that the different of raw material structure resulted in the pyrolysis product yield [12]. The used engine oil had the lowest pyrolysis oil yield. However, when used engine oil was mixed with polypropylene, polyethylene, and polystyrene plastic waste, the pyrolysis oil yield was increased.

Table 4. The pyrolysis of various raw materials results.

Raw materials	Yield (%)		
	Gas	**Pyrolysis oil**	**Solid**
Polyethylene	15.58	75.71	8.71
Polystyrene	1.19	93.13	5.68
Used engine oil	8.14	65.72	26.14
Mixed PP : PE : PS : Oil (1:1:1:1)	16.98	78.67	4.35

The results of pyrolysis oil properties analysis are shown in Table 5. The color of all substrates met the diesel fuel standard, which were not higher than 7.5. The used engine oil produced the pyrolysis oil having the highest color value up to 3.5. For specific gravity, the pyrolysis oil from the used engine oil and the mixed raw materials passed the standard criteria of diesel fuel. Only the used engine oil produced the pyrolysis oil passing the viscosity value standard of the diesel fuel. The distillation value at 90% passed the diesel standard excluding the oil from the polystyrene and the mixed substrates. The foaming occurred while the distillation value measurement of the pyrolysis produced from the foam food container, thus the value could not be determined.

Table 5. The properties of pyrolysis oil from the pyrolysis of different raw materials.

	Poly ethylene	**Poly styrene**	**Used engine oil**	**Mixed**
Color	1	2.5	3.5	2
Specific gravity	0.79	0.94	0.81	0.83
Flash point (°C)	35	40	31	28
Viscosity (cSt)	1.68	1.78	2.63	1.45
Distillation (°C)	277.5	-*	318.1	-*
Cetane index	59.5	-*	65.5	-*

*Distillation could not be measured, so cetane number were not able to be calculated.

From GC-MS analysis, the main hydrocarbon compound of the pyrolysis oil from the polyethylene and used engine oil was the straight chain paraffin approximately 60%. Other components in the oil from the polyethylene and the used engine oil were the double bonding compounds and the alicyclic compounds. Whereas the polystyrene plastic waste provided the oil containing only the aromatic compounds owing to the styrene ring in the polystyrene. These pyrolysis conditions cannot crack the aromatic structure of raw material. For the co-pyrolysis between the used engine oil and the plastic waste namely polypropylene, polyethylene, and polystyrene, it produced the oil consisting of 6 groups of hydrocarbon compounds that are straight chain paraffin, branched chain paraffin, double bonding compounds, triple bonding compounds, alicyclic compounds and aromatic compounds.

4 Conclusion

The pyrolysis process of polypropylene plastic waste with and without the molybdenum modified alumina-silica catalysts produced over 80% yield of pyrolysis oil. The increasing molybdenum content resulted in the enhancing amount of pyrolysis oil. The 10% molybdenum on alumina-silica catalyst presented the highest yield of pyrolysis oil at 86.63%. The carbon compounds in the pyrolysis oil were mostly double bonding compounds and alicyclic compounds. The use of molybdenum modified alumina-silica catalyst can decompose the polypropylene plastic waste structure to shorter chain carbon compound comparing to the pyrolysis process without the catalyst. The product yields were different when the different sorts of raw materials were applied. Polystyrene presented the highest pyrolysis oil yield, but the oil properties did not qualify the diesel standard. The used engine oil provided the pyrolysis oil qualifying the standard of color, specific gravity, viscosity, distillation, and cetane index despite the lowest pyrolysis oil yield.

The authors would like to acknowledge the Rajamangala University of Technology Thanyaburi (RMUTT) annual government statement of expenditure in 2018 for the financial support. The Energy from Waste for Sustainable Environment Research group at Chemical and Materials Engineering department, RMUTT is also grateful for the instrumental and laboratory supports.

References

1. S. Sukganta, P. Srichamni, P. Fuengauksorn, L. Komkoontod, W. Kullasiri, P. Kreablumyai, J. Mookprom, S. Niyomthong, *Municipal solid waste situation in Thailand in 2017* (Polution Control Department, Thailand, 2018)
2. K. Tanakwang, N. Tangtinchai, *Overview on Municipal solid waste management in Thailand* (Polution Control Department, Thailand, 2013)
3. I. Macro, B.M. Cabellero, A. Lopez, M.F. Laresgoiti, A. Torres, M.J. Chomon, J. Anal. Appl. Pyrol. **85**, 384-391 (2009)
4. T. Kan, V. Strezov, T.J. Evans, Renew. Sust. Energ. Rev. **57**, 1126-1140 (2016)
5. M.S. Hossain, M.R. Islam, M.S. Rahman, M.A. Kader, H. Haniu, Energy. Proced. **110**, 453-458 (2017)
6. F. Abnisa, W. Daud, Energy. Convers. Manag. **87**, 71-85 (2014)
7. P. Pan, C. Hu, W. Yang, Y. Li, L. Dong, L. Zhu, D. Tong, R. Qing, Y. Fan, Bioresource. Technol. **101**, 4593-4599 (2010)
8. R. Miandad, M. Rehan, Abdul-Sattar, Nizami, M.A. Barakat, I.M. Ismail, *Recycling of solid waste for biofuels and bio-chemicals* (Springer Science and Business Media, Singapore, 2016)
9. T. Chanchuey, C. Autthanit, B. Jongsomjit, J. Chem-ny. **2016**, 1-8 (2016)
10. N. Tungsupanich, *Characteristics and quality of diesel fuel* (Thai Department of Energy Business, Thailand, 2015)
11. M.Z.H. Khan, M. Sultana, M.R. Al-Mamun, M.R. Hasan, J. Environ. Public. Health. **2016**, 1-6 (2016)
12. N. Phetyim, S. Pivsa-Art, Energies **11**, 1-11 (2018).

https://doi.org/10.1051/e3sconf/201912201006

Dew, Fog and Rain Collector in a Hyper-arid Climate: Case Study in Abu Dhabi

Banu Sizirici[1,*]

[1]Khalifa University of Science and Technology, Civil, Infrastructure, and Environmental Engineering Department, Abu Dhabi, P.O Box: 127788, United Arab Emirates

Abstract. Dew condensers collect dew via thermal radiation and cooling affect without the use of external energy. The aim of this project was to design the 1m x 1m with 30° angle square funnel dew, fog and rainwater collector to calculate the dew yield in Abu Dhabi urban area. In addition, studies were conducted to determine whether the collected dew quantity is significant for practical use and to verify the cooling effect in condenser surface using temperature sensors. Aluminum was chosen as dew collector material due to its high emissivity. Square funnel was chosen as condenser shape due to easiness of the assembly and feasibility of its design. The quantity of the collected water depended on the weather condition. The average collected water was 0.016 ml/day (0.016 mm/m^2/day). Temperature sensor analysis showed that there was a 3.5 °C temperature difference between condenser top and bottom indicating the cooling effect of square funnel shape.

1 Introduction

The increases in industrialization and population put heavy pressure on water supplies in our planet. In addition, there is a high demand on the energy-efficient and environmentally-friendly methods to obtain water for house-hold and agriculture purposes. Since the UAE does not have any surface waters, the main sources of water are seawater and groundwater. However, groundwater is of poor quality, and the desalination of the sea water is quite expensive. Therefore, alternative water production systems are in demand. Dew has been proposed as an alternative water resource in arid to sub-tropical regions. It helps plants and animals thrive in arid and semi-arid climates. Dew is atmospheric water vapor which condenses on surfaces cooler than dew point temperature (the temperature at which water droplets form). Radiation (the loss of heat to the sky) leads to cooling and this process is the main contributor to the formation of dew droplets [1]. Moreever, meteorological conditions such as ambient air temperature, relative humidity, wind speed, cloud cover, and sky radiation also play significant roles in condensation process. The dew formation is notable during nighttime with a clear sky due to high radiant heat loss from the exposed surface. When the warmer moist air pass over the cool surface, the air is cooled and the water vapor condense into droplets on the surfaces [2]. Dew collectors can collect not only dew but also rain water. There are two types of dew water collectors: radiative (passive) and active dew water condensers. Radiative condensers rely on collecting dew water via radiative exchange with sky which results zero energy input [2]. Dew collectors have been placed upon the sloped roofs of the buildings, terraces, and directly on the ground[3-6]. The maximum theoretical dew condensation is limited to 0.8 mm (i.e. 0.8 L.m^{-2}) per night [7]. According to the available radiative energy for condensation, the maximum dew water yield is about 0.3–0.6 L/day/m^2 of surface area in arid and semi-arid climates [8-10]. The maximum dew yield measured was almost 0.6 L/m^2/day [11]. Dew collectors in humid climates showed lower dew water yields. The collector in a grassland area in the Netherland collected 0.19 L/day/m^2 water [12]. Dew collector in humid tropical island in French Polynesia collected 0.23 L/day/m^2 water [13]. Another collector in an agricultural area close to an urban site in SainteAnne-de-Bellevue, QC, Canada collected 0.37 L/day/m^2 water [2]. Plain, uninsulated, and corrugated galvanized iron roofs yielded 113.5 L water (6.3 mm) in Kothara, India [4]. A dew harvesting system collected 0.28 L /m^2/day of dew compared to 0.17 L/ m^2/day of rainfall in Morocco [5]. In this project, we designed 1m^2 planar dew, fog and rainwater collector to calculate the dew water yield and to determine the cooling effect in dew collector using temperature sensors. The UAE's climate is hot and hyper arid, and humidity can reach up to 90% with low wind speed which are in favor of dew condensation. Square funnel shape was chosen as dew collectors' shape. The quantity and frequency of collected dew water were measured to determine if the quantity is significant for practical use.

2 Materials and Methods

* Corresponding author: banu.yildiz@ku.ac.ae

2.1 Weather and Geography

The city of Abu Dhabi is located on the Arabian Peninsula between latitudes 24° and 26° North and between 54° and 38° East with 7 m of altitude. The dew collector prototypes were built and tested in Khalifa University, Abu Dhabi Campus (24°26′52"N and 54°23′41"E), which is 15 km away from the sea. The location was an urban area and dew collector was set on the stone pavement. Abu Dhabi is considered to have a desert climate. Köppen and Geiger classified this climate as hot dessert climate (BWh: B: dry climate, BW: arid climate, h: warm) [14]. The average annual temperature is 26.8 °C, and the average annual rainfall is 75 mm in Abu Dhabi. The greatest amount of precipitation occurs in February with an average of 29 mm. There are two main seasons: winter lasts from November to March (average 26°C during the daytime and 15°C during the nighttime), and summer lasts from April to September. Summers are very hot, and temperatures can rise to about 48°C with humidity levels reaching as high as 90% as shown in Table 1. The annual average pan evaporation is around 7.5-8 mm/day which makes the area hyper arid [15, 16].

Meteorological data such as ambient temperature, relative humidity, wind speed, wind direction, and cloud cover were obtained from a nearby weather station located 7 km northeast of the campus to assess the influence of local meteorological conditions on dew and fog formation as shown in Table 1. In addition, humidity and temperature sensors connected with ARDUINO with data logger were placed on top of the dew collector to collect the real time ambient temperature and humidity data with the ±5% accuracy.

Table 1. Weather data in Abu Dhabi in 2018

Month	Av. Temperature °C	Av. Humidity %	Av. Wind km/h	Av. Rainfall mm
January	20	66	13.8	9.1
February	21	66	15	4.2
March	23	64	14.7	8.8
April	27.5	59	13.5	6.5
May	31.5	57	13	0
June	33	62	13.7	0
July	35.1	61	13.7	0
August	35.6	60	13.9	0
September	33.8	63	13.3	24.8
October	30.9	63	12.8	22.8
November	26.7	61	13.5	23.7
December	22.5	66	13.6	23.4

2.2 Condenser surface design

Square funnel design (1m x 1m) with 30 degree angle was chosen due to easiness of the assembly, feasibility of its design and commonality. The square funnel (inverted pyramid) shape of the dew collector and its effect on water collection capacity have been studied in different studies. A square funnel dew collector with an angle of 30^0 in a grassland area in the Netherland collected 20% more water than a standard 1 m^2 planar dew collector [12]. Another study showed that a funnel-shaped condenser with a half-angle of 30° showed 40% more water than 1 m^2 and 30° inclined planar condenser [17, 18]. The shape of funnel reduced the flow of warm air and held the heavier cold air at the bottom. Also the square funnel design protected condenser from the wind and shielded from infrared emission of the lower layers of the atmosphere [12, 13]. Another study stated that the ''best'' tilt angle for the passive dew drop recovery was 30^0 due to the gravity flow of the dewdrops[17, 19].

2.3 Material Selection for the Condenser Surface

Aluminum 1060 (aluminum content of 99.1%) was used as condenser material which has high corrosion resistance and workability [20]. Aluminum has been selected because it is cheap, strong (density: 2.7 g/cm3), high in emissivity (0.03-0.77 depend on surface) and thermal conductivity (237 kW/(m. K) @ 300K); it can be folded and machined easily: it is readily available in the local market. Square funnel was chosen as dew condenser shape. Underside of the aluminum condenser was insulated with the white coating to prevent heat gain from the ground below as shown in Figure 1. Another study tested three materials (aluminum, glass and polyethylene) for dew condenser surface, and found that aluminum had the highest dew yield because its polished surface enables dew water easily slide along the surface [21]. Dew water was collected in the plastic bottles connected to the tip of the condenser, and the volumes were measured each morning at 08:00 a.m at the local time.

Figure 1. Dew, fog and rain collector

2.4 Condenser surface temperature

Cooling effect of the condenser surface were analyzed using temperature sensors. 16 DHT11 temperature and humidity sensors (4 sensors for each side) were placed at the inner side of the condenser surfaces with different elevations as shown in Figure 2. Sensor 1 was placed at the top of the the condenser and sensor 4 was placed at the bottom/tip of the condenser. DHT11 is a basic temperature and humidity sensor which uses capacitive humidity sensor and a thermistor to measure nearby air. The sensor sends digital signal on the data pin (1s and 0s) and it collects new data every 2 seconds [22]. DHT11

https://doi.org/10.1051/e3sconf/201912201006

sensors were programmed to collect daily data. Arduino compatible ESP8266 based board with a micro-controller (Wi-Fi module) was used as an access point to allow receiving and sending data instantaneously (receive from: DHT11, send to: ThingSpeak.com). Wi-Fi module sent the data to the website Thingspeak to be viewed online. Thingspeak.com is an "open data platform that allows its users to easily document and share the data collected" [23]. The data was collected with 10-minute intervals from 7.00 pm to 7.00 am twice a week from September 1, 2017 to May 1, 2018.

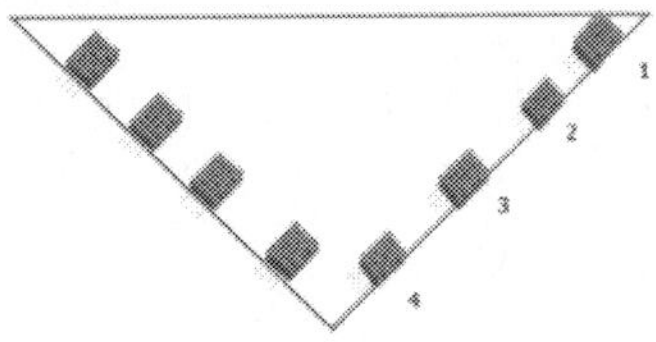

Figure 2. Positions of the sensors on the condenser surface

3 Results and Discussion

3.1 Temperature analysis

The weekly average temperature obtained from the sensors showed that there were 0.98 °C to 3.3 °C temperature differences between the top (sensor 1) and the bottom of the condenser (sensor 4), and 0.2 °C to 1.4 °C temperature difference between top of the condenser and ambient temperature as shown in Figure 3.

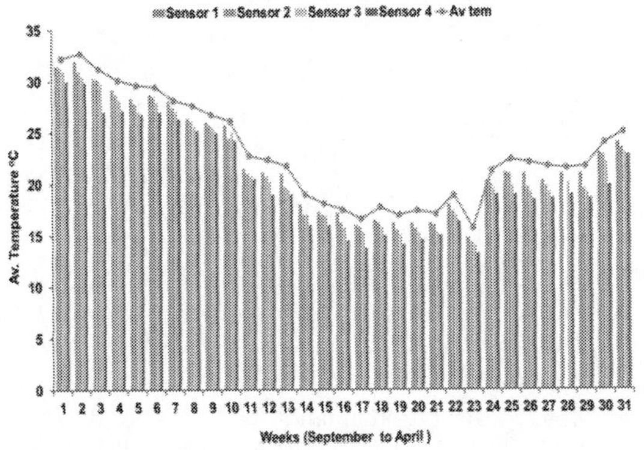

Figure 3. Weekly average measured temperatures by the sensors together with the ambient temperatures.

Hourly data revealed that cooler temperature values were observed at sensor 4. The maximum temperature differences were observed between 1.00 pm and 5.00 pm as shown in Figure 4.

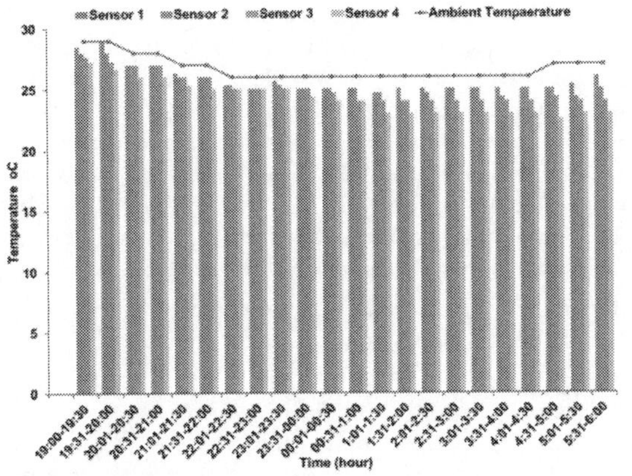

Figure 4. Hourly temperature sensor data together with ambient temperature recorded for April26, 2018

According to a study, plain, uninsulated, corrugated, and galvanized iron roofs yielded maximum 2 °C of cooling temperature, while a condenser that was thermally insulated had maximum 3.4–3.7 °C of cooling temperature in North West India [2]. Another study reported that on the nights when large amounts of dew condensed, the condenser surface would cool below the ambient temperature by 4 °C or more [10] . Our results confirm the cooling effect of the square funnel. In addition, it is in parallel with the results of computational fluid dynamics simulations performed in other studies. According to these simulations air gets cooler inside the funnel and the cooler air slips to the center of funnel. [17, 24].

3.2 Water collection

The analysis of the data collected during a one-year study showed that the dew occurred from September to April. Dew, rain and fog events occurred approximately 100 days in 2018, but water collection occurred only 51 days because not all droplets formed on the condenser slid to the collection container. Similar results were reported in other studies as well [8, 10, 11, 25]. The dew events were observed when the relative humidity was at least 80% or more as shown in Figure 5.

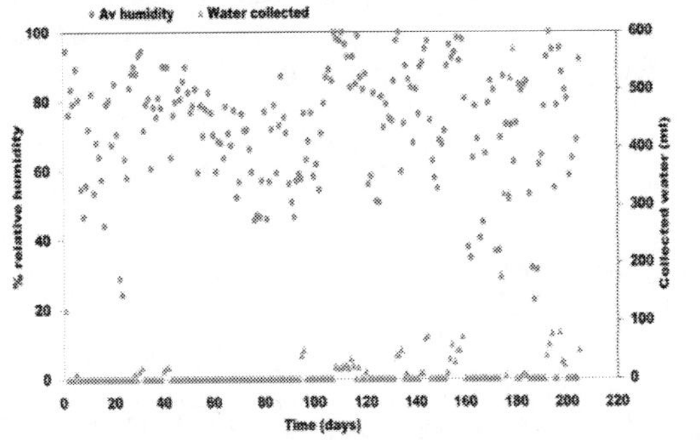

Figure 5. Water collection from September to April 2018

Similarly, a study in the southwest Morocco showed that when the relative humidity was between 74%, and

92%, water condensed on the condenser surface [10]. The average collected water was 0.016 ml/day (0.016 mm/m^2/day). The daily dew yield in Jerusalem was 0.2 mm on average with a maximum of 0.6 mm and the dew yield in Zadar and Beirut was 0.04 mm on average [18]. As another study stated, the water collection per day for dew collectors is relatively low and difficult to predict [2]. Some other studies stated that following effects could lead to lower water collection such as:

1. Buildings in urban areas obstruct the sky view resulting in higher nocturnal temperatures which lead to urban heat island effect and inhibit dew formation [26-28].
2. During warmer months, urban areas have lower relative humidity than nearby rural areas, therefore less humidity correlated with urban heat island effects could reduce dew formation [29, 30].
3. Increased evaporation in urban areas due to urban heat island effect inhibits the condensation which leads to less dew formation [31].
4. Increased atmospheric pollution in urban environment may inhibit the atmospheric transmittance and leads to less dew formation [2, 18, 32]
5. Dust accumulation is a serious issue in Abu Dhabi, and the accumulation of the dust on the condenser surface may inhibit the gravity slide of the formed dew to the collection container [33].

4 Conclusion

A passive dew, fog, and rain water collector made from square-funnel-shaped aluminum was used as an additional water source in Abu Dhabi. The temperature sensors placed on the condenser with different elevations revealed that there was a maximum of 3.5 °C temperature difference between the top and bottom sensors and a maximum of 1.4 °C difference between the ambient temperature and the top surface of condenser indicating the cooling effect of the condenser. One-year data collection showed that the dew occurred in 100 days in which 51 of them yielded water collection. The average collected water was 0.016 ml/day (0.016 mm/m2/day). The lower dew yields might be as a results of urban heat island effect. To be able to make dew water as a useful supplementary water resource in Abu Dhabi, more analysis is necessary related to the shape, size, and material of the dew collector as well as the location where it should be placed.

References

1. Middleton, W.E., *History of the Theories of Rain and other forms of precipitation.* (1966).
2. Khalil, B., et al., *A review: dew water collection from radiative passive collectors to recent developments of active collectors.* Sustainable Water Resources Management, (2016). **2**(1): p. 71-86.
3. Beysens, D., et al., *Collecting dew as a water source on small islands: the dew equipment for water project in Bis˘evo (Croatia).* Energy, (2007). **32**(6): p. 1032-1037.
4. Sharan, G., D. Beysens, and I. Milimouk-Melnytchouk, *A study of dew water yields on Galvanized iron roofs in Kothara (North-West India).* Journal of Arid Environments, (2007). **69**(2): p. 259-269.
5. Clus, O., et al., *Dew, fog and rain water collectors in a village of S-Morocco (Idouassksou).* Desalination and Water Treatment, (2013). **51**(19-21): p. 4235-4238.
6. Sharan, G., et al., *A very large dew and rain ridge collector in the Kutch area (Gujarat, India).* Journal of hydrology, (2011). **405**(1): p. 171-181.
7. Monteith, J.L., *Dew.* Quarterly Journal of the Royal Meteorological Society, (1957). **83**(357): p. 322-341.
8. Muselli, M., et al., *Dew and rain water collection in the Dalmatian Coast, Croatia.* Atmospheric Research, (2009). **92**(4): p. 455-463.
9. Maestre-Valero, J., et al., *Comparative analysis of two polyethylene foil materials for dew harvesting in a semi-arid climate.* Journal of hydrology, (2011). **410**(1): p. 84-91.
10. Lekouch, I., et al., *Rooftop dew, fog and rain collection in southwest Morocco and predictive dew modeling using neural networks.* Journal of hydrology, (2012). **448**: p. 60-72.
11. Berkowicz, S.M., et al., *Urban dew collection under semi-arid conditions: Jerusalem.* (2004).
12. Jacobs, A.F.G., B.G. Heusinkveld, and S.M. Berkowicz, *Passive dew collection in a grassland area, The Netherlands.* Atmospheric Research, (2008). **87**(3): p. 377-385.
13. Clus, O., et al., *Study of dew water collection in humid tropical islands.* Journal of hydrology, (2008). **361**(1): p. 159-171.
14. Peel, M.C., B.L. Finlayson, and T.A. McMahon, *Updated world map of the Köppen-Geiger climate classification.* Hydrology and earth system sciences discussions, (2007). **4**(2): p. 439-473.
15. Ministry of Presidential Affairs, N.C.o.M. (2018) [cited 2018 25/5/2018]; Available from: https://www.ncm.ae/en/index.html#!/Radar_UAE_AD_ReflectivityAD-DP/26.
16. data.org, c. 2018 25/05/2018]; Available from: https://en.climate-data.org/asia/united-arab-emirates/abu-dhabi/abu-dhabi-3217/.
17. Clus, O., et al., *Comparison of various radiation-cooled dew condensers using computational fluid dynamics.* Desalination, (2009). **249**(2): p. 707-712.
18. Tomaszkiewicz, M., et al., *Dew as a sustainable non-conventional water resource: a critical review.* Environmental reviews, (2015). **23**(4): p. 425-442.
19. Beysens, D., et al., *Using radiative cooling to condense atmospheric vapor: a study to improve*

water yield. Journal of hydrology, (2003). **276**(1): p. 1-11.

20. Avallone, E., I. Baumeister, and A. Sadegh, *Marks' Standard Handbook for Mechanical Engineers. 10.* (2006): New York: McGraw-Hill.
21. Alnaser, W. and A. Barakat, *Use of condensed water vapour from the atmosphere for irrigation in Bahrain.* Applied Energy, 2000. **65**(1): p. 3-18.
22. Adafruit Industries, U.f.D.e.a.k. *DHT11 basic temperature-humidity sensor + extras ID: 386* 2017 [cited 2017 25/12/2017]; Available from: https://www.adafruit.com/product/386.
23. ThingSpeak, , *The open IoT platform with MATLAB analytics.* 2018.
24. Edouard Cabay, E.P., *Dew harvesting devices and their application in architecture.* (2014), Ecole Spéciale d'Architecture, Paris.
25. Muselli, M., D. Beysens, and I. Milimouk, *A comparative study of two large radiative dew water condensers.* Journal of Arid Environments, (2006). **64**(1): p. 54-76.
26. Spronken-Smith, R. and T. Oke, *Scale modelling of nocturnal cooling in urban parks.* Boundary-Layer Meteorology, 1999. **93**(2): p. 287-312.
27. Kim, Y.-H. and J.-J. Baik, *Maximum urban heat island intensity in Seoul.* Journal of applied meteorology, (2002). **41**(6): p. 651-659.
28. Gedzelman, S., et al., *Mesoscale aspects of the urban heat island around New York City.* Theoretical and Applied Climatology, (2003). **75**(1-2): p. 29-42.
29. Unkašević, M., O. Jovanović, and T. Popović, *Urban-suburban/rural vapour pressure and relative humidity differences at fixed hours over the area of Belgrade city.* Theoretical and Applied Climatology, (2001). **68**(1-2): p. 67-73.
30. Fortuniak, K., K. Kłysik, and J. Wibig, *Urban–rural contrasts of meteorological parameters in Łódź.* Theoretical and applied climatology, (2006). **84**(1-3): p. 91-101.
31. Kuttler, W., et al., *Urban/rural atmospheric water vapour pressure differences and urban moisture excess in Krefeld, Germany.* International Journal of Climatology: A Journal of the Royal Meteorological Society, (2007). **27**(14): p. 2005-2015.
32. Beysens, D., et al., *Measurement and modelling of dew in island, coastal and alpine areas.* Atmospheric Research, (2005). **73**(1-2): p. 1-22.
33. Sizirici, B., *Modified biosand filter coupled with a solar water pasteurizer: Decontamination study.* Journal of Water Process Engineering, (2018). **23**: p. 277-284.

A comparative study regarding to physical and calorific properties of briquettes from vegetable biomass

Cosmin Spirchez[1*], *Aurel* Lunguleasa[1], and *Cătălin* Croitoru[2]

[1]Transilvania University of Brasov, Wood Processing and Design Wooden Product Department, 29 Street Eroilor, Brasov, Romania
[2] Transilvania University of Brasov, Engineering Materials Department, 29 Street Eroilor, Brasov, Romania

Abstract. The paper makes a comparison between four types of briquettes: first was wheat and rapeseed briquettes, second was wheat briquettes (batch 2018), third was wheat briquettes (batch 2017), and fourth was lucernes briquettes, from the point of view of the physical and energetic characteristics, in order to choose the best ones and their corresponding market. The main properties studied are the density and moisture of the briquettes, physical properties, and also calorific power and ash content as energetic properties. Research results show that wheat and rape briquettes have had high calorific value but low density, compared to other types of briquettes. A general conclusion rise from whole paper, respectively the vegetable biomass is a renewable material and briquettes from it remains the best option of combustible materials.

1 Introduction

The biomass is the biodegradable part of the vegetal products, wastes and agricultural residues, including the vegetal and animal substances, as well as the biodegradable part of industrial and / or urban wastes.
The biomass represents the most abundant renewable resource on the planet. This includes absolutely all the organic matter, obtained through metabolic processes of live organisms.

The biomass is a renewable energetic source by the fact that, it increases year by year, it is widely spread worldwide and it has low costs in comparison to the fossil fuels. The biomass resources, the fuel material is made of, may include the wood and wooden wastes, agricultural cereals and wastes resulted from their production, municipal wastes, animal dejections, wastes from food processing, aquatic biomass and algae. The biomass is one of the renewable sources forms which may be converted in solid, liquid and gaseous fuel, and which may generate energy as heat through its burning, as well as electrical energy through the industrial conversion processes, individually or in combination with fossil coal [1].

The biomass in vegetal form is a complex compound and differs from a vegetal species to other. This comprises all the forms of vegetal materials, which grow on the terrestrial surface, in water or on water, as well as other substances by biological development or degradation.

The biomass is environmentally friendly and a neutral energy against the CO_2 emissions. The carbon dioxide is absorbed by the plants during growth and forms a closed circle of entering and continuity, because the quantity of CO_2 which was absorbed by the vegetal part (either the one formed by vegetal plants, shrubs or trees) during the growth will be equal the same quantity which will be released during the burning out processes or of discomposure in nature under the action of bio-pests. The biomass may be used in the combustion processes and in most of cases it does not require very high investments as it is necessary for the production of hydro-electrical, solar, wind and geothermal energy[2].

At the worldwide level, according to the latest research in the field, there are highlighted preoccupations in the field of biomass use on energetic purposes. Romania has a potential of 60% in energy production from the existing sources of biomass. Also, in the same manner, it might be said that Romania has a surface of 6.3 million ha, which represents 27% of the surface of the whole existing territory. The resources from biomass, at the present, represents the raw material resulted from the wood processing, from agriculture, municipal wastes and animal. All this raw material resulted from the wood processing, from agriculture, municipal wastes and animal dejections. All this raw material recycled and used becomes a valuable material to produce energy, heat and a bio-fuel [3].

The biomass differs from the other forms of renewable sources forms by the fact it represents o resourceful raw material which can be easily transported due to its solidity, and it can be processed through different processes of conversion in gaseous, liquid or solid fuels. The biomass is divided in 4 categories described in the normative SR EN 14961-1:

- Forest production: wood, wastes from wood cutting, sawdust, trees, shrubs, splints, bark, resulted from the forest exploitation and trimming;

* Corresponding author: cosmin.spirchez@unitbv.ro

- Wastes: resulted from the agricultural production, from agricultural processes, cereal wastes, organic urban wastes;
- Energetic cereals: crops from short-term processing, crops of starch (corn, wheat and barley), sugar crops (sugar cane and sugar beetroot), fodder crops (grass, alfalfa and clover), oligeneous crops (sun flower, soy, saffron);
- Aquatic plants: algae, water weeds, water hyacinth, reed and rush;

The wooden biomass comes from the wood resulted after trimming the tree branches, which cannot be used industrially, from wood quantities which is due to be processed, from forests and wood processing. Practically the forest exploitation starts by cutting the wood in the forest, trimming the branches and removing the roots, after which there is the transportation operation and the storage within the processing plants, the place where the real sorting and classification on quality groups of the wood is performed, all of this representing approximately 10% wooden wastes [4].

Briquettes and pallets are products made from wooden wastes such as mince, sawdust or / and agricultural wastes. The quality of the briquettes depends on the compression endurance and their density, having a caloric power higher as they are more compressed. The briquettes are the most advantageous heating methods of dwellings, because they are cheap, they may be used in stoves and central heating, with an intense burning without leaving much ash, they are easily transported and stored.

At the present, it is noticed that the used of biomass is in a continuous increase. This increase is materialized by the usage directions of biomass and the products resulted by biomass processing, separated in material to produce biogas and biofuel.

The wooden crops are differentiated, according to the research, in net biomass quantity of material available through the following four factors: quantity of solar energy, quantity of biomass intercepted by biomass, the efficiency of the photosynthetic conversion in energy, loss of biomass by exploitation. Besides the solar energy, obtaining some higher crops of primary biomass and more qualitative, is conditioned by several factors, such as: internal – genetic factors (species, variety, hybrid), which have a critical role in the accumulation capacity of biomass, external factors (environment factors, technological factors) [5].

Using the wood as energy source has increased very much during the last decades in all the counties in developing process, but this resource has not been exploited as durable, but through massive and uncontrollable exploitations of the forest areal. The biomass potential could be doubled only by systematic recovery of all the organic origin waste (household, industrial, agricultural, forest).

Romania is a country with high potential in the renewable energy sources field and that is in producing energy from biomass, one of the reasons for the increase of this field being the raising of prices for fossil fuels. Obtaining energy from renewable energy sources becomes thus a valuable investment, for consumers as well as for industrial plants of high capacity. The favorable conditions for producing energy from alternative sources are determined by the legislation in force and the European funds which encourages the production of renewable energy from biomass, as well as the use of solar energy, wind energy, geothermal and hydro energy.

The use rhythm of the renewable sources will continue to increase, so that the production of electrical energy from these sources in 2020 will represent 38%. The biomass is regarded as alternative energy source of wide scope, widely spread worldwide and presents different forms which may be used for green energy production.

The wood covers nowadays more than 10% of the demand of primary energy in many countries from Asia, Africa and Latin America and in some countries in Europe (Sweden, Finland, Austria)[6].

In Europe, the renewable energy has a rate of annual increase of primary energy with an average of 3,4% during 1990-2005. At the worldwide level, the biomass is the sole energy source which can be used in future as an accessible renewable source.

The chemical composition of the biomass, respectively for plants is 25% lignin and 75 % carbohydrates (celluloses and hemi-celluloses) or saccharides. One of the most important carbohydrates is cellulose. Lignin appears in plants as something like glue having a role of adhesion between the usual structures of resistance, which ties the cellulosic molecules between them [7].

During the last centuries the humanity has learnt to obtain fossil types of biomass, especially, as coal. Fossil fuels present the result of the very slow chemical reaction of transformation the poly saccharides in chemical compounds resembling the lignin fraction.

In some developed countries, the biomass is used intensively enough. Sweden plans for the future the increase of the biomass volume used concomitant with the closing of nuclear – electrical and thermo – electrical plants, which uses fossil fuel coal type or methane gas. In U.S.A there are installations with a total capacity of 90000 MW which function, where the biomass burns with the purpose of obtaining electrical energy. The biomass may assure over 20% from the energetic necessities of the country. At the beginning of the 1980s, almost 1.3 billion people provided the necessity in fuels based on decreasing the forest reserves, by massive cuttings[8].

The biomass is the assembly of non-fossil organic materials, where there are registered: wood, vegetal oils and wastes in the forest, agricultural and industrial field, but also the cereals and fruits, from which ethanol may be produced. Similar to the energy obtained from fossil fuels, the energy produced from biomass comes from the solar energy stored in plants, through the photosynthesis process with the help of light.

The main difference between the two forms of energy is that the fossil fuels could not be extracted and transformed in energy useable only after thousands of years, while the biomass energy is renewable, being used each and every each (if some clear rules are observed regarding their sustainability)[9].

Supplying energy from biomass took off starting the 2000, and in 2010 this field has already supplied 280 TWh of electrical energy at the global level, which equals 1.5 % of electrical energy supplied worldwide. Presently the biomass continues to play an important role in the countries which are not OECD member (The Organization for Economic Co-operation and Development) and represents the main source which provides the energetic necessary. Presently, the countries in developing process assure their energy necessary through sources such as biomass using central heating with a low rating of 10 -15%. Taking into consideration these aspects, it becomes necessary to supply the energetic market with fuels and thermal installations with high rating, thus the biomass should become a competitive source on the energetic market.[10]

Romania must encourage the companies and citizens in order to invest in alternative energy sources. Therefore, the electrical energy balance produced from renewable resources of energy, in comparison to the gross natural consumption of electric energy must reach 33% until 2020. The main renewable energy sources in Romania could be: biomass, solar, wind and geothermal energy [11].

The main purpose of the paper is to make a comparative study between the physical and energetic proprieties of the four types of briquettes (wheat and rape briquettes, 2017 wheat briquettes, 2018 wheat briquettes, lucerne briquettes). Among the physical proprieties, the individual density and humidity was analyzed, among the energetic proprieties the caloric power and ash content were analyzed[12].

2 Materials and methods

There were analyzed four types of briquettes: wheat and rape briquettes, 2017 wheat briquettes, 2018 wheat briquettes, lucerne briquettes, in order to determine which are better to be used in combustion.

In order to determine the density, the classical method to determine their weight and volume was used. The samples, over 20 pieces of each type, were taken from test tubes and the ends were cut perpendicularly lengthwise, in order to obtain a clear length. The cylindrical shape of briquettes was chosen because, only for this type the method to determine the perpendicular compression is applied, and the volume is easily determined, as volume of a smooth circular cylinder. The briquettes were weighted using an electronic scale TP KERN EW 1500-24 with precision of 0,1 grams. As well, it was determined the average diameter of the briquettes, as average mean in three points (at the ends and middle), for two perpendicular diameters. The determining relation of density was as follows:

$$\rho = \frac{4m}{\pi D_m^2 l} \quad [\text{g/cm}^3]$$

where: m-briquette weight, in g;

D_m- average diameter of the briquettes, in mm;

l- test sample length, in mm.

The moisture content determined for the four types through the classical method of weighting-drying-weighting for briquettes was 8% [13-15].

The caloric power represents the quantity of heat resulted while burning a weight unit of fuel. There were performed 2 determinations for the caloric power for briquettes: superior caloric power (when the water vapors condensate, relieving evaporation heat) and the inferior caloric power (when a part of the heat for water evaporation is lost). The inferior caloric power is the difference between the superior caloric power and the quantity of heating relieved for the water evaporation from the burning gas the devices used to determine the caloric power of the four types of briquettes is the calorimetric bomb XRY-1C (made in China) (Fig 1)[16-18].

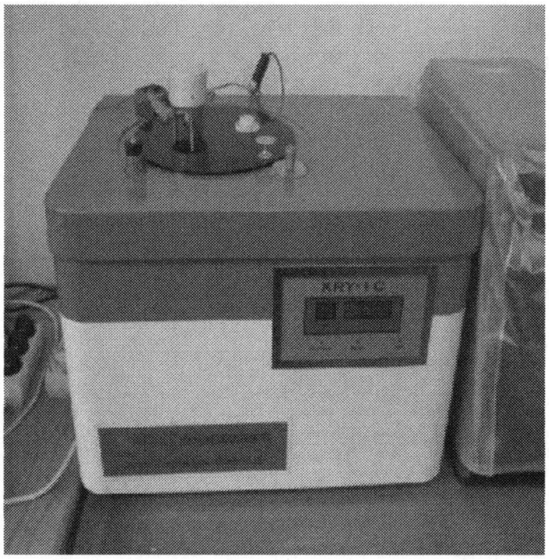

Fig. 1. Calorimetric bomb XRY-1C

Determining the caloric power of fuels is researched worldwide. Until the present there are a series of manners to determine the caloric power for wooden species, on an experimental manner as well as empiric equations. For the first time, the calorimeter Junkers type was used, which determines the caloric power for gaseous fuels. The method consists in burning out a known quantity of gas and the practical transmission of the heat losses relieved in the burning process, a water flow which circulates through the calorimeter. For this purpose, a quantity of fuel measured is burnt with the help of a gas burner or burner. The gas resulted from burning is cooled to the temperature of the environment by a powerful water flow. The weight of water, the temperature variation, the weight of condensed vapors from the burning products are measured and thus the burning superior heat is determined, as well as the inferior one.

For combustible materials with high content of water and hydrogen, such as wooden biomass, two types of caloric powers may be identified, respectively the superior caloric power and the inferior caloric power. The superior caloric power cannot be used in practical

https://doi.org/10.1051/e3sconf/201912201007

manner. Because the water vapors evacuate outside the vent and only the inferior caloric power remains, which is the most important and leads to the assurance of the energetic necessary.

Determining the caloric power implies going over three phases: initial phase, main phase, final phase. The initial phase represents the determination of temperature variation of water in the interior of the calorimetric container, because of the heat exchange with the exterior before the burning. At the end of the initial phase the burning of the samples extracted from briquettes begins.

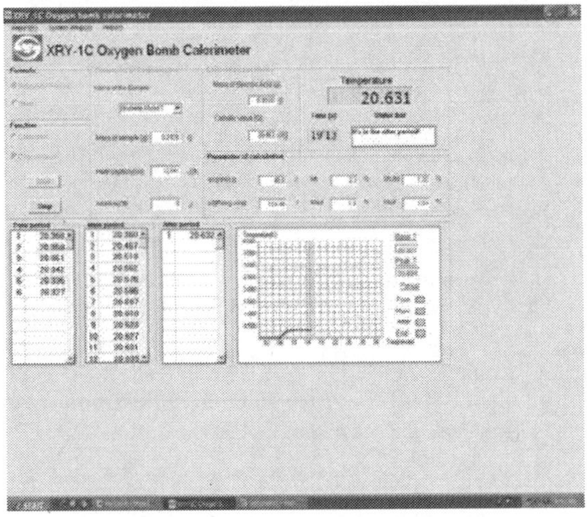

Fig. 2. The image of the computer soft for the calorimeter with explosive burning for the determination of the caloric power

The superior caloric power is determined with the relation:

$$Q_s = \frac{K \cdot (t_f - t_i)}{m} - q_s \quad [\text{kJ/kg}]$$

where:

K is the calorimetric factor, expressed in kJ/degree Celsius;

t_f – final value of the temperature, expressed in °C;

t_i – initial value of temperature, in °C;

q_s- heat consumed for the burning of incandescent wire in the gersdorffite and the cotton wore, for the initiation of burning, expressed in kJ/kg;

m – fuel sample weight, in kg.

In order to determine the ash content of briquettes, it was used the general method of standardized determination (ASTM D2866-11, 2012). According to this method, material is minced and dried until 0% humidity at a temperature of 650 °C in a laboratory stove, at least for 3 hours. The advanced burning operation is performed on a high temperature proof metallic pot, and the weighting is performed on an analytical scale with a precision of 3 decimals.

When determining the ash content, it will be taken into account that the sample is completely dried and the pot weight clean and empty.

3 Results and discussion

The density values for the four types of briquettes are: 937 kg/m^3 – wheat and rape briquettes, 1094 kg/m^3-2018 wheat briquettes, 1052 kg/m^3- 2017 wheat briquettes, 983 kg/m^3- lucerne briquettes.

In figure 3 there are presented individual density values for the four types of studied briquettes.

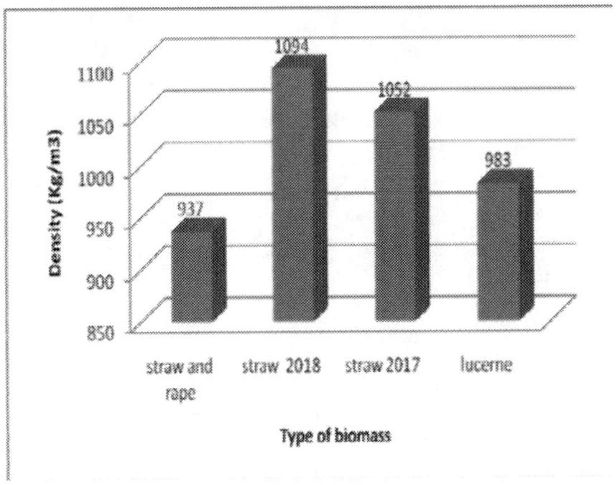

Fig. 3. The values of individual density for the four types of briquettes

The values of the superior caloric power for the four types of briquettes are: 19458 kJ/kg wheat and rape briquettes, 17689 kJ/kg 2018 wheat briquettes, 17661 kJ/kg 2017 wheat briquettes, 17989 kJ/kg alfalfa briquettes.

In figure 4 it is presented the values of the superior caloric power for the four types of briquettes.

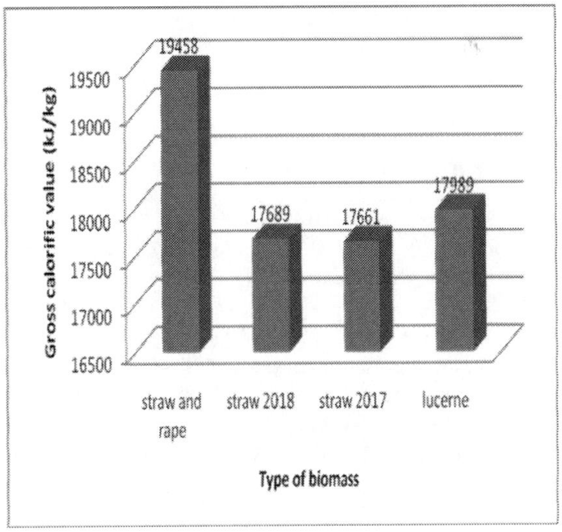

Fig. 4. The values of superior caloric power for the four types of briquettes

In figure 5 there are presented the values of the ash content for the four types of briquettes.

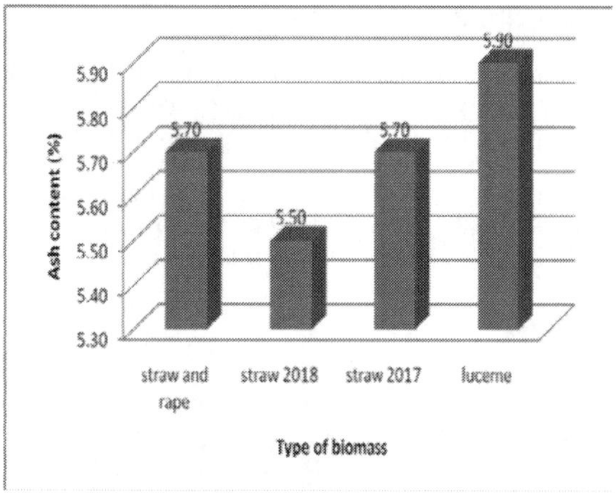

Fig. 5. The values of the ash content for the four types of briquettes

4 Conclusions

Briquettes have an exceptional thermal capacity; therefore, they retain heat for a much higher period of time and keep the temperature high within the focal point, allowing a facile burning of the new introduced briquettes.

The biomass represents a renewable source of energy. This does not contribute to the problem of environment changing, because it recycles the carbon dioxide in the atmosphere.

Wheat and rape briquettes have the best caloric capacity, while the 2017 wheat briquettes have the lowest caloric capacity, among the four types of analyzed briquettes. The main conclusion of the paper refers to the use of these briquettes and there is stated that they are used as well as other briquettes resulted from wood or other wooden materials. This thing is possible due to the good characteristics of briquettes, equivalent to the ones made of other materials.

References

1. J.P. Boutin, G. Gervasoni, R. Help, K. Seyboth, P. Lamers, M. Ratton et al. Alternative Energy Sources in Transition Countries. The Case of Bio-energy in Ukraine. Environ Engineering and Management Journal , vol.6, nr.1, pp.3-11 (2007)
2. A.V. Bridgwater, Review of Fast Pyrolysis of Biomass and Product Upgrading, Biomass bioenergy, vol. **38**, pp. 68-94 (2012)
3. C.Ciubota-Rosie, M. Gavrilescu, M. Macoveanu Biomass-an Important Renewable Source of Energy in Romania, Environ Engineering and Management Journal, vol.7, pp.559-568 , (2009)
4. A. Demirbas Resource Facilities and Biomass Conversion Processing for Fuels and Chemicals, Energy Conversion Management, vol. **42**, nr. 11, pp. 1357-1378, (2011)
5. A.M. Garcia, B.M.J. Barcia, J.A. Hernandez Preparation of active Carbon from a Comercial Holm-Oak Charcoal: Study of Micro and Meso-porosity, Journal Wood Science and Technology, vol. **37,** nr. 5, pp. 499-509 (2008)
6. D. Gavrilescu Energy from biomass in pulp and paper, Environmental Engineering and Management Journal, vol. **7(5)**, pp.537-546, (2008)
7. M. Gavrilescu Biomass Power for Energy and Sustainable Development, Environmental Engineering and Management Journal, vol. **7(5)**, pp.617-640, (2008)
8. N. Kaliyan, R.V. Morey Factors Affecting Strength and Durability of Densified Biomass Product, Biomass Bioenerg, vol. **33**, nr. 3, pp. 379-359, (2009)
9. T. Kuhlman, V. Diego, E. Koomen Exploring the potential of reed as a bioenergy crop in the Netherlands, Biomass Bioenergy, vol. **55**, pp.41-52, (2013)
10. J. Lako, J. Hancsok, T. Yuzhakova, G. Marton, A. Utsai, A. Redey, Biomass and Bioenergy , vol. **40,** pp. 63-73, (2012)
11. A. Lunguleasa Compaction coefficient of wooden briquettes used as renewable fuel, Environmental Engineering and Management Journal, vol.**10,** nr. 9, pp.1263-1268, (2011)
12. A. Lunguleasa The compressive strength of wooden briquettes used as renewable fuel, Environmental Engineering and Management Journal, vol. **9,** pp.977-982, (2010)
13. C. Okello, S. Pindozzi, S. Faugno, L. Boccia Bioenergy potential of agricultural and forest residues in Uganda, Biomass Bioenergy, vol. **56**, pp. 515-525, (2013)
14. N. A. Pambudi, K. Itaoka, A. Chapman, N. D. Hoa, N. Yamakawa Biomass energy in Japan: Current status and future potential, International Journal of Smart Grid and Clean Energy, vol. **6**, no. 2, pp. 119-126 (2017)
15. J. Swithenbank, Q. Chen, X. Zhang, V. Sharifi, M. Pourkashamiani Wood would burn, Biomass and bioenergy, vol.**3**, (2011)
16. J.Tanchuling, R. De Leon Solid fuel characterization of torrefied coconut shells in an oxidative environment, International Journal of Smart Grid and Clean Energy, vol. **7**, no. 2, pp. 145-151 (2018)
17. O. Teuch A. Hofeanuer, F. Troger, J. From Basic properties of specific wood based materials carbonised in a nitrogen atmosphere, Wood Science Technology Journal, vol.**38,** nr.3, pp.323-333, (2004)
18. V.K. Verna, S. Bram, J. de Rucky Small Scale Biomass Systems: Standards, Quality, Labeling and Market Driving Factors-An Outlook Biomass Bioenerg, vol.**33**, nr.10, pp. 1393-1402, (2009)

https://doi.org/10.1051/e3sconf/201912202001

Potential assessment of hybrid PV-Wind systems for household applications in rural areas: Case study of Morocco

Lhoussaine Tenghiri[1,*], *Yassine* Khalil[1], *Farid* Abdi[2], and *Anas* Bentamy[1]

[1]School of Science and Engineering, Al Akhawayn University, Ifrane, 53000, Morocco
[2]Electrical Engineering Department, Faculty of Sciences and Technologies, P.O. box 2202 – Route d'Imouzzer, Fez, Morocco

Abstract. Proper combination of wind and solar photovoltaic (PV) systems can result in optimal configurations that maximise the Annual Energy Production (AEP) while being economically attractive. This paper presents a typical approach to the design of a hybrid PV-Wind system for household applications in rural areas. Based on the capacity factor of the hybrid system components, a design methodology was developed to maximize the AEP and to minimise the investment cost. The electricity generated will be used to meet the load requirements of the user while the potential excess of the power will be stored in the battery system or dissipated in a dump resistor. This design methodology is suitable for household applications presenting limitations in the available roof area where the PV panels will be installed. Implementing the PV panels on the house's roof intends to eliminate the shadow over the panels, to avoid the material damage, and to prevent vandalism. Four different cities were selected to conduct the study. These are Tetouan, Essaouira, Dakhla, and Ouarzazate. For the present design approach, the installed cost of the hybrid system includes the price of the storage system.

1 Introduction

Hybrid PV-Wind Systems (HPWSs) are becoming an attractive energy sources that can be used for decentralized electrification in remote areas. Several methods and optimisation approaches have been developed for the design of HPWSs in a cost effective way. Erdinc and Uzunoglu have presented a summary of several sizing approaches that have been applied for the design of hybrid renewable energy systems [1]. Their literature overview includes methods based on the commercially available software (such as HOMER and HYBRID2). These tools are based on optimisation methods such as Genetic Algorithm (GA), Simulated Annealing (SA), Linear Programming, Particle Swarm Optimisation (PSO), Neutral Network, and iterative and probabilistic approaches. Predominant constraint in these methods is that the required PV power calculated does not take into consideration the available surface on the user's roof. The design methodology used in this paper is based on the Capacity Factor (CF) which is defined as the ratio of the actual annual energy production to the theoretical annual energy production. The optimum percentage of the hybrid system components will be determined such that the HPWS meets the load requirements while being able to adequately exploit the available resources of the given locations. The potential excess of the power will be stored in the battery bank or dissipated in a dump resistor (water heater or heating system of the house). The analysis in this paper is based on the following steps: at first, site-specific wind conditions of the selected locations was performed. Then, typical daily electricity consumption of a Moroccan household as well as the models used for the calculation of PV and wind systems are presented. In a third part, the hybrid system sizing procedure is given. Finally, the results are presented and discussed.

2 Wind and solar potential assessment

To ensure an appropriate level of efficiency and reliability for the HPWS design, wind and solar energy potential must be evaluated for the selected sites. Figure 1 shows the geographical locations of the sites that were selected for the analysis.

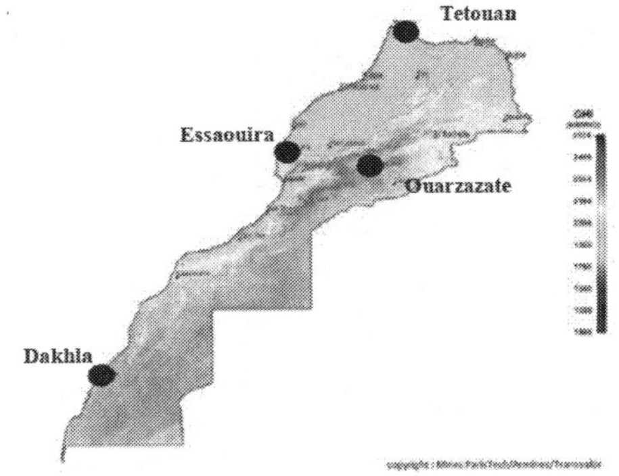

Fig. 1. Geographical location of the studied locations

* Corresponding author: lhoussaine.tenghiri@usmba.ac.ma

2.1 Wind potential assessment

Various mathematical tools have been used for the analysis of wind speed data. A convenient mathematical distribution function that has been found to fit well with wind data is the Weibull probability density function, whose probability density match in most of the times the actual wind speed measurements [2]. This function is mainly described using two parameters: the shape parameter, k, and the scale parameter, c. The general form of the Weibull distribution, for a specific wind speed, V, is given by the following equation [3]:

$$f(V) = \frac{k}{c} \times \left(\frac{V}{c}\right)^{k-1} \times e^{-\left(\frac{V}{c}\right)^{k}} \quad (1)$$

The value of the shape parameter varies usually between 1.5 and 3. If the values of these parameters, at the anemometer height, z_a, are c_a and k_a, then their corresponding values, at a higher level z, can be determined using the following equations [4].

$$C_z = c_a \times \left(\frac{z}{z_a}\right)^{n} \quad (2)$$

$$k_z = \frac{k_a \times \left[1 - 0.088 \ln\left(\frac{z_a}{10}\right)\right]}{\left[1 - 0.088 \ln\left(\frac{z}{10}\right)\right]} \quad (3)$$

The exponent, n, is given as:

$$n = \frac{[0.37 - 0.088 \ln c_a]}{\left[1 - 0.088 \ln\left(\frac{z_a}{10}\right)\right]} \quad (4)$$

Different methods have been developed to estimate the Weibull distribution parameters. The most widely used are the method of moments which was selected for wind data analysis in this study. Based on this method, the following equations give the values of the Weibull parameters [3]:

$$k = \left(\frac{\sigma}{V_{ave}}\right)^{-1.086} \quad (5)$$

$$c = \frac{V_{ave}}{\Gamma\left(1 + \frac{1}{k}\right)} \quad (6)$$

Where Γ is the gamma function, which is given by [4]:

$$\Gamma(x) = \int_0^{\infty} t^{x-1} e^{-t} dt \quad (7)$$

To estimate wind speeds at various elevations, a vertical wind profile was used. The common power law that can be used to obtain the extrapolated values of wind speed at different heights is given by the following equation [4]:

$$V_{z2} = V_{z1} \times \left(\frac{z_2}{z_1}\right)^{m} \quad (8)$$

Where V_{z1} is the wind speed at the measurement height, usually taken as 10 m. V_{z2} is the height at which the wind speed estimation is required, and the parameter, m, is the power law exponent of the wind speed. The power law exponent, also called roughness coefficient, varies according to the roughness of the terrain. If there is no specific information, the International Electrotechnical Commission (IEC) standard for the design of small wind turbines (IEC 61400-2) recommends using a roughness coefficient of 0.2 [5]. This value corresponds to a roughness length of approximately 0.10 m (taking into account agricultural land with some houses and 8 m high hedges with spacing of approximately 1250 m).

The wind power density function is a parameter that gives the distribution of wind energy at different wind speeds. It is well known that the power of the wind that flows at speed, V, through a blade sweep area, A, increases as the cube of its velocity and is given by the following equation [4]:

$$P(V) = \frac{1}{2} \times \rho \times A \times V^3 \quad (9)$$

The parameter ρ is the air density which is taken as 1.225 Kg/m^3. The parameter A represents the wind turbine rotor swept area. Wind power density per unit area for a specific region (based on the Weibull probability density function) can be determined using the following equation [4]:

$$P = \int_0^{\infty} P(V)\, f(v)\, dv = \frac{1}{2} \times \rho \times c^3 \times \Gamma\left(1 + \frac{3}{k}\right) \quad (10)$$

For wind data analysis, a time series of daily measured wind speeds, recorded over a period of 20 years, were used for the statistical analysis of the wind speed and the wind power density. These wind data are given by the climatic data center of the National Oceanic and Atmospheric Administration (NOAA) [6]. Climate data provided by this center constitutes a reliable source for wind energy community, especially for pre-feasibility study of wind farm projects development. Table 1 summarizes the results of the wind potential assessment.

Table 1. Annual mean wind speed and Weibull distribution parameters of the four studied locations

Station	V_{ave} (m/s) 10 m	c (m/s)	k	V_{ave} (m/s) 24 m	Power density (W/m) 10 m
Tetouan	6.4	7.27	1.96	7.62	318
Essaouira	5.6	6.31	1.85	6.67	223
Dakhla	6.0	6.75	2.22	7.14	227
Ouarzazate	4.0	4.51	1.72	4.77	90

2.2 Solar potential assessment

The potential of solar energy is mainly determined by the solar irradiation which is defined as the amount of energy that reaches the ground surface over a stated

 https://doi.org/10.1051/e3sconf/201912202001

period of time, and expressed as Wh/m .day. The solar irradiation was assessed using the Photovoltaic Geographical Information System (PVGIS) [7]. This geographical information system, which has been developed by the Joint Research Centre (JRC), is the most commonly used GIS-based web application for estimating the performance of PV generation systems. Table 2 shows the annual average daily irradiation, I_{rr}, on an optimum tilted surface for the four studied locations (In°). The inclination of the solar PV panels is chosen such that power extraction from solar irradiations will be maximized. The solar panels will be mounted at an angle to point directly at the sun. The PV panels will be also oriented towards the south with a zero azimuth angle. The south facing modules increases the irradiance falling on the panels compared to the other orientations.

Table 2. Annual average solar irradiation for the four studied stations

	Tetouan In°(31°)	Essaouira In°(30°)	Dakhla In°(25°)	Ouarzazate In°(31°)
	I_{rr} (Kwh/m .day)			
Average	5.430	6.720	6.540	6.690

3 Daily electricity consumption

The residential electricity consumption was determined based on the needs of the household appliances that include water pumping systems for irrigation. The load demand was assessed using the following steps:

a) Determination of typical household appliances in rural areas: this is based on the acquisition rate of household appliances. This is defined as the ratio of the number of households owning the appliance at the total number of households involved in the survey. This step is based on the results of the national survey of households' standard of living (HCP 2014) which is the latest available report on the household's standard of living in Morocco.

b) Determination of the daily electricity consumption of a typical household: the dimensioning of the HPWS is strongly dependent on the overall daily energy consumption of the selected appliances. Based on this statistical approach, the daily electricity consumption of a typical Moroccan household in rural areas is 103.637 kWh/day. This value is relatively high due to the fact that the needs of the majority of the pumping systems in Morocco have an average Total Dynamic Head (THD) of 80 m with a daily water flow of 300 m^3. This need corresponds to the water required to irrigate 5 hectares of agricultural area in a typical farm.

4 Wind/PV power assessment and capacity factors

4.1 Wind power assessment and capacity factor

Based on a typical household's daily electricity consumption of 103.637 kWh, equation (11) gives the required electrical power of the wind power system [8].

$$P_{wa} = \frac{E_{Load}}{C_{ptot} \times 24hour} \quad (11)$$

C_{ptot} is the total efficiency of the wind turbine conversion system which is taken as 40% [8]. This total efficiency includes the efficiency of the drive train, the generator (assuming an overall transmission efficiency of 90%), and the aerodynamic losses. From equation (11), an electrical power generation of 10.8 kW is deduced. This power represents the required power output to meet all the load requirements when the wind generation system is used alone. This is because the power calculation given above assumes the wind power system will meet all the user's daily needs.

Productivity of the wind power system is usually evaluated using a quantity known as capacity factor:

$$AEP_w = C_{fw} \times 8760 \times P_{rated} \quad (12)$$

P_{rated} is the rated power of the wind turbine. The capacity factor of the wind power system is determined using equation (13) [9]. This model describes the capacity factor as a function of the Weibull distribution and the wind turbine parameters.

$$C_{fw} = \frac{e^{\left[-\left(\frac{V_c}{c}\right)^k\right]} - e^{\left[-\left(\frac{V_r}{c}\right)^k\right]}}{\left(\frac{V_r}{c}\right)^k - \left(\frac{V_c}{c}\right)^k} - e^{\left[-\left(\frac{V_f}{c}\right)^k\right]} \quad (13)$$

Parameters V_c, V_r and V_f are respectively the cut-in wind speed, the rated wind speed, and the cut-out wind speed of the wind turbine. Typical values of 3 m/s for V_c, 12.5 m/s for V_r, and 25 m/s V_f were used. The wind energy system will be assessed for a hub height of 24 m. This typical tower height is usually proposed by the wind turbine manufacturers for commercially available small wind turbines.

4.2 PV power assessment and capacity factor

The required power output of the PV generation system is obtained by the following mathematical equation:

$$P_{va} = \frac{E_{Load}}{I_{rr} \times \eta_{tot}} \quad (14)$$

Where I_{rr} is the annual average daily irradiation of the geographical location and η is the total efficiency of the PV system. By considering the PV panels reduction factor and the efficiencies of the cables, the controller, the battery, and the inverter, the total system efficiency is found to be 60% [10]. With an annual average daily irradiation of 6.34 kwh/m .day (average value of the four studied stations), a required power of 27.24 kW_p is deduced for the PV generation system. Similarly, the calculated power represents the required power output to meet all the load requirements when the PV generation system is used alone. The capacity factor of PV generation systems is defined as the ratio of the actual annual energy production to the annual energy the PV system would produce if it operates at its full rated power. This can be determined using the following equation:

$$C_{fv} = \frac{AEP_v}{8760 \times P_{rated}} = \frac{H_{tot} \times PR}{8760 \times P_{rated}} \quad (15)$$

AEP_v is the actual annual energy production of the PV generation system. H_{tot} is the total amount of solar energy received by the PV panels in one year which is given by:

$$H_{tot} = 365 \times (I_{rr} \times S_v \times \eta_P) \quad (16)$$

S_v is the total area covered by the solar PV panels (in m) and η_p is the PV panel efficiency. P_{rated} is the rated power of the PV panel (peak power). The parameter PR is the performance ratio which defines the total losses in the system (optical losses, array losses, and system losses). A typical value of 80% was used for PR in the present analysis [11].

5 Optimal sizing methodology

The objective function of this nonlinear optimization problem is defined as the ratio of initial investment cost to the total AEP of the hybrid system which is multiplied by the unit price of kilowatt hour (kWh) of electricity. The flow chart given in figure 2 summarizes of the steps of design methodology.

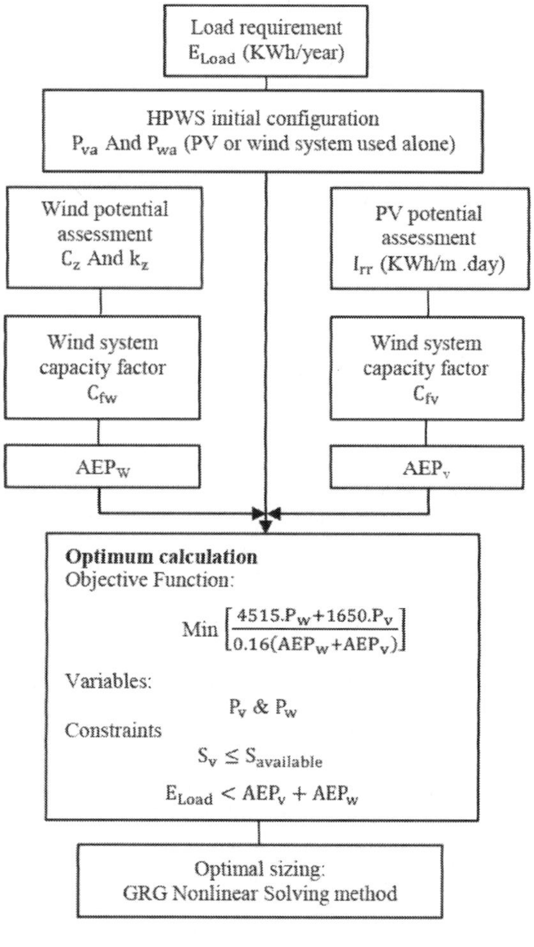

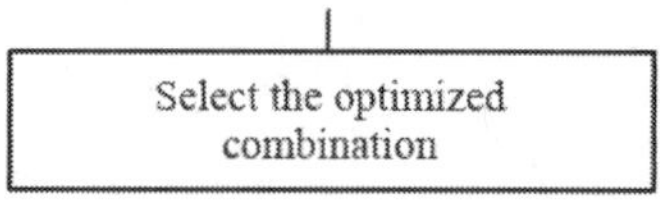

Fig. 2. Flow chart of the design methodology

The optimal hybrid configuration corresponds to the minimum of the objective function with the adequate values of installed powers of PV and wind turbine systems. For the wind generation system and according to the World Wind Energy Association (WWEA), the average worldwide installed cost of small wind turbines, in 2016, is about 4515 USD$/kW. The average installed cost of PV systems in 2016, is about 1650 USD$/kW$_p$. The installed cost of both wind and PV systems include the price of the battery bank. Therefore, the initial cost of an optimum design of the hybrid PV-wind system is given by:

$$Cost_H\ (USD\$) = 4515 \times P_w + 1650 \times P_v \quad (17)$$

The analysis was conducted for surface areas of 50 m , 100 m , and equal or more than 200 m . The contribution of the system to the payback period corresponds to the net cash flow saved from not being connected to the grid utility. For the calculation of the pay-back periods, the price of unit kilowatt (kWh) of electricity is fixed to 0.16 USD$. The optimal sizing results for the considered available surface, $S_{available}$, are given in table 3.

Table 3. Results of the design methodology

	Tetouan		
$S_{available}$	**50 m**	**100 m**	**≥200 m**
Need (kWh/year)	37820		
Total AEP (kWh/Year)	37 820	37 820	37 820
P_w (kW)	8.16	4.87	0
P_v (kWp)	6.86	13.73	23.91
Capital cost (USD$)	31 419.19	31 419.19	31 419.19
Cash-flow-20 years (USD$)	108 364		
Payback period (year)	8	7	6
Percentage of contribution in total AEP (%)	**Wind : 71%** **Solar: 29%**	**Wind : 43%** **Solar: 57%**	**Wind : 0%** **Solar: 100%**
	Essaouira		
$S_{available}$	**50 m**	**100 m**	**≥200 m**
Need (kWh/year)	37820		
Total AEP (kWh/Year)	37 820	37 820	37 820
P_w (kW)	9.07	4.07	0
P_v (kWp)	6.86	13.73	19.32
Capital cost (USD$)	35 133.71	33 298.48	31 816.42
Cash-flow-	108 364		

20 years (USD$)			
Payback period (year)	9	7	5
Percentage of contribution in total AEP (%)	**Wind : 64%** **Solar: 36%**	**Wind : 29%** **Solar: 71%**	**Wind : 0%** **Solar: 100%**
	Dakhla		
$S_{available}$	**50 m**	**100 m**	**≥200 m**
Need (kWh/year)	37820		
Total AEP (kWh/Year)	37 820	37 820	37 820
P_w (kW)	9.19	4.33	0
P_v (kWp)	6.86	13.73	19.85
Capital cost (USD$)	52856.78	42231.96	32763.69
Cash-flow-20 years (USD$)	108 364		
Payback period (year)	9	7	5
Percentage of contribution in total AEP (%)	**Wind : 65%** **Solar: 35%**	**Wind : 30%** **Solar: 70%**	**Wind : 0%** **Solar: 100%**
	Ouarzazate		
$S_{available}$	**50 m**	**100 m**	**≥200 m**
Need (kWh/year)	37820		
Total AEP (kWh/Year)	37 820	37 820	37 820
P_w (kW)	16.41	7.42	0
P_v (kWp)	6.86	13.73	19.41
Capital cost (USD$)	70 758	56 196.62	32 029
Cash-flow-20 years (USD$)	108 364		
Payback period (year)	15	9	5
Percentage of contribution in total AEP (%)	**Wind : 65%** **Solar: 35%**	**Wind : 30%** **Solar: 70%**	**Wind : 0%** **Solar: 100%**

The design methodology gives the optimal combination of the PV and the wind power systems that meet the load requirements of the users for all the studied stations.

For all studied cases, the hybrid system tends to exploit the available surface for PV system. This finding is not surprising given that the average installed cost per kilowatt of the PV system (1650 USD$/kW$_p$) is very low compared to that of the wind generation system (4515 USD$/kW).

The capacity factor of the HPWS components (PV and wind systems) has a significant effect on the economic performance (payback period) of the hybrid system. This can be easily noticed in the results obtained for the case of Ouarzazate. This location presents the highest investment cost which is mainly due to the lower capacity factor of the wind system component. Therefore, the HPWS in this location is not economically viable because of the relatively high payback periods. On the contrary, the case of Tetouan and Essaouira presents the lowest investment cost, and hence relatively short payback periods. This is mainly due to the highest capacity factors of the both PV and wind generation systems.

The available surface for the PV panels has a significant effect on the cost effectiveness of the HPWS. For the case of Ouarzazate, the hybrid system is not economically effective as the pay-back period is more than 10 years. The minimum required surface that makes the HPWS cost effective in Ouarzazate is 142 m . However, for the sites of Tetouan and Essaouira, $S_{available}$ does not have a major effect on the cost effectiveness of the system. This is mainly due to high wind potential at these locations. Therefore, the system tends to compensate the limitations in the available surface with small wind power systems.

When $S_{available}$ is equal or more than 200 m , all the electricity needed to meet the load requirements comes from the PV system. This is normal since the design approach tends to minimize the initial cost of the HPWS by exploiting all the available surface area for PV generation which offers the advantage of very low costs per kilowatt installed.

6 Conclusion

An optimal design methodology for hybrid PV–wind systems was presented in this paper. This practical approach present a solution for complex nonlinear problems by taking into account the surface availability for the PV panels as a design constraint. The variables of this optimization exercise are the installed power of both the PV and the wind turbine system. The objective function consists in minimizing the ratio of the initial investment cost to the total AEP multiplied by the price of unit kilowatt hour (Kwh). It was found that the sites with high wind and solar potentials (Tetouan and Essaouira) provide configurations with less investment costs and short payback payments. However, a decrease in the wind energy potential (low wind capacity factors) in the site of Ouarzazate affects the economic feasibility of the HPWS (high investment costs and very long payback periods). For all studied locations, the design methodology demonstrates its ability to provide an optimal HPWS configuration without using commercial software.

This work was funded by the «Institut de Recherche en Énergie Solaire et Energies Nouvelles-IRESEN».

References

1. O. Erdinc, M. Uzunoglu, *Optimum design of hybrid renewable energy systems: Overview of different approaches*, Renew. Sustain. Energy Rev., 16 (2012)
2. A. K. Azad, M. G. Rasul, M. M. Alam, S. M. Ameer Uddin, S. K. Mondal, *Analysis of wind energy*

conversion system using Weibull distribution, Procedia Eng., 90 (2014)

3. S. Farhan Khahro, K. Tabbassum, A. M. Soomro, L. Dong, X. Liao. *Evaluation of wind power production prospective and Weibull parameter estimation methods for Babaurband, Sindh Pakistan*, Energy Conversion and Management, 78 (2013)
4. B. Safari, J. Gasore, *A statistical investigation of wind characteristics and wind energy potential based on the Weibull and Rayleigh models in Rwanda*, Renewable Energy, 35 (2010).
5. IEC 61400-2 (International Electrotechnical Commission standard). *Wind turbines-Part 2: Requirements for small wind turbines*, (2013)
6. National Oceanic and Atmospheric Administration (NOAA), *National Climatic Data Center* [Online]. Available: http://www.ncdc.noaa.gov/. (2016)
7. A. Bocca, L. Bottaccioli, E. Chiavazzo, M. Fasano, A. Macii, P. Asinari, *Estimating photovoltaic energy potential from a minimal set of randomly sampled data*. Renew. Energy, 97 (2016)
8. D. Wood, *Small Wind Turbines: Analysis, Design, and Application*. New York: Springer. doi: 10.1007/978-1-84996-175-2, (2011)
9. M. F. Akorede, M. I. Mohd Rashid, M. H. Sulaiman, N. B. Mohamed, S. B. Ab Ghani, *Appraising the viability of wind energy conversion system in the Peninsular Malaysia*, Energy Convers. Manag., 76 (2013).
10. T. Kulworawanichpong, J. J. Mwambeleko, *Design and costing of a stand-alone solar photovoltaic system for a Tanzanian rural household*, Sustain. Energy Technol. Assessments, 12 (2015)
11. M. Shravanth Vasisht, J. Srinivasan, S.K. Ramasesha, *Performance of solar photovoltaic installations: Effect of seasonal variations*. Sol. Energy, 131 (2016)

Assessment of various parametric effects on the performance of a commercial grade photovoltaic solar panel

Susanta K. Das[*] and *David* Frank

Department of Mechanical Engineering, Kettering University, 1700 University Avenue, Flint, MI 48504, USA.

Abstract. This paper presents the implementation of an integrated photovoltaic solar panel model which includes all the necessary parameters in order to simulate and analyze using Matlab/Simulink software. The integrated photovoltaic model is validated with Matlab/Simulink simulation for a commercially available 100W photovoltaic solar panel module. Using the commercial solar panel specifications, such as solar irradiation, panel temperature, etc. the integrated model solved with Matlab/Simulink returns an I-V and P-V performance characteristics under various conditions which enabled detailed analysis, assessment of parametric effect on the performance of the solar panel. According to the model simulation results for a commercial RNG-100D solar panel, it was found that panel performance was mostly affected by the variations of solar radiation/insolation, panel surface temperature, series resistance, shunt resistance and band gap energies of the semiconductor materials. The commercial solar panel simulation results show that a photovoltaic panel output power reduces as module temperature decreases. Taking the effect of sunlight irradiance and cell temperature into consideration, the output current and power characteristics of PV model are simulated and results can be optimized using the proposed model. The integrated model enables the dynamics of PV power system to be easily simulated, analyzed, and optimized.

1 Introduction

Increasing world population and rapid expansion of industrialization have put an excessive strain on the fossil fuel based energy supply chain. Fossil fuels such as gas, coal, oil and derivatives of oil account for over 80% of the world's primary energy consumption and out of which 58% is consumed by the transportation sector and rest are used to meet the electrical energy demand [1-3]. Continuous burning of fossil fuels resulted gradual depletion of natural fossil fuel reserve and pollute global climate due to emission of greenhouse gases which compelled policy makers to take action for reduced dependency on fossil fuels and look for renewable energy resources to meet the growing energy demand.

Solar energy is a renewable, environment friendly and inexhaustible source of energy. Moreover, solar energy has numerous direct or indirect advantages such as (i) it does not deplete natural resources, (ii) does not release harmful greenhouse gas emission into air and water or generates liquid or solid waste products, (iii) can accelerate rural or remote areas electrification in developing countries, (iv) may reduce transmission lines from electricity grids, (v) may rescue degraded land, (vi) may increase regional/national energy independence, (vii) may enhance diversification and security of energy and (viii) provide healthy environment by harnessing usable energy. Furthermore, solar ray or radiation is available virtually at any location on the entire earth surface [2-4]. The sun's average radiation intensity to the earth surface is about 1.35kW/m^2. The solar power density within the diameter of the exposed surface area of the earth is 178x10^6 MW. The amount of solar energy can be harnessed from the entire surface of the earth is 1.22x10^{14} tons coal equivalent or 0.814x10^{14} tons of oil equivalent in one year. That is to say, the amount of incoming solar energy on the earth surface in one year is fifty times the known reserves of coal and 800 times the known oil reserves [8]. Hence, if solar energy can be harnessed and used in a proper way, it has the capacity to fulfil multiple times the energy demands of the world population. Since the solar being thousands of time larger as energy source than the present energy consumption rate in the world and hence it enables more and more research in the field of solar energy so that the present and future energy needs of the world can be met.

The solar energy is a dilute source of energy and its availability varies significantly with time and earth's surface location. Now the primary question arises as how to harness solar energy? One important way to convert solar radiation into useful electrical energy through the photovoltaic (PV) effect which is the emergence of an electric voltage between two electrodes attached to a solid or liquid system upon shining light onto the system [3-8]. Photovoltaic (PV) solar cell is basically a p-n junction fabricated in a thin wafer or layer of semiconductor. Usually semiconductor materials absorb a large part of the solar spectrum. As PV cell exposed to the sunlight, photons with energy greater than the band-gap energy of the semiconductor are absorbed and create

[*] Corresponding author: sdas@kettering.edu

some electron-hole pairs proportional to the incident irradiation. Under the influence of the internal electric fields of the p-n junction, these carriers are swept apart and create a photocurrent which is directly proportional to the incident radiation or solar insolation [4-6]. When the PV cell is short circuited, the current flows in the external circuit; when open circuited, the current is shunted internally by the intrinsic p-n junction diode. The characteristics of this diode therefore set the open circuit voltage characteristics of the PV cell [3-6]. Since a typical PV cell produces less than 2W at 0.5V, the PV cells must be connected in series configuration to form a PV module to produce required amount of power. A PV solar panel is a group of several PV modules those are electrically connected in series and parallel circuits to generate the required current and voltage to meet the power demand. Since a PV solar panel integrate current electronics, it is difficult to simulate and analyze the PV computational model which. motivates to develop an integrated PV solar panel model. In this study, an integrated PV solar panel model has been developed and simulated the model using Matlab/Simulink package and conducted detailed assessment of the effects of various physical parameters on the performance of the PV solar panel.

2. PV solar panel model and governing equations

The actual photovoltaic solar panel is the result of combination of a group of photovoltaic cells arranged in series and parallel orientations with their protection devices and energy conversion unit as shown in the Figure 1.

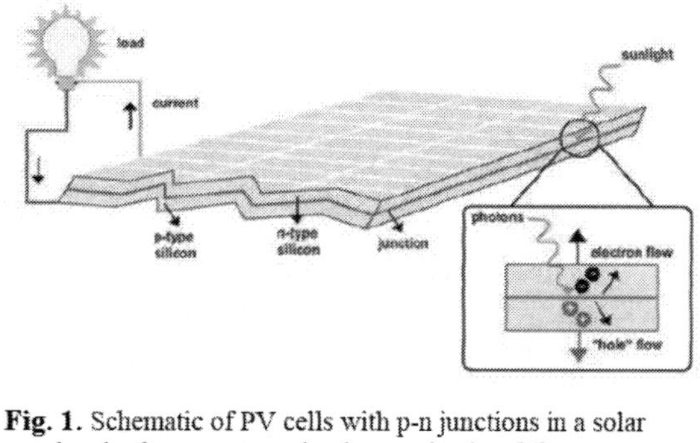

Fig. 1. Schematic of PV cells with p-n junctions in a solar panel and solar energy production mechanism [9].

Hence, the equivalent circuit approach with resistances arranged in parallel and series can be used to develop an integrated computer model for simulation of the actual PV solar panel performance as shown in Fig. 2.

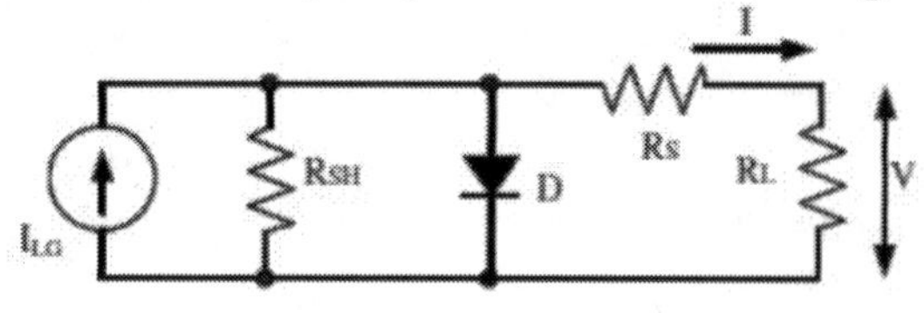

Fig.2. Schematic of equivalent circuit diagram representing PV solar panel [10].

Mathematical model equations of a PV cell has been studied for over past several decades to understand the PV cells I-V output characteristics [1-5] using equivalent circuit approach. As shown in Fig. 2, the equivalent circuit model consists of a photo current, a diode, a parallel resistor expressing a leakage current, and a series resistor describing an internal resistance to the current flow. The voltage-current characteristic equation of the PV solar cell can be written as [3-6]:

$$I_{PV} = I_{PH} - I_S\left[\exp\left(\frac{q(V_{PV}+I_{PV}R_S)}{kT_cA}\right)-1\right]-\frac{(V_{PV}+I_{PV}R_S)}{R_{SH}} \quad (1)$$

where I_{PV} is the cell current, I_{PH} is the photocurrent, I_S is the cell saturation current, $q\left(=1.6\times10^{-19}C\right)$ is an electron charge, $k\left(=1.38\times10^{-23}J/k\right)$ is the Boltzmann's constant, T_C is the cell's operating temperature, A is the field factor, R_{SH} is the shunt resistance, and R_S is the series resistance. The photocurrent mainly depends on the solar radiation and cell's operating temperature, which is described as

$$I_{PH} = \left[I_{SC}+k_I\left(T_C-T_{ref}\right)\right]\cdot\frac{\beta}{1000} \quad (2)$$

where I_{SC} is the cell's short-circuit current at a reference temperature and unit intensity of solar radiation per square meter, k_I is the cell's short-circuit current temperature coefficient, T_{ref} is the cell's reference temperature, and β is the solar radiation or insolation in kW/m^2. The cell's saturation current varies with the cell temperature as [2-6]

$$I_S = I_{RS}\left(T_C/T_{ref}\right)^3\exp\left[\left(\frac{1}{T_{ref}}-\frac{1}{T_c}\right)\frac{qE_g}{kA}\right] \quad (3)$$

where I_{RS} is the cell's reverse saturation current at a reference temperature and a solar radiation, E_g is the bang-gap energy of the semiconductor used in the PV cell. The field factor A is dependent on PV technology [3-6]. The PV cell reverse saturation current can be described as [2-6]

$$I_{RS} = \frac{I_{SC}}{\left[\exp\left(\frac{q\,V_{oc}}{kAT_C}\right)-1\right]} \quad (4)$$

where V_{oc} is the open-circuit voltage which can be obtained by assuming the output current is zero and I_{SC} is the short-circuit current.

If the PV solar panel is made by arranging N_p number of PV cells in parallel and N_s number of PV cells in series then the equivalent circuit equation for representing the solar panel's current and voltage characteristics can be written as [2-7].

$$I_{PV} = N_p I_{PH} - N_p I_S \left[\exp\left(\frac{q\left(V_{PV}/N_s + I_{PV} R_S / N_p\right)}{kT_c A} \right) - 1 \right] - \frac{\left(N_p V_{PV}/N_s + I_{PV} R_S\right)}{R_{SH}} \quad (5)$$

For an ideal PV solar cell, there is no series loss and no leakage to ground, i.e., $R_s = 0$ and $R_{SH} = \infty$. Furthermore, normally photo current is much larger than the saturation current $(I_{PH} \gg I_S)$ and ignoring the small diode and ground-leakage currents under zero-terminal voltage, then the short-circuit current I_{SC} is approximately equal to the photocurrent I_{PH}, i.e., $I_{PH} = I_{SC}$. The maximum power of the PV cell can be expressed as

$$FF = \frac{P_{\max}}{V_{oc} \cdot I_{SC}} = \frac{V_{\max} \cdot I_{\max}}{V_{oc} \cdot I_{SC}} \quad (7)$$

$$P_{\max} = V_{\max} \cdot I_{\max} = FF \cdot V_{oc} \cdot I_{SC} \quad (6)$$

where $V_{\max}$ and $I_{\max}$ are the terminal voltage and output current of the PV cell at maximum power point and FF is the cell fill factor which is a measure of the cell quality. The PV cell's efficiency can then be calculated as

$$\eta = \frac{P_{\max}}{P_{in}} = \frac{V_{\max} \cdot I_{\max}}{A_{area} \cdot G_{in}} \quad (8)$$

where G_{in} is the solar radiation flux occurs on the PV cell.

3. Simulation results of PV solar panel and discussions

In this study, we integrated all the parameters needed to consider in the mathematical model through equations (1)-(8).

Table 1. RNG-100D Specifications (1kW/m2, 25°C).

Characteristics	Specification
Maximum Power at STC (Pmax)	100W
Voltage at Max Power (Vmax)	18.9V.
Current at Max Power (Imax)	5.29A
Open circuit voltage (Voc)	22.5V
Short circuit current (Isc)	5.75A
Maximum system voltage	600VDC
Temperature Coefficient of Pmax	-0.44%/°C
Temperature Coefficient of Voc	-0.30%/°C
Temperature Coefficient of Isc (Ki)	0.04%/°C
Nominal Operating Cell Temperature (NOCT)	47±2°C
Operating Module Temperature	-40°C to +80°C

The integrated mathematical model was simulated using a refined Matlab/Simulink block-diagram program [11]. To simulate and validate the integrated model equations (1)-(8), we used a high performance commercial Renogy RNG-100D [12] 100W monocrystalline PV solar panel. The parametric specifications of the RNG-100D solar panel are shown in Table 1. The specifications given in Table 1 are used to simulate the integrated model using Matlab/Simulink program to study the effect of various model parameters on the characteristics performance of the RNG-100D solar panel.

Figure 3 represents the *I-V* performance curve of a commercially available solar panel, RGN-100D, for different band gap energies. Band gap energy describes jump in potential between a higher and lower electron state. In essence, it is the 'gap' in which no electron states can exist. Thus, for any given semiconductor material, higher and lower electron states differ only by an integer multiple of a constant. That constant, in eV, is a material's band gap energy. In the context of PV cells, a lower band gap energy correlates to a higher short circuit voltage according to equations (1) and (3). As far as performance is concerned, the results in Figure 3 suggest that lower band gaps are ideal, however the limited effect on performance could easily be considered trivial.

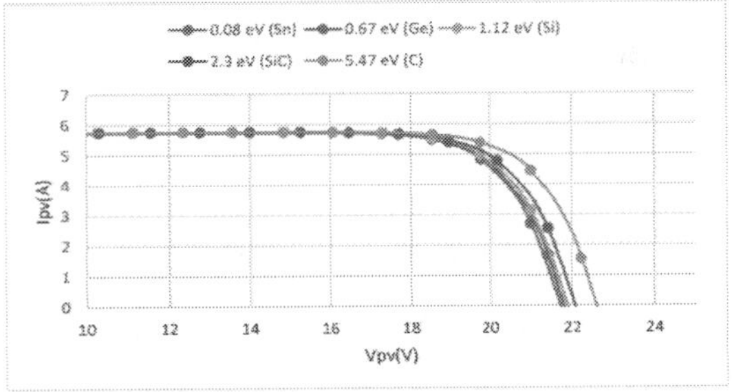

Fig. 3. *I-V* Curves for diffèrent band gap énergies of RNG-100D solar panel

Figure 4 represents the power output of a commercially available RNG-100D solar panel for different band gap energies. The band gaps' effect on power output is directly proportional to its effect on short circuit voltage, and its effects are noticeably small as can be seen from Fig. 4. This comes as no surprise since equation (6) defines power as the product of voltage and current. Yet again, the results from Fig. 4 suggested that lower bandgaps are ideal although the effect on the solar panel performance remains small.

Figure 5 shows performance curves for different amount of solar radiations/flux intensities on a commercially available RNG-100D solar panel. From

Figure 5 it can be seen that the different solar intensities produce the most significant effects on the *I-V* characteristics of the solar panel. This should come as no surprise as sunlight is the 'prime mover' as it were, in PV cells and panels alike. Moreover, equations (1) and (2) affirm this. And as those equations imply, the effect is almost perfectly scalar with the effect seen in an individual PV cell.

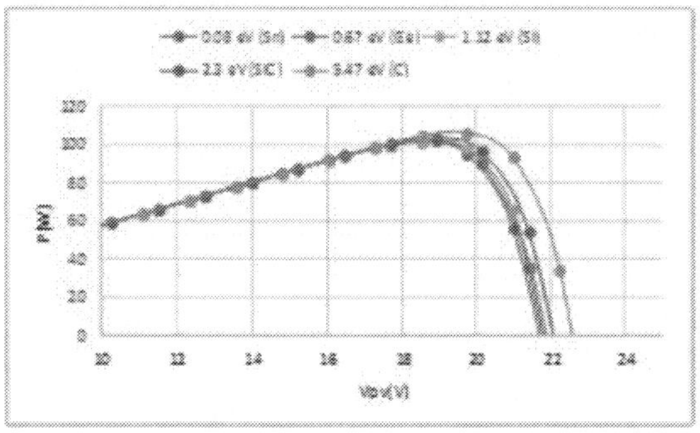

Fig. 4. *P-V* Curves for different band gap energies of RNG-100D solar panel.

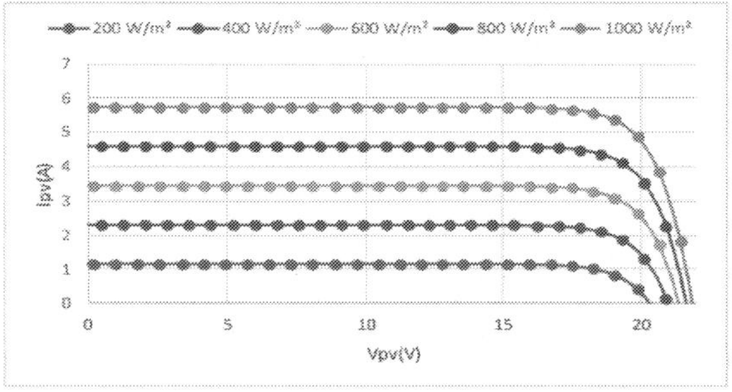

Fig. 5. *I-V* Curves for different solar radiations on RNG-100D solar panel.

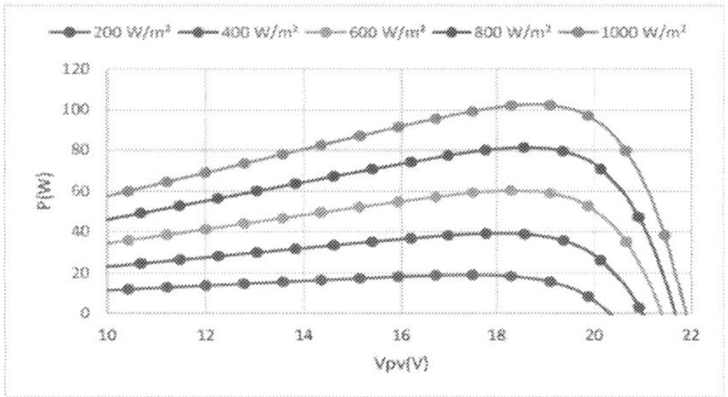

Fig. 6. *P-V* Curves for different Solar Radiations of RNG-100D solar Panel.

Figure 6 represents power curves for different solar radiations for a commercially available RNG-100D solar panel. In accordance with equations (1) and (2), the voltage is unaffected by solar radiation/intensity. Be that as it may, power still suffers as it is a function of current via equation (6). The figure clearly illustrates these predictions.

Figure 7 represents *I-V* performance curves for different temperatures for the commercially available RNG-100D solar panel. From Figure 7 it can be seen that initially the solar panel produce about 5.9A constant current as voltage increases up to 16V. As voltage increases the current production drops sharply with decreased temperature. Figure 8 represents power curves for different temperatures for a commercially available solar panel. Although current is only marginally increased, the subsequent deficit in voltage has a significant effect on power. From Figure 8 it is noted that the wide variation in maximum power for each temperature. Equations (2) and (3) define this effect. This figure reaffirms the conclusion that temperature has a near identical effect on performance regardless of scale. The effect is as significant as it is consistently observed in other solar panel.

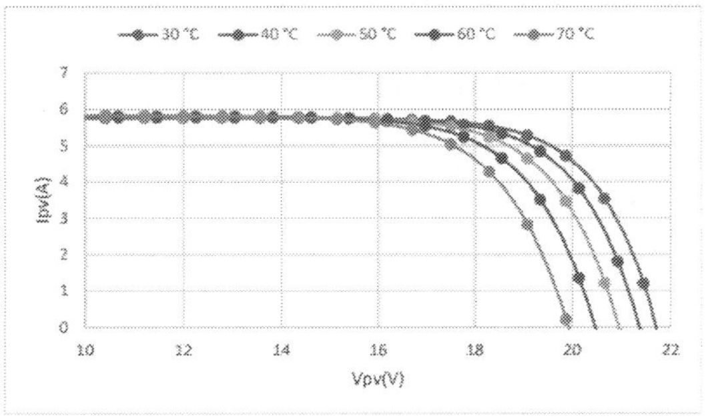

Fig. 7. *I-V* Curves for different surface temperature of RNG-100D solar panel.

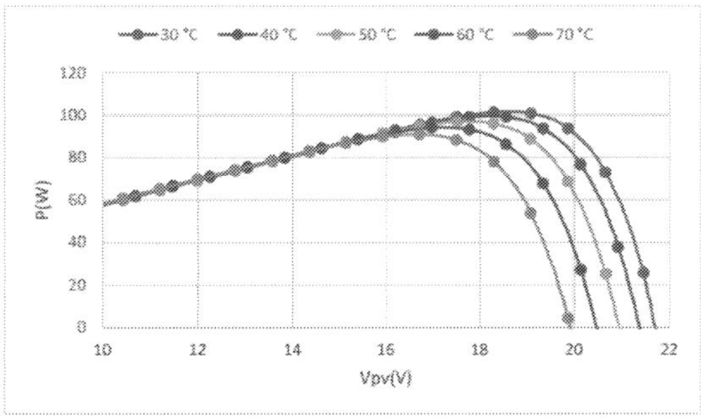

Fig. 8. *P-V* Curves for different surface temperature of RNG-100D solar panel.

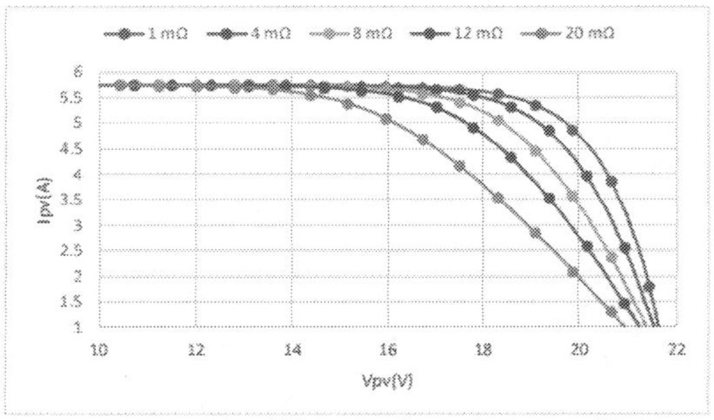

Fig. 9. *I-V* Curves for different series resistances of RNG-100D solar panel.

Figure 9 represents *I-V* performance curves for different series resistances for a commercially available

RNG-100D solar panel. According to Ohm's Law, it was assumed that the total series resistance of the panel would affect its performance in a perfectly linear fashion. The results in Figure 9 do affirm that prediction.

Figure 10 represents *I-V* performance curves for different shunt resistances for a commercially available RNG-100D solar panel. Shunt resistance is an undesired short circuit between the front and back contacts of a solar cell. It is an unavoidable consequence of PV cell manufacture and usually worsens over a cell's lifetime due to wafer damage. For that reason it can be seen as an indicator of quality in a brand new solar panel as well as an indicator of condition in older ones. The results presented in Figure 10 displays how as the shunt resistance decreased, performance of the solar cell dwindled. This can be explained as having increased the number of short circuits, effectively wasting cell current. A higher resistance better resists the flow of this wasted current.

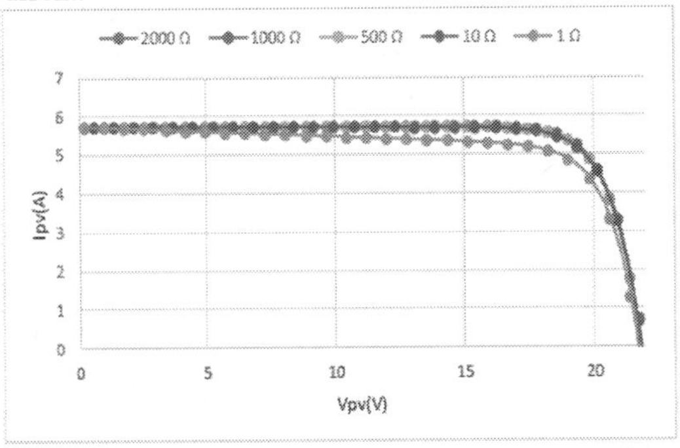

Fig. 10. *I-V* Curves for different shunt resistances of RNG-100D solar panel.

4. Conclusions

An accurate photovoltaic solar panel module is presented based on equivalent circuit approach and demonstrated in Matlab/Simulink for a commercially available Renogy RNG-100D 100W photovoltaic solar panel. The outlined integrated PV solar model is easy to implement on proposed refined Matlab/Simulink modeling and simulation platform. Using the refined Matlab/Simulink block diagram there is now an integrated PV model which can be used for the model and analysis in the filed of PV panel solar energy conversion system. Using the commercial solar panel manufacturer specifications, such as solar irradiation, panel temperature, etc. the integrated model solved with Matlab/Simulink returns an I-V and P-V characteristics under various conditions which enabled detailed analysis, assessment of various parametric effect on the performance of the solar panel. Based on the model simulation results for a commercial RNG-100D solar panel it was found that panel performance was mostly affected by the variations of solar radiation/insolation, panel surface temperature, series resistance, shunt resistance and band gap energies of the semiconductor materials. The commercial solar panel simulation results show that a photovoltaic panel output power reduces as module temperature decreases. The integrated model results obtained by simulation with Matlab/Simulink can be used to optimized solar panel performance and manufacturing process.

References

1. M. Buresch, McGraw-Hill, NY, (1983).
2. S. W. Angrist, Allyn and Bacon, Inc, 4th edition (1982).
3. A. Kumar, K. Kumar, N. Kaushik, S. Sharma, S. Mishra, Renewable and Sustain. Energy Reviews, **14**(8), (2010).
4. J. C. H. Phang, D. S. H. Chan, and J. R. Philips, Electronics Letters, **20**(10), (1984).
5. S. Rustemli and F. Dincer, J. Electronics and Electrical Engineering, **35** (2011).
6. S. Nema, R. Nema, and G. Agnihotri, Int.l J. Energy Environ., **1**(3) (2010).
7. A. Mellora, D. A. Alvareza , I. Guarracinob, A. Ramosb, A. R. Lacastac, L. F. Llind, A.J. Murrelle, D.J. Pauld, D. Chemisanac, C.N. Markidesb, N.J. Ekins-Daukes, J. Solar Energy, **174** (2018).
8. A. Ramos, M. A. Chatzopoulou, I. Guarracino, J. Freeman, C. N. Markides, J. Energy Conversion and Management **150** (2017).
9. B. Miedzinski, Electronics and Electrical Eng. – Kaunas: Technologija. **5**(101), (2010).
10. M. Demirta|, e-Journal of New World Sciences Academy, **4**(3), (2009).
11. J. B. Dabney and T. L. Harman, Mastering Simulink, Pearson Education,Inc. (2004).
12. Renogy. (n.d.). *RNG-100D*. Retrieved from www.renogy.com (2019)

Modeling and Thermal Analysis of a CCP Collector System Based on Fractal Architecture: Receiver Proposal

Angélica Palacios [1], *Darío* Amaya [2,*], *Olga* Ramos [3]

[1]u1801712@unimilitar.edu.co, Universidad Militar Nueva Granada, Colombia
[2]dario.amaya@unimilitar.edu.co, Universidad Militar Nueva Granada, Colombia
[3]olga.ramos @unimilitar.edu.co, Universidad Militar Nueva Granada, Colombia

Abstract. Solar concentrator technology has been researches in different fields in order to enhancement efficiency and energy storage of these systems. Some changes in fluid and numerous components of the collectors has been proposed in recent years. In this context, this paper present results associates to the modeling and thermal analysis of a concentrating system based on parabolic through collector with receiver piper, designed with fractal geometry with the aim of improve system heat transfer. Founded on thermal modeling of heat transfer phenomena as radiation, convection and conduction, physical and mathematical relations between fractal geometrical parameters and transfer heat coefficient were established to find the influence of a chaotic structure on thermal behavior. Proposed designs were simulated through Solid works ® Flow Simulation tool. Was possible obtain maximum temperatures in air with fractal and cylindrical pipes of de 89°C, 86°C y 81°C respectively. The gap between fractal geometry results and cylindrical geometry is 10°C around.

1 Introduction

Within technology with high development in solar energy, we found first of all, photovoltaic and concentrating collectors. The last, count with a system of thermal energy storage, wherewith is possible improve the controlling capacity of the concentrating system output [1]. For the year 2050 it hopes that concentrating stations by parabolic cylindrical collectors (CCP), satisfices global demand in a 6%, [2]. At the same, energy system will benefit by this kind of technology, contaminate emissions decrease, electrical trade reliability and renewable energies, will be evident across de world. [3]–[6]. Within advantages it's found accessible cost, high concentration, thermal storage wherewith the system works full power, even if solar radiation is not available for hours [7].

Some studies and researches globally, about solar collection systems by parabolic cylindrical collectors, are related with heat transfer enhancement in receiver pipe under different guidelines, [8]–[10]. Some of them directly interact with pipe and other ones with transfer fluid. In the last one, Nano-fluids and Nano-particles concepts are involves in collector thermal efficiency, [11]–[13] . Although presently this kind of fluids are not applied in CCP stations, use of them involve specific requirements in hydraulic elements as pumps, valves, connection pipes, among others, [14]. Nano-fluids potential has been experimented in [15], where heat transfer with forced convection was studied, inside the receiver of a parabolic cylindrical collector. From there, was possible identify a 28% increase of thermal coefficient with Al2O3 – H2O combination and a 35% increase with CuO- H2O combination regard to pure water.

Furthermore, receiver pipe modification has been achieved in different researches, some works present mechanical design where pipe has inside fin kind incrustations. At the same, receiver size has been studied, with the purpose of enhancement system heat transfer, [16], with the thermal resistance decrease, turbulence intensity increase and increase of effective thermal conductivity. As presented in [17], [18], where thermal and thermodynamic efficiency of a CCP was analyzed with a perforated plate, geometrical parameters related to the plate as slope angle, were modified in the research, a 8% increase was obtained in system efficiency.

Within modifications developed in receiver to enhancement the transference, is proposed convex pipes, symmetrical and asymmetrical. In papers as [19], [20], was possible obtain varied in efficiency with Reynolds number increase in some geometrical parameters. Finally, a 135% increase was obtained in heat transfer for a symmetrical geometry and 148% increase for asymmetrical geometry. Use of fractal geometry has been developed in some researches to enhancement flow conditions. In [21] modelling and experimental study, was developed about flow distribution and heat transfer of a system conformed with hollow fibers in a fractal array structure. Likewise, friction factor and irregularity according to fractal dimension was studied. Wherewith

* Corresponding author: dario.amaya@unimilitar.edu.co

results it's found, that a minor fractal dimension the uniformity in flow distribution decrease.

This paper present the results obtained with a modelling and thermal analyses of a concentrating system by parabolic cylindrical collector with a pipe receiver, designed based on fractal geometries. With the purpose of enhancement in heat transfer modifying hydraulic diameter and transversal area where heat fluid flow.

2 Methods

2.1 Solar Resource

A high quantity of energy equivalent to 1.74×10^7W , is radiated by Sun to Earth atmosphere. However, this energy don't achieved in a 100% the surface, is attenuated by reflection, absorption and dispersion phenomena involved in chemistry components which conform air planet. Thus, almost 51% of radiated energy achieved the surface, despite the percentage represent a high energy [22], [23]. On a parabolic cylindrical collector, as shown in Figure 1, radiated energy by Sun and focused on Earth is concentrated on an area, known as aperture area.

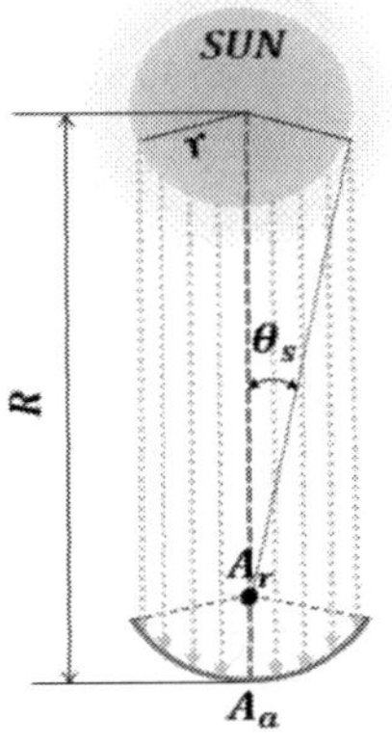

Figure 1. Solar-collector relation.

2.2 CCP Construction by Parabolic Geometry

A construction of parabolic cylindrical collector, is based on parabolic equation relation (Eq.(1), which parable aperture depends of the height and focal distance.

$$X^2 = 4fY \quad (1)$$

Under rays concentrating theory as of parabolic surface to receiver pipe, when focusing rays by reflection in parable geometry of focal distance. Related parameters in concentrating CCP theory, are illustrated in Figure 2.

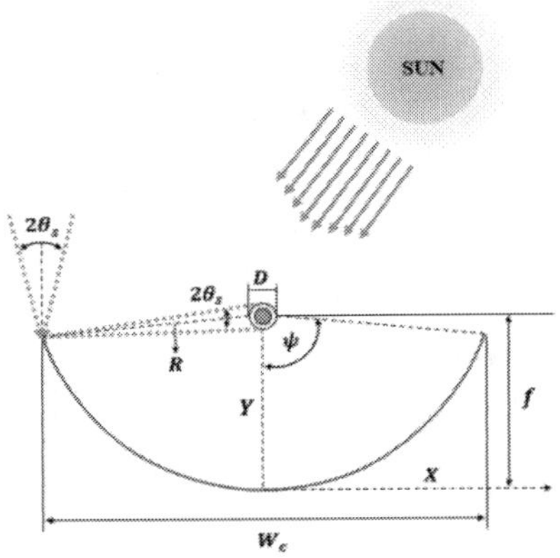

Figure 2. Parabolic Cylindrical Collector geometry.

To calculated keys parameters of collector design, as receiver diameter (D), parable width (W_c), focal distance (f) and finally concentration ratio (C_n) are stablish geometrical relations presented down below. Receiver diameter (Eq.(2), is related with admissible incident angle (θ_s=16'), through trigonometric association with distance between concentrating point [R] and parabolic surface.

$$D = 2 \times R \times \sin(\theta_s) \quad (2)$$

Similarly, the aperture width of parable is obtained with trigonometric relations, between rhim angle and distance (R), as shown in (Eq.3).

$$W_c = 2 \times R \times \sin(\psi) = \frac{4f \sin\psi}{1 + \cos\psi} = 4f \times \tan\left(\frac{\psi}{2}\right) \quad (3)$$

Finally, concentrating ratio of a parabolic cylindrical collector (C_n), is realted between aperture area of collector (A_a) and focus area o receiver (A_r). This parameter is described by (Eq.4).

$$C_n = \frac{A_a}{A_r} = \frac{W_c}{\pi D} \quad (4)$$

To obtain design parameters, a concentration factor was set of ($C_n = 10$) and a diameter of ($D = 0.0254$ m). Other surface and receiver dimensions are presented in Table 1.

Table 1 .CCP Dimensions

Surface Dimensions		
Parameter	**Value**	**Unity**
W_c	0.798	$[m]$
f	0.199	$[m]$
X_{max}	-0.4	$[m]$
Y_{max}	0.4	$[m]$
ψ	90	$[°]$
R	0.4	$[m]$
L_c	0.92	$[m]$
Receiver Dimensions		
D	0.0254	$[m]$
L_r	0.92	$[m]$

https://doi.org/10.1051/e3sconf/201912202003

2.3 Proposed Design

Once conventional parabolic cylindrical collectors were analyzed, as well as enhancement elements to improve heat transfer. In this paper is proposed a design of a pipe based on fractal geometry with the purpose of enhance area and transfer coefficients. Pipe geometries are presented down below in Figure 3.

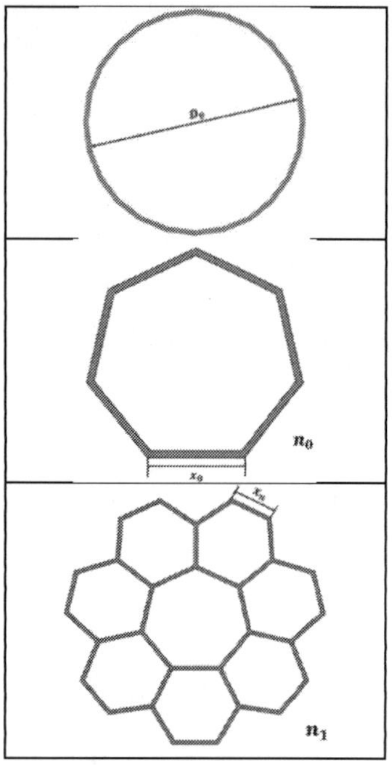

Figure 3. Receiver geometry.

Sides number (k_n) of fractal depends of initial value ($k_0 = 7$), in next iterations ($\{n_1, n_2, \ldots, n_n\}$), the representation is given by (Eq.(5). Geometry scale factor (f_s) consist in number of elements involve on geometry, (n_0 $f_{s=1}$). For next iterations factor can be defined as (Eq.(6).

$$k_n = 3 \times k_{n-1} \tag{5}$$

$$f_s = \frac{k_n}{3} + 1 \tag{6}$$

Fractal geometry analysis on receiver thermal study is based on transfer area and the relation with convection coefficient, through hydraulic diameter ($D_{h[n]}$). The last depends of transversal area (A_t) and perimeter (P) of seven sides polygon (n) as (Eq.7).

$$D_{h[n]} = \frac{4 \times A_{[n]}}{P_{[n]}} \tag{7}$$

2.4 Study Conditions

CFD Model

Laminar model of Flow Simulation tool in SolidWorks software, Works with Navier Stokes equations [24], where conservation of mass, momentum and energy laws are related on a cartesian coordinate system [25], [26]. These equations are presented down below [Eq.(8 - Eq.(10].

$$\frac{\partial \rho}{\partial t} + \frac{\partial}{\partial x_i}(\rho u_i) = 0 \tag{8}$$

$$\frac{\partial \rho u_i}{\partial t} + \frac{\partial}{\partial x_j}(\rho u_i u_j) + \frac{\partial p}{\partial x_i} = \frac{\partial}{\partial x_j}(\tau_{ij} + \tau^R{}_{ij}) + S_i \tag{9}$$

$$i = 1,2,3$$

$$\frac{\partial \rho H}{\partial t} + \frac{\partial \rho u_i H}{\partial x_i} = \frac{\partial}{\partial x_i}(u_j(\tau_{ij} + \tau^R{}_{ij}) + q_i) + \frac{\partial p}{\partial t} - \tau^R{}_{ij}\frac{\partial u_i}{\partial x_j} + \rho\varepsilon + S_i u_i + Q_H, H = h + \frac{u^2}{2} \tag{10}$$

3 Analysis and Results

Study results for each proposed receiver are presented independently. First is described outlet temperature in fluid and concentrator temperature increasing, with cylindrical pipe for each fluid (air, refrigerant and ethylene).

3.1 Cylindrical Receiver

Maximum temperature in air, refrigerant and ethylene inside of cylindrical pipe was 81.56°C, 83.79°C and 80.216°C respectively. The maximum delta of temperature was found in refrigerant as fluid, where initial temperature was increased around 62°C. The results are related in Table 2.

Table 2. Cylindrical Receiver

Fluid	*Solid*		*Fluid*	
	T_{max} [°C]	ΔT [°C]	T_{max} [°C]	ΔT [°C]
Air	81.57	59.76	81.57	59.75
R4010	83.79	61.97	83.79	61.97
Ethylene	80.22	58.39	80.21	58.39

Inside and outside temperature distribution of cylindrical pipe are shown in the image of Figure 4. Temperature vary between 64°C and 77°C on pipe zone, was the highest temperature on solid.

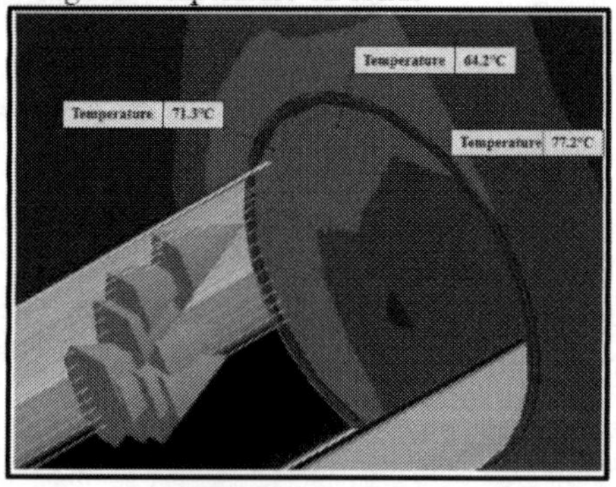

Figure 4. Thermal results of receiver with air.

Curves in Figure 5 present the temperature of each fluids inside cylindrical pipe. As shown refrigerant was the highest temperature followed by air and ethylene.

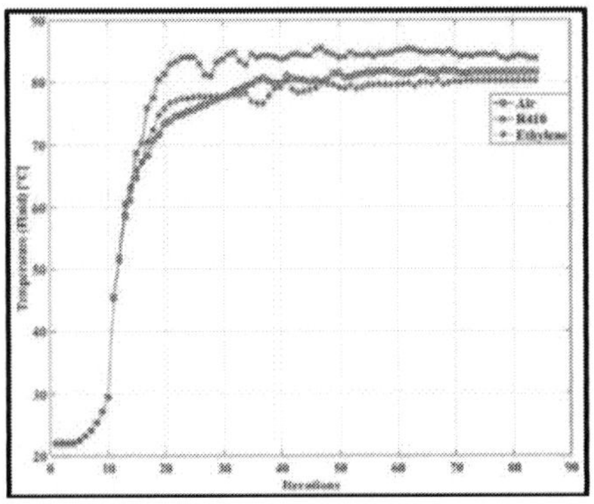

Figure 5. Thermal graph of cylindrical pipe.

3.2 Fractal n_0 Receiver

Maximum temperature and delta temperatures obtained on receiver and each analyzed fluids are presented on Table 3. Highest temperature in receiver was 86.54°C with air as fluid, while refrigerant achieved a maximum temperature of 80.95°C , finally ethylene present a temperature of 6.57°C less than the maximum obtained. Similar variation to the calculated with refrigerant as fluid.

Table 3. Receiver-Fractal N0

Fluid	*Solid*		*Fluid*	
	T_{max} [°C]	ΔT [°C]	T_{max} [°C]	ΔT [°C]
Air	86.55	63.87	86.55	64.36
R4010	80.96	58,91	80.96	59.06
Ethylene	79.97	58	79.97	58.34

Figure 6 shows temperature distribution on pipe and fluid which flow through receiver. The highest temperature on this section was in pipe surface with a 84.46°C, furthermore air achieved a temperature of 76.67°C, while envirement remain in 22°C.

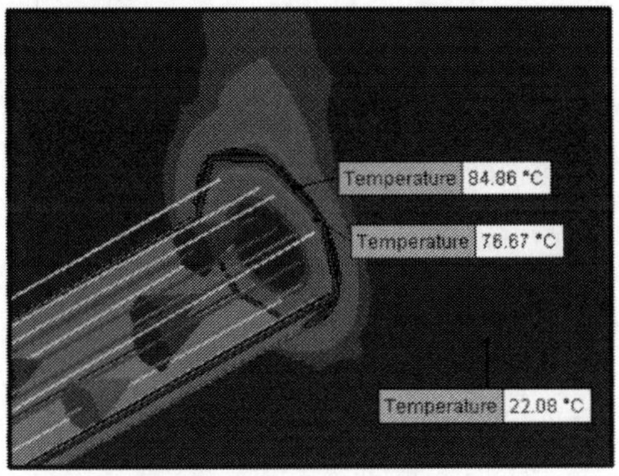

Figure 6. Thermal results of receiver with air.

As shown in graph of Figure 7, air was the fluid with the highest temperature, followed by refrigerant and ethylene. Each of the fluids present a temperature increase of 64.35°C, 59.06°C y 58.33°C respectively.

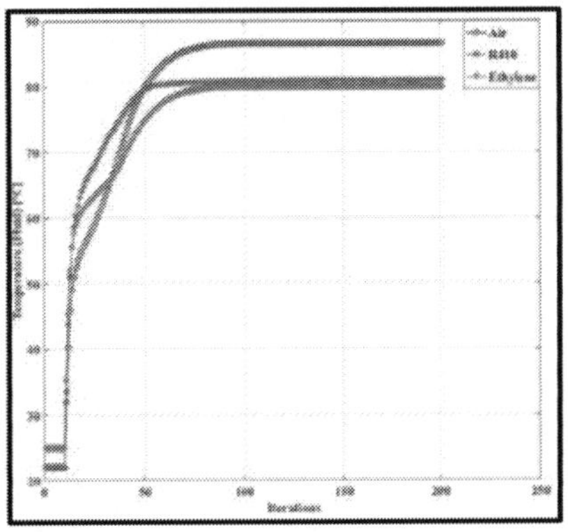

Figure 7. Thermal graph of pipe n0.

3.3 Fractal n_1 Receiver

Obtained results in the analysis of solar concentrator with Fractal pipe n_1, are describes in Table 4. Receiver temperature and different fluids are presented in this table. In contrast to receiver n_0 fluid which system achivied the highest temperature in receiver n_1 was refrigerant with a temperature of 88.84°C , however difference between air is only 0.19°C, thus values are really nearby to compare. Furthermore, with ethylene the solid achieved a maximum temperature of 83.74°C and a delta of 62.68°C.

Table 4. Receiver-Fractal N1

Fluid	*Solid*		*Fluid*	
	T_{max} [°C]	ΔT [°C]	T_{max} [°C]	ΔT [°C]
Air	88.65	67.47	88.66	67.65
R4010	88.84	67.76	88.85	67.76
Ethylene	83.75	62.69	83.75	62.69

Fluid temperature (R410) inside the fractal pipe n_1, is ilustred in Figure 8. Through the image is possible evidence the temperature distribution from the inside to exterior pipe, where areas with a maximum heat concentration find in receiver pipe.

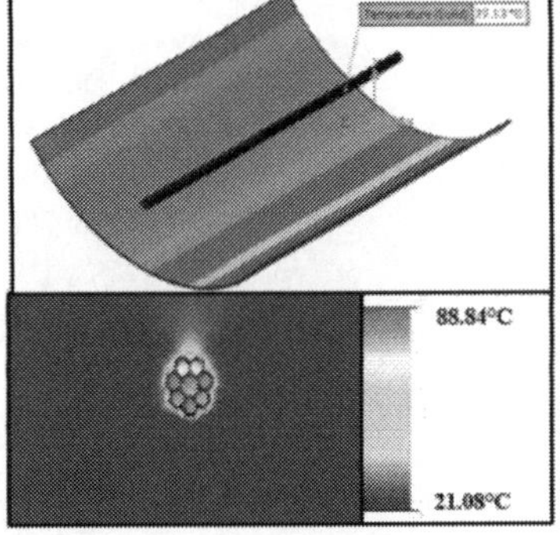

Figure 8. Receiver n_1 thermal results with R410.

Finally, graph in Figure 9 illustrated the temperature in different fluids (air, refrigerant and ethylene) studied in geometry of receiver n_1. Based on the figure is possible to identify closeness between thermal performance in air and refrigerant, while ethylene show a difference regard to the others fluids around 5°C.

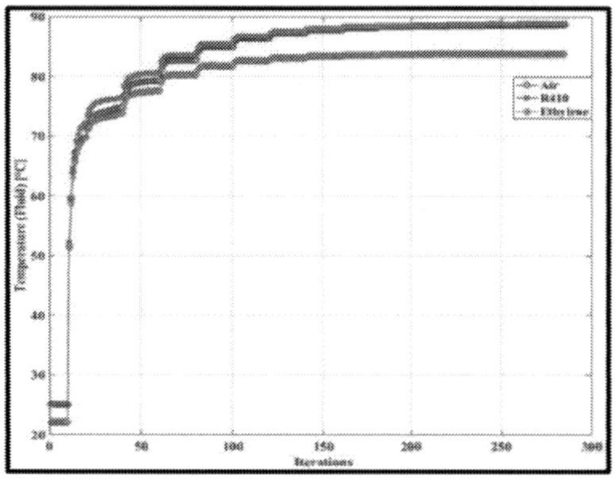

Figure 9. Thermal graph of pipe n_1.

As was evident the geometry with the highest temperature in fluid was the concentrator with fractal receiver n_1, achivied a maximum value of 88.84°C with refrigerant. The system with the lowest temperature both fluid and solid was the concentrator with cylindrical receiver, which obtain a maximum temperature of 83.79°C with refrigerant as fluid. Table 5 present the results unified for analyzed systems.

Table 5. Receiver results.

Pipe	Fluid	T max [°C]
Cylindrical	Air	81.57
	R410	83.79
	Ethylene	80.21
Fractal n_0	Air	86.55
	R410	80.96
	Ethylene	79.97
Fractal n_1	Air	88.66
	R410	88.85
	Ethylene	83.75

4 Conclusions

Different geometries for pipe receiver of a parabolic cylindrical collector were studied in this work. Both conventional cylindrical pipe, heptagonal pipe and next iteration of lineal fractal, general heat transfer coefficient or convection coefficient, vary according to hydraulic diameter and transfer area, which are related with Nusselt and Reynolds number, by reducing in each fractal iteration the geometry diameter is increase the transfer heat.

The fractal tube n_1 was the system that reached the highest temperature in simulation, under the same conditions and the different fluids used in the study. The maximum temperature reached in this tube was approximately 89°C with an increase in the initial temperature of 68°C, with R410 refrigerant as the transfer fluid. Furthermore, the conventional cylindrical pipe reached the lowest temperatures of the study with an average of 81.85°C in the fluids.

For the cylindrical geometry and the fractal n_1 in the receiver pipe, the fluid with the best thermal behavior was the air, while the fractal geometry n_0 presented better results with air. The difference between this and R410 refrigerant was 6°C, with a maximum temperature in the fluid of 86.55°C. In turn, it was possible to identify that the difference in the results obtained by the fractal geometries was approximately 2°C, while the difference with the cylindrical geometry was approximately 10°C.

Acknowledgment

The authors would like to thank the Nueva Granada Military University research center for financing this work (research project IMP-ING-2656, 2019).

References

1. S. Zhao Y. Fang, and Z. Wei, "Stochastic optimal dispatch of integrating concentrating solar power plants with wind farms," *Int. J. Electr. Power Energy Syst.*, vol. **109**, pp. 575–583, (2019).
2. S. Izquierdo, C. Montanes, C. Dopazo, and N. Fueyo, "Analysis of CSP plants for the definition of energy policies: The influence on electricity cost of solar multiples, capacity factors and energy storage," *Energy Policy*, vol. **38**, no. 10, pp. 6215–6221, (2010).
3. K. Dallmer-Zerbe, M. Bucher, A. Ulbig, and G. Andersson, "Assessment of capacity factor and dispatch flexibility of concentrated solar power units," in *IEEE Grenoble Conference*, (2013), pp. 1–6.
4. P. Denholm and M. Hummon, "Simulating the value of concentrating solar power with thermal energy storage in a production cost model," *Contract*, vol. **715**, no. 6, pp. 1–4, (2013).
5. S. Madaeni, R. Sioshansi, and P. Denholm, "Estimating the capacity value of concentrating solar power plants with thermal energy storage: A case study of the southwestern United States," *IEEE Trans. Power Syst.*, vol. **28**, no. 2, pp. 1205–1215, (2013).
6. G. He, Q. Chen, C. Kang, and Q. Xia, "Optimal offering strategy for concentrating solar power plants in joint energy, reserve and regulation markets," *IEEE Trans. Sustain. Energy*, vol. **3**, no. 3, pp. 1245–1254, 7AD.
7. R. Sioshansi and P. Denholm, "The value of concentrating solar power and thermal energy storage," *IEEE Trans. Sustain. Energy*, vol. **1**, no. 3, pp. 173–183, (2010).
8. H. Ghaebi, T. Parikhani, H. Rostamzadeh, and B. Farhang, "Thermodynamic and thermoeconomic analysis and optimization of a novel combined cooling andz power (CCP) cycle by integrating of

ejector refrigeration and," *Energy*, vol. **139**, pp. 262–276, (2017).

9. H. Rostamzadeh, M. Ebadollahi, H. Ghaebi, M. Amidpour, and R. Kheiri, "Energy and exergy analysis of novel combined cooling and power (CCP) cycles," *Appl. Therm. Eng.*, vol. **124**, pp. 152–169, (2017).
10. M. Pan, I. Bulatov, and R. Smith, "Improving heat recovery in retrofitting heat exchanger networks with heat transfer intensification, pressure drop constraint and fouling mitigation," Appl. Energy, vol. 161, pp. 611–626, 2016.
11. B. Balakin, O. Zhdaneev, A. Kosinska, and K. Kutsenko, "Direct absorption solar collector with magnetic nanofluid: CFD model and parametric analysis," *Renew. Energy*, vol. **136**, pp. 23–32, (2019).
12. M. Dehaj and M. Cells, "Experimental investigation of heat pipe solar collector using MgO nanofluids," *Sol. Energy Mater. Sol. Cells*, vol. **191**, (2019).
13. K. Farhana et al., "Improvement in the performance of solar collectors with nanofluids—A state-of-the-art review," *Nano-Structures & Nano-Objects*, vol. **18**, (2019).
14. A. Kasaeian, A. Eshghi, and M. Sameti, "A review on the applications of nanofluids in solar energy systems," *Renew. Sustain. Energy Rev.*, vol. **45**, pp. 584–598, (2015).
15. S. Ghasemi and A. Ranjbar, "Thermal performance analysis of solar parabolic trough collector using nanofluid as working fluid: a CFD modelling study," *J. Mol. Liq.*, vol. **222**, pp. 159–166, (2016).
16. E. Baysal, A. R. Dal, and N. Şahin, "Investigation of heat transfer enhancement in a new type heat exchanger using solar parabolic trough systems," *Int. J. Hydrogen Energy*, vol. **40**, no. 44, pp. 15254–15266, Nov. (2015).
17. A. Mwesigye, T. Bello-Ochende, and J. P. Meyer, "Multi-objective and thermodynamic optimisation of a parabolic trough receiver with perforated plate inserts," *Appl. Therm. Eng.*, vol. **77**, pp. 42–56, (2015).
18. A. Mwesigye, T. Bello-Ochende, and J. P. Meyer, "Heat transfer and thermodynamic performance of a parabolic trough receiver with centrally placed perforated plate inserts," *Appl. Energy*, vol. **136**, pp. 989–1003, (2014).
19. W. Fuqiang, L. Qingzhi, H. Huaizhi, and T. Jianyu, "Parabolic trough receiver with corrugated tube for improving heat transfer and thermal deformation characteristics," *Appl. Energy*, vol. **164**, pp. 411–424, (2016).
20. W. Fuqiang, T. Zhexiang, G. Xiangtao, T. Jianyu, H. Huaizhi, and L. Bingxi, "Heat transfer performance enhancement and thermal strain restrain of tube receiver for parabolic trough solar collector by using asymmetric outward convex," *Energy*, vol. **114**, pp. 275–292, (2016).
21. L. Z. Zhang, "Heat and mass transfer in a randomly packed hollow fiber membrane module: a fractal model approach," *Int. J. Heat Mass Transf.*, vol. **54**, pp. 13–14, (2011).
22. T. Li, R. Wang, J. Kiplagat, and Y. Kang, "Performance analysis of an integrated energy storage and energy upgrade thermochemical solid–gas sorption system for seasonal storage of solar thermal," *Energy*, vol. **50**, pp. 454–467, (2013).
23. X. Meng, X. Xia, C. Sun, and X. Hou, "Adjustment, error analysis and modular strategy for Space Solar Power Station," *Energy Convers. Manag.*, vol. **85**, pp. 292–301, (2014).
24. X. J. Yang, D. Baleanu, and J. A. Tenreiro Machado, "Systems of Navier-Stokes equations on Cantor sets.," *Math. Probl. Eng.*, (2013).
25. G. Galdi, "An introduction to the mathematical theory of the Navier-Stokes equations: Steady-state problems," (2011).
26. J. Ferziger and M. Peric, *Computational methods for fluid dynamics*. (2012).

E3S Web of Conferences **122**, 02004 (2019) https://doi.org/10.1051/e3sconf/201912202004
REEE 2019

Economic feasibility of developing large scale solar photovoltaic power plants in Spain

Javier Menéndez[1,*] and *Jorge* Loredo[2]

[1]HUNASER ENERGÍA, 33005 Oviedo, Spain
[2]UNIVERSITY OF OVIEDO, Mining Exploitation Department, 33004 Oviedo, Spain

Abstract. In 2017, electricity generation from renewable sources contributed more than one quarter (30.7%) to total EU-28 gross electricity consumption. Wind power is for the first time the most important source, followed closely by hydro power. The growth in electricity from photovoltaic energy has been dramatic, rising from just 3.8 TWh in 2007, reaching a level of 119.5 TWh in 2017. Over this period, the contribution of photovoltaic energy to all electricity generated in the EU-28 from renewable energy sources increased from 0.7% to 12.3%. During this period the investment cost of a photovoltaic power plant has decreased considerably. Fundamentally, the cost of solar panels and inverters has decreased by more than 50%. The solar photovoltaic energy potential depends on two parameters: global solar irradiation and photovoltaic panel efficiency. The average solar irradiation in Spain is 1,600 kWh m^{-2}. This paper analyzes the economic feasibility of developing large scale solar photovoltaic power plants in Spain. Equivalent hours between 800-1,800 h $year^{-1}$ and output power between 100-400 MW have been considered. The profitability analysis has been carried out considering different prices of the electricity produced in the daily market (50-60 € MWh^{-1}). Net Present Value (NPV) and Internal Rate of Return (IRR) were estimated for all scenarios analyzed. A solar PV power plant with 400 MW of power and 1,800 h $year^{-1}$, reaches a NPV of 196 M€ and the IRR is 11.01%.

1 Introduction

The use of renewable energies in the generation of electric power has increased considerably in recent years. Gross electricity generation in Organisation for Economic Co-operation and Development (OECD) from renewable energy sources (excluding generation from pumped storage plants) reached 2,588.3 TWh in 2016, a 3.8% increase from the 2015. This represents 23.8% of total OECD electricity production in 2016 (Fig. 1), which is the largest share of renewables in gross electricity production for any year in the renewables time series beginning from 1990 [1].

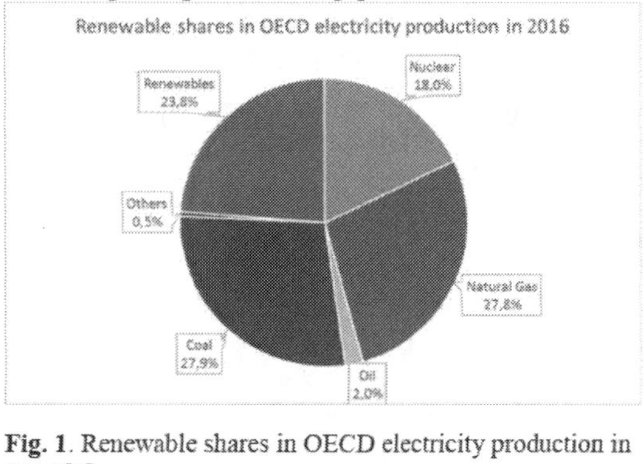

Fig. 1. Renewable shares in OECD electricity production in 2016 [1].

The increase in electricity production from renewables was mainly caused by wind and solar photovoltaic (PV). Solar PV increased by 35.0 TWh, again mainly driven by the US, which had increased electricity production from solar PV by 18.0 TWh, followed by Japan (8.0 TWh), UK (2.7 TWh) and Chile (1.3 TWh) [1].

Solar energy is currently the lowest cost power generation source in many regions of the world – and its cost continues to decrease rapidly. Solar has the potential to play a major role in the European Union meeting its 32% renewables target by 2030. Bloomberg NEF, in its New Energy Outlook 2018, anticipates that renewables will cover 87% of Europe's electricity generation by 2050 in Europe, in which 1.4 TW of solar is installed and contributes to 36% of total power generation.

The increase of renewable energy sources requires flexible energy storage systems. Pumped-storage hydropower (PSH) is the most mature and efficient technology that currently exists. Other systems, such as Compressed Air Energy Storage (CAES) plants, can also be used. Underground pumped storage power plants can be an alternative using underground space, such as closed mines. [2-6].

Fig. 2 shows the renewable energy sources in electricity in EU'28 in 2017. The share of renewable energy sources has increased from 16.97% to 30.75% in

* Corresponding author: javiermenendezr@gmail.com

https://doi.org/10.1051/e3sconf/201912202004

the period 2008-2017. It highlights, Austria, Sweden, Denmark, Latvia and Portugal, with a share in the electric mix higher than 50% in 2017. The share of renewable energies in electricity in Spain was 36.34%.

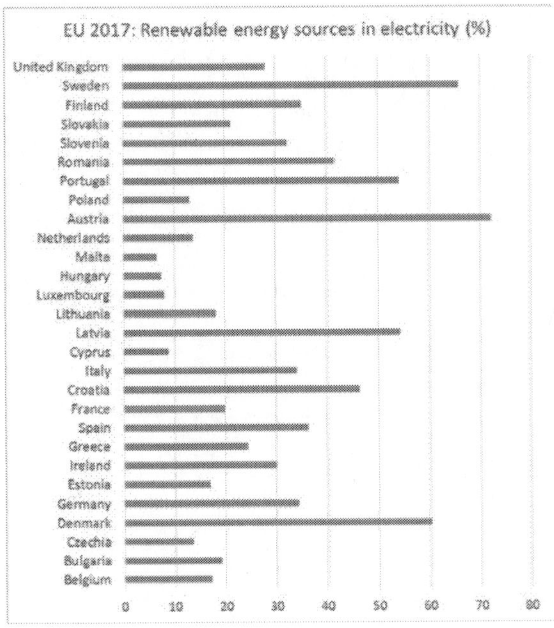

Fig. 2. Renewable energy sources in electricity in EU'28 (2017).

This paper analyzes the economic feasibility of developing large scale solar PV projects in Spain. Different output power (100-400 MW) and equivalent hours per year, depending on the situation of the solar PV power plant (800-1,800 h year^{-1}) have been considered in this study. A profitability analysis has been carried out for different prices of the electricity produced in the daily market (50-60 € MWh^{-1}). Net Present Value (NPV) and Internal Rate of Return (IRR) were obtained in all scenarios.

2 Materials and methods

2.1 Solar energy in Spain

Spain offers optimal conditions for the installation of solar PV power plants [7]. Fig. 3 shows a detailed map of the annual accumulated solar irradiation level (kWh m^{-2}) for EU considering optimally-inclined PV modules [8, 9]. In Spain, the average values vary from 1,000 hours year^{-1} (equivalent hours) on the northern and 1,700 hours year^{-1} (equivalent hours) on the southaestern zone [10]. The national average is 1,600 kWh m^{-2} [11].

Electricity generation in Spain from renewable energy spurces reached 101,6 TWh in 2018, a 40.1% of total electricity production in 2018. The total installed power in Spain reached 98.5 GW in 2018. The total power of solar PV accumulated in 2018 reached 4,436 MW (4.5% of the total installed power).

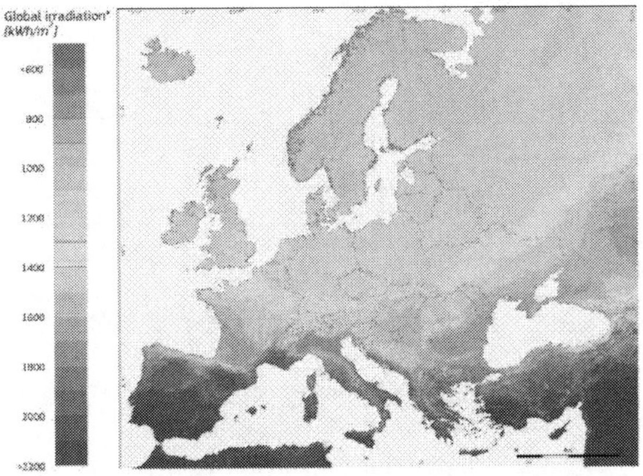

Fig. 3. Map of the annual accumulated solar irradiation level for EU. Optimally-inclined PV modules [8].

2.2 Solar energy calculator

The solar energy potential depends on two parameters: solar irradiation and PV panel efficiency. The power output of the PV is given by Eq. (1).

$$P_{PV} = \sum Y_{PV} D_f \left[1 + \alpha(T_o - T_{C,STC})\right] P_{VN} \qquad (1)$$

where:

P_{PV} : Annual PV solar power output
Y_{PV} : PV panel power output
D_f : Degradation factor (0.5% year^{-1})
α : Temperature loss coefficient
T_o : Operating temperature
$T_{C,STC}$: Temperature at Standard Test Conditions (25°C)
P_{Vn} : Number of PV panels
The panel power output, Y_{PV} (kW) is given by Eq. (2). Finally, the total power produced by PV power (P_{PV}) plant is given by Eq. (3)

$$Y_{PV} = R_{GHI} A\, d_m \eta_{PV} \qquad (2)$$

$$P_{SP} = P_{PV} L_{SP} \qquad (3)$$

where:

R_{GHI} : Monthly average solar irradiation (kWh m^{-2})
A : Area of PV panel (m^2)
d_m : Number a days per month
η_V : PV conversion efficiency
P_{SP} : Power output of solar plant (kWh)
L_{SP} : Solar plant losses

Table 1 shows the system losses in a solar PV power plant. Considering a useful life of 25 years, the PV module degradation reaches a value of 12.5% of the total energy produced in year 25. The solar PV modules must be periodically cleaned to avoid losses in the generation of electricity. The typical power of a PV module (peak power) is 315 W, with a dimensions of 1,956x992 mm.

Table 1. Solar PV power plant. System losses

Solar PV plant system losses	
PV module degradation (D)	0.5% year^{-1}
PV module performance due to temperature (α)	0.41% °C^{-1}
Inverter efficiency loss	2.5-5%
Loss due to soiling of module surface	0.5-10%
AC wiring	1-3%
DC wiring	1-2.5%
Connectors	1-2.5%
System availability	1-3%

2.3 Economic evaluation: NPV and IRR

The economic evaluation consists in the analysis of the profits obtained from the solar PV power plant invested by a promoter on the basis of NPV (€) and IRR (%). NPV is calculated in the way of subtracting the investment cost (current) from the future cash flow obtained during the operation phase. IRR is used to analyze what discount rate would cause the NPV of a project to be '0'€ and then comparing the value with expected rate of return. NPV and IRR are given by Eq. (4) and Eq. (5) as follows:

$$NPV = \sum_{t=0}^{n} \frac{B_t}{(1+r)^t} = \sum_{t=0}^{n} \frac{CF_t}{(1+r)^t} \quad (4)$$

$$IRR = \sum_{t=0}^{n} \frac{B_t}{(1+r)^t} = \sum_{t=0}^{n} \frac{CF_t}{(1+r)^t} \quad (5)$$

where:

B_t : Benefit from the perspective
n : Project period cash flow is received
r : Discounted rate
CF_t: Net after-tax cash inflow-outflows during a period t

3 Results and discussion

3.1 Solar energy production

Table 2 shows the energy production (MWh year^{-1}) in a solar PV power plant depending on the equivalent hours per year (obtained by subtracting the system losses) and the installed power (MW). A typical solar PV power plant in Spain, with a power of 200 MW and 1,600 effective hours of operation, could reach up to 320 GWh year^{-1}.

Table 2. Solar energy production (MWh year^{-1})

Equivalent hours [h year^{-1}]	Power [MW]			
	100	200	300	400
800	80,000	160,000	240,000	320,000
1,000	100,000	200,000	300,000	400,000
1,200	120,000	240,000	360,000	480,000
1,400	140,000	280,000	420,000	560,000
1,600	**160,000**	**320,000**	**480,000**	**640,000**
1,800	180,000	360,000	540,000	720,000

Fig. 4 shows the energy production in GWh year^{-1} for solar PV power plants with 100, 200, 300 and 400 MW of output power, considering 1,600 equivalent hours per year (typical value in the south of Spain) and a PV panel degradation of 0.5% year^{-1}. In the year 25, a reduction of income of 12.5% is produced due to this issue.

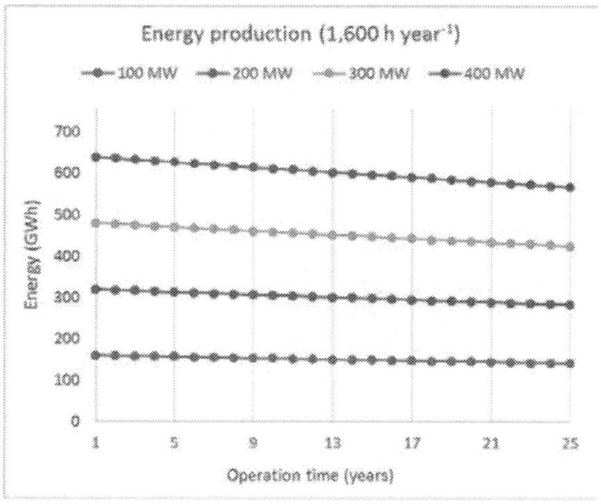

Fig. 4. Energy production (GWh year^{-1}) considering 1,600 h year^{-1} and a degradation of 0.5% year^{-1}.

3.2 Investment and O&M costs and land-take requirements

Fig. 5 shows the summary of investment cost of a solar PV power plant with 100 MW of output power. The main cost is supply and installation of the PV panel, with a 37.9% of the total cost.

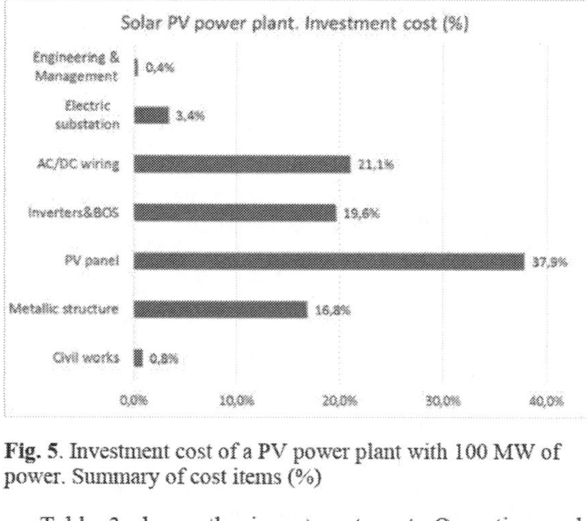

Fig. 5. Investment cost of a PV power plant with 100 MW of power. Summary of cost items (%)

Table 3 shows the investment cost, Operation and Maintenance (O&M) costs and land-take requirements for solar PV power plants for output power between 100-400 MW. A power typical power plant with a power of 200 MW has an investment cost of 141.05 M€ and requires more than 190 ha of land. The land is usually rented during the period of operation of the PV power plant (25 years). A cost of 1,500 € ha^{-1} $year^{-1}$ has been considered.

Table 3. Investment and O&M costs and land-take requirements (ha)

Output power (MW)	CAPEX (M€)	Land-take (ha)	O&M (M€ $year^{-1}$)
100	70.80	190.00	0.49
200	141.05	380.00	0.87
300	210.44	570.00	1.26
400	276.05	760.00	1.64

3.3 Profitability analysis

In this section, the results obtained in the economic model are shown. Table 4 shows the main parameters that have been considered in the economic model.

Table 4. Economic model parameters

Economic model parameters	
Daily market price (€ MWh^{-1})	50-60
Operation time (years)	25
Equity (%)	20%
Debt (%)	80%
Loan interest rate (%)	3%
Loan term (years)	15
Depreciation (years)	25
Anual CPI (%)	2.0%

Fig. 6 shows the profitability analysis, NPV and IRR, for a PV power plant with a power of 100 MW and a daily price in the electricity market of 50 € MWh^{-1}. In a typical PV power plant in the South of Spain, with an average of 1,600 h $year^{-1}$, NPV is 17.4 M€ and IRR reaches 6.91%.

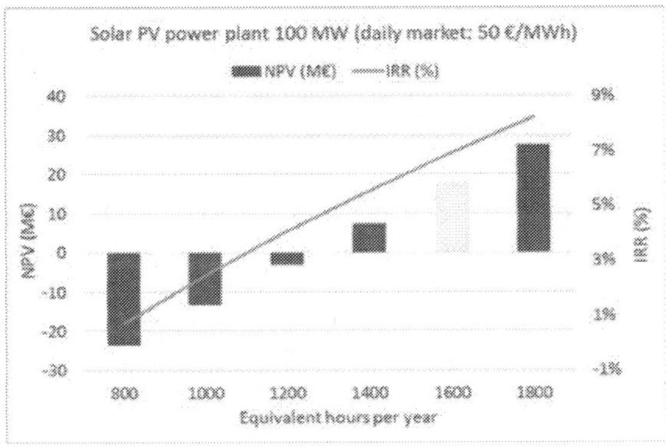

Fig. 6. Profitability analysis. PV power plant with 100 MW of power and daily market price of 50 € MWh^{-1}.

Fig. 7 shows NPV and IRR for a PV power plant with a power of 100 MW and a daily price in the electricity market of 60 € MWh^{-1}. In this scenario, NPV increases to 34 M€ and IRR increases to 9.01%.

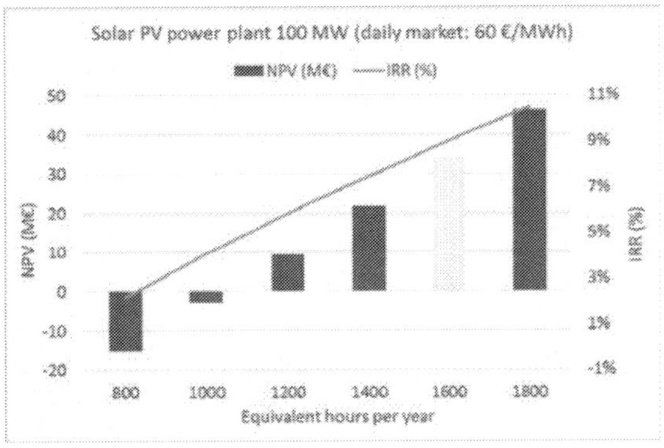

Fig. 7. Profitability analysis. PV power plant with 100 MW of power and daily market price of 60 € MWh^{-1}.

Finally, Fig. 8 shows NPV and IRR values for a solar PV power plant with 400 MW of power and a daily price in the electricity market of 60 € MWh^{-1}. In this scenario, NPV reaches 147 M€ and IRR is 9.49% for 1,600 h $year^{-1}$.

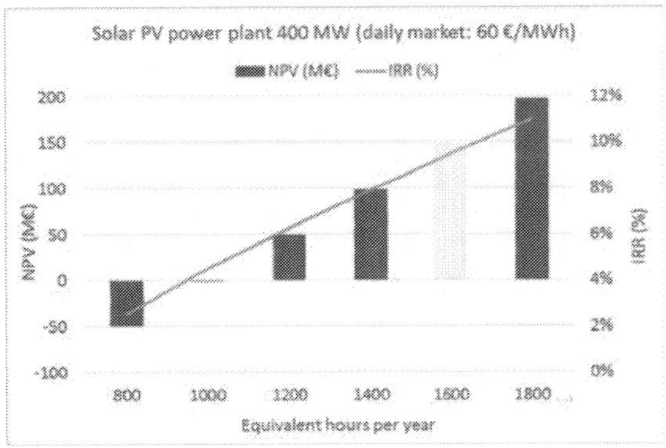

Fig. 8. Profitability analysis. PV power plant with 400 MW of power and daily market price of 60 € MWh^{-1}.

4 Conclusions

Spain offers optimal conditions for the installation of solar PV power plants. The average solar irradiation in Spain is 1,600 kWh m^{-2}. In this paper, the economic feasibility of large-scale solar PV power plants has been studied. PV power plants with power between 100-400 MW, with a number of equivalent hours between 800-1,600 h $year^{-1}$ have been considered.

The economic feasibility depends on the number of equivalent hours per year and the daily price in the electricity market. The results obtained show that the economic viability of these plants is reached for a number of hours exceeding 1,400 h $year^{-1}$, fundamentally in PV power plants with powers below 200 MW. In the north of Spain, with a number of equivalent hours less than 1,200 h $year^{-1}$, there is no profitability for the current investment cost.

In the south of Spain, with an average of 1,600 h year^{-1}, the profitability increases, reaching a NPV of 147 M€ and an IRR of 9.49% for a solar PV plant of 400 MW of power and a daily price in the electricity market of 60 € MWh^{-1}. If the power of the plant is lower, the profitability is reduced, mainly because the O&M costs are higher for each MWh produced.

References

1. International Energy Agency (IEA). Key world energy trends, except from: renewables information.
2. Menendez J, Loredo J, Fernandez-Oro J, Galdo M. Underground pumped-storage hydro power plants with mine water in abandoned coal mines. In: *Proceedings of the IMWA 13th International Congress*, (2017); pp. 6-13.
3. Menendez J, Loredo J, Fernandez-Oro JM, Galdo M. Energy storage in underground coal mines in NW Spain: Assessment of an underground lower water reservoir and preliminary energy balance. *Renew. Energy* (2019); 134:1381-1391.
   https://doi.org/10.1016/j.renene.2018.09.042
4. Menendez J, Ordóñez A, Álvarez R, Loredo J. Energy from closed mines: Underground energy storage and geothermal applications. *Renew Sustainable Energy Rev* (2019);108:498–512. https://doi.org/10.1016/j.rser.2019.04.007.
5. Menendez J, Loredo J. Use of closured open pit and underground coal mines for energy generation: Application to the Asturias Central Coal Basin (Spain). *E3S Web of Conferences* (2019); 80: 01005. https://doi.org/10.1051/e3sconf/20198001005
6. Menendez J, Fernandez-Oro JM, Galdo M, Loredo J. Pumped-storage hydropower plants with underground reservoir: Influence of air pressure on the efficiency of the Francis turbine and energy production. Renew. Energy (2019); 143:1427-1438. https://doi.org/10.1016/j.renene.2019.05.099
7. Pablo-Romero M, Sánchez Braza A, Pérez M. Incentives to promote solar thermal energy in Spain. *Renew Sustain Energy Rev* (2013);22:198–208 http:// dx.doi.org/10.1016/j.rser.2013.01.034.
8. Huld T, Müller R, Gambardella A. A new solar radiation database for estimating PV performance in Europe and Africa. *Solar Energy* (2012); 86: 1803-1815.
9. Montoya F, Aguilera MJ, Manzano-Agugliaro F. Renewable energy production in Spain: A review *Renew Sustainable Energy Rev* (2014);33:509-531.
10. Moreno F, Martínez-Val JM. Collateral effects of renewable energies deployment in Spain: impact on thermal power plants performance and management. *Energy Policy* (2011);39(10):6561–74.
11. Ruiz-Arias J, Terrados J, Pérez-Higueras P, Pozo-Vázquez D, Almonacid G. Assessment of the renewable energies potential for intensive electricity production in the Province of Jaén ,Southern Spain. *Renew Sustain Energy Rev* (2012);16(5):2994–3001 http://dx.doi.org/10.1016/j.rser.2012.02.006.

Using Life Cycle Assessment to Determine the Environmental Impacts Caused by Solar Photovoltaic Systems

Anushka Pal[1*] and *Jeff* Kilby[1]

[1]School of Engineering, Computing and Mathematical Sciences, Auckland University of Technology, Auckland, New Zealand

Abstract. The paper presents research that investigated the Life Cycle Assessment of multi-crystalline photovoltaic (PV) panels, by considering environmental impacts of the entire life cycle for any solar PV systems. The overall manufacturing process of a solar PV panel ranging from silica extraction, crystalline silicon ingot growth, wavering to module fabrication and packing of the solar PV panel. The results from this research showed that the module assembly and cell processing of the manufacturing process contributed towards the main environmental impacts of the life cycle of solar PV systems.

1 Introduction

The initial concept of Life Cycle Assessment (LCA) began in 1960's with concerns over energy resources and raw materials of a products life and finding ways to minimise the energy use [1]. In 1963, Harold Smith reported his initial calculations at the World Energy Conference for calculation of cumulative energy requirements for the manufacturing of chemical intermediates and products [1].

LCA is a tool for environmental management, which is not applicable to all decision-making process, to solve a problem. It considers environmental; problems such as resource consumption, human health, ecological environment and does not consider, social or economic effects such as costs, public image, quality, performance etc. Hence, the decision-making process should involve another type of assessments and information. LCA methodology does not consider environmental law and regulations, which is an important aspect for any organisation dealing with environmental policy and decision-making process [2].

LCA analysis is a method that calculates the environmental inputs and outputs of a process or product from which the results are extensively used in energy technologies to evaluate the research and development funding and in formulating energy policies. The European community publication [3], was used by decision makers and reflects when compared to any other technology, PV have higher environmental impact e.g., Greenhouse gas (GHG) emissions of 180g carbon dioxide/kWh in Germany, 10 times higher GHG emissions of nuclear fuel cycle and 45% of combined cycle (CC) natural gas power generation [4]. This indicates that fossil fuel consumed during the production of material used for the manufacturing of multi-crystalline solar PV. The data used in this study was outdated and few assumptions were invalid and the LCA methodology is used to calculate the environmental impact of multi-crystalline PV technology based on the latest data. Since 1970 major research has taken place in the field of solar PV with few instances where the amount of energy required in the production of solar cells to the finished products is calculated [5].

Technically, LCA is not a scientific methodology as it considers both subjective and objective components. The subjective components include assumption, choice and value 18 judgements such as the selection of data sources, the selecting of environmental damage types, system boundaries, the selection of calculation procedures, evaluation process etc. Circulatory effects are the common problem in the boundary definition. It means one must know the complete LCA of all the components of related material and process to analyse the LCA for any material or process, which is next to impossible. The subjective components consist of trade-offs, hypothesis and value judgments. Therefore, LCA results in consists of detailed explanations of the assumption made, information obtained and scientific methods behind the subjective judgements. Time and geographic location also play an important role in LCA results. Environment data might change with different geographic scope and time, so the calculated results are limited to a specific location and time-period.

2 Methodology

This research which is related to Life Cycle Assessment (LCA) of solar PV systems and will be conducted covering the following points:

- Determination of system boundaries (goal and scope).
- SimaPro software will be used to analyse the data to conduct LCA on solar PV systems.
- Interpreting the results obtained from the SimaPro Software.

* Corresponding author: anushka.pal@aut.ac.nz

Life Cycle Assessment (LCA) is a highly recognised method to evaluate environmental impacts of a system or product [6]. The structure of the LCA methodology has been well established by the International Standard 14040 (ISO 2006b) [7]. The LCA framework as shown in
Figure 1 is done in four phases, which are:

- Goal and Scope Definition
- Inventory Analysis
- Impact Assessment
- Interpretation

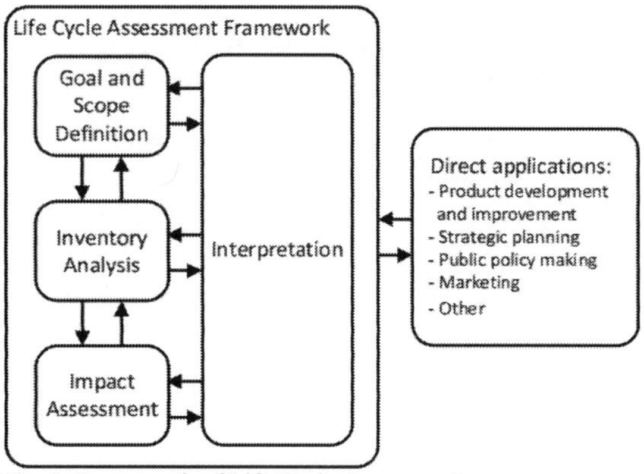

Fig. 1. Framework of Life Cycle Assessment

2.1 Goal and Scope

The most important step in LCA is to determine the goals, scope, and functional unit, which provides a basis to decide the relevant global, regional, or local environmental metrics. Wherever needed researchers must determine assumptions and requirements of data before setting the boundaries.

The goal choices for LCA experts need to be determined based on the intended viewers, application, and study scope. The functional unit is a key element of LCA which must be clearly defined. The functional unit is a measure of the function of the studied system and it provides a reference to which the inputs and outputs can be related.

2.2 Inventory Analysis

Life Cycle Inventory (LCI) is the collection of study data pertaining to the flow of material and energy. It is the process of quantifying the raw material requirement, energy, solid waste, atmospheric and waterborne emissions and other releases for the whole life cycle of a process, product or activity. This data is obtained from the available literature, general databases, LCI databases, or a combination of these sources.

2.3 Impact Assessment

This phase evaluates the environmental and potential human health impacts of the environmental resource. It addresses the resource depletion, ecological and human health effects. It attempts to create a relation between the process or product and its potential environmental impacts.

Also in this phase, interpretation of inventory and impact assessment is drawn to recommendations and conclusions. For this, sensitivity, uncertainty analysis and peer review would be performed as recommended by the International Organisation Standard ISO 14040.
So, system boundaries were defined as follows:

- Quartz mining
- Metallurgical smelting
- Silicon purification
- Polycrystalline ingot
- Wafer slicing
- Cell processing
- Module assembly

These are shown in Figure 2 along with the inputs and outputs.

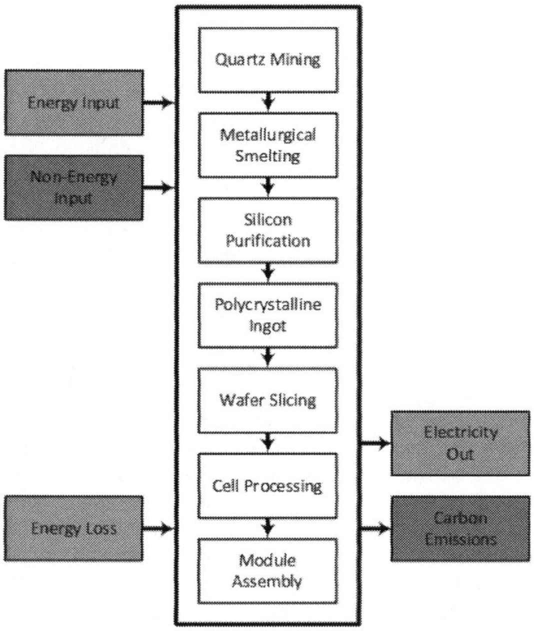

Fig. 2. System Boundaries for solar PV module

To analyse the LCA it was necessary to create a material flow structure for the solar PV module as shown in Figure 3. The material flow was divided into four stages from beginning to the end process such as (a) metal extraction, (b) pre-manufacturing, (c) manufacturing and (d) module assembly.

3 Results

Once the material flow was fixed, materials were set up in the SimaPro software to perform the analysis. The flows as shown in Figure 3 between all the constituents of the solar panel, a small cut-off was made to have a better overview of the most important process and material flows.

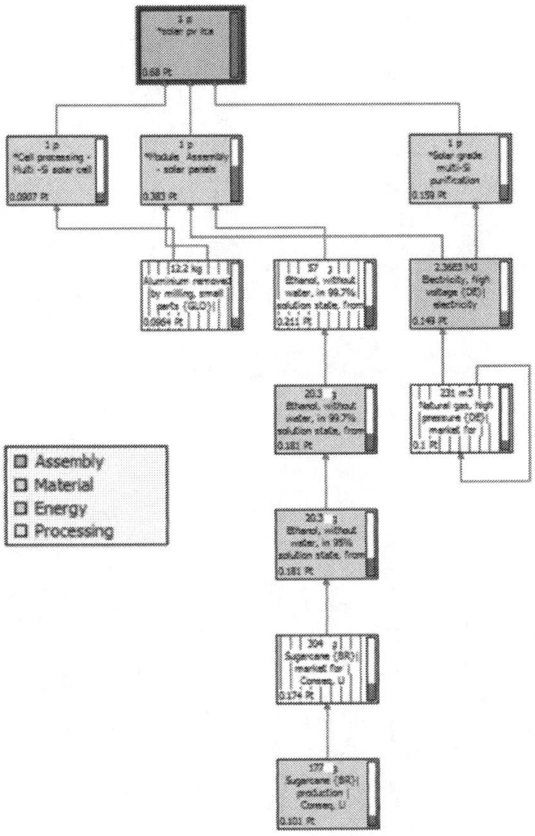

Fig. 3. A scheme of LCA for multi-crystalline solar cell

This scheme determines all the material used in the manufacturing process of solar PV from cradle to grave. All the materials used in the process are fed into the software to calculate the environmental effects. The impact assessment method IMPACT 2000+ was used for this analysis where the impacts are generally categorized by damage to human health, ecosystem quality and resources. It is an end damage assessment method and easily understandable to all through its various impact categories. The major part of the environmental impacts during the whole lifecycle of a solar module is expected to stem from the production and processing of the different components. A categorization of these components obtained from SimaPro, according to their individual contribution to the environmental impact is shown in Figures 4, 5 and 6.

The results obtained in Figure 4, shows human health, resources and climate change have highest impact coming from the cell processing and module assembly. However, category ecosystem quality has the least impact on the environment.

The results in Figure 5 shows that the module assembly and cell processing have the highest impact on human health, ecosystem quality, climate change and resources. However, ingot casting, metallurgical silicon smelting and wafer slicing have considerably less impact on the environment.

The results in Figure 6 shows, emissions that are harmful to human health, such as carcinogens and respiratory inorganic, Aluminium constitutes the highest contribution. This can be explained by the aggressive emissions (SO_2, carbon dioxide, and PFC, PAH etc.) related to each stage of Aluminium production starting from extraction to processing. Health risks related to module assembly and cell processing, however, from using different chemicals as toluene, hydrochloric acid, methyl chloroform at various production stages as shown in Figure 6. It was found, electricity consumption during solar grad Poly-Si production was one of the factors that had the most effect on the primary energy demand, Acidification Potential and Eutrophication Potential, followed by electricity consumption during cells processing, steam consumption during solar grad Poly-Si production, aluminium and glass consumption during modules assembly. Electricity consumption during solar grad Poly-Si production was also the factor that had the most influence on the Global Warming Potential and Photochemical Ozone Creation Potential, followed by electricity consumption during cells processing, aluminium consumption during modules assembly, steam

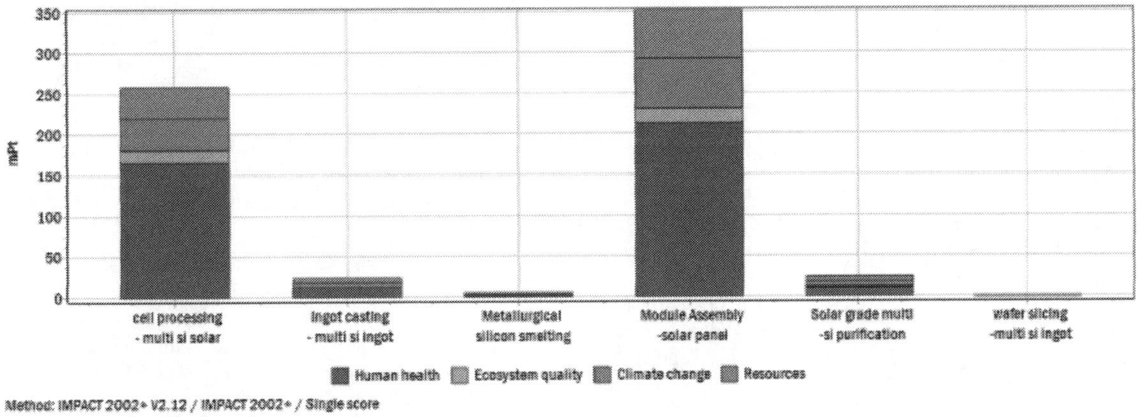

Fig. 4. Single score graph comparing processes

https://doi.org/10.1051/e3sconf/201912202005

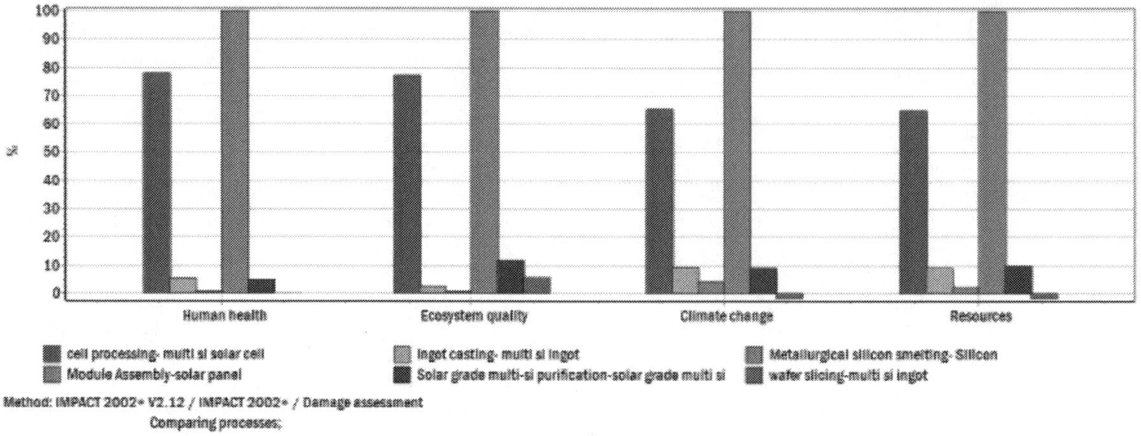

Fig. 5. Damage assessment for solar module life-cycle

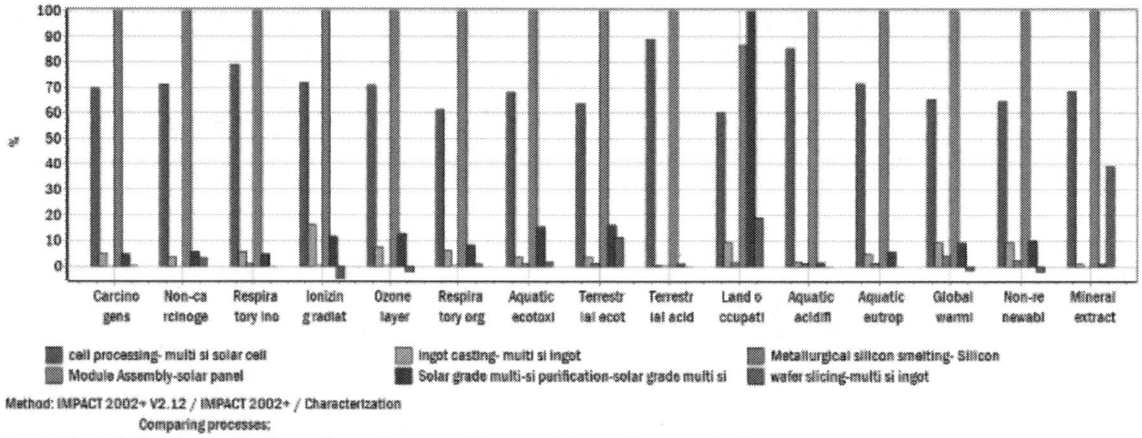

Fig. 6. The individual contribution to the environmental impact of the production of different components of a solar cel

consumption during solar grad Poly-Si production, and glass consumption during modules assembly.

3 Conclusion

Using SimaPro software, the environmental impacts and energy demand were compared with other forms of energy generating systems as per kWh of electricity generated. It was found that the environmental impacts and primary energy demand were far less when compared with coal-fired electricity production. Therefore, using coal-fired plants with PV plants will improve the primary energy demand tremendously. PV technology has improved in terms of efficiency and cost in last few decades, so with the help of subsidies and government incentives, implementation of PV technology in conjunction with other energy production systems is more feasible. Most of the impacts come from the production and the processing of its different components. The LCA showed that the module assembly and cell processing constitute the major contribution to most of the impact categories. It also showed the obvious correlation between energy consumption and environmental impacts from the fact that the most energy consuming processes were also those with the highest negative impacts both to the environment and human health. The module assembly and cell processing were identified as the most energy consuming processes. To reduce the energy required for solar cell production, the efforts should therefore mainly be concentrated on making improvements in these two processes.

4 References

1. S. A. I. Corporation and M. A. Curran, *Life-cycle* assessment*: principles and practice*. National Risk Management Research Laboratory, Office of Research and Development, US Environmental Protection Agency Cincinnati, Ohio, (2006).
2. Remmen, " Limitation of current LCA approaches," *CALCAS seminar,* (2007).
3. European_Commission, "Directorate-General for Research, 2003. External costs. research results on socio-environmental damages due to electricity and transport. Office for Official Publications of the European Communities," vol. 92-894-3353-1, no. EUR 20198, (2003).
4. C. Lamnatou and D. Chemisana, "Photovoltaic/thermal (PVT) systems: A review

with emphasis on environmental issues," *Renewable Energy,* vol. **105**, pp. 270-287, (2017).

5. L. P. Hunt, " Total energy use in the production of silicon solar cells from raw materials to finished product.," *12th IEEE PV Specialists Conference. New York;* , pp. 347–52., November 15–18, (1976).
6. M. A. Curran, "Goal and Scope Definition in Life Cycle Assessment," in *Goal and Scope Definition in Life Cycle Assessment*: Springer, 2017, pp. 1-62.
7. ISO, "Environmental management–life cycle assessment–requirements and guidelines, German and English version EN ISO 14044," (2006).

Synthesis of Phosphorus Solution for n^+ Si Selective Emitter Solar Cell by Spin on Doping

Thipwan Fangsuwannarak*, *Supanut* Laohawiroj and *Kamonchanok* Mekmork

School of Electrical Engineering, Suranaree University of Technology, Nakhon Ratchasima, Thailand

Abstract. A strong requirement in manufacturing of high-efficiency solar cells is its cost reduction. One approach of aim is to merge several steps of n^+ Si selective emitter processing into one step without degrading the performance of solar cells. By varying the doping level in the selective area, intrinsic fields can be built into solar cells with potential benefits long recognized. In this paper, the spin-on doping (SOD) method was used for the purpose of important tasks, different phosphorus diffusion to form n^+ Si selective area consisting of the lightly and heavily doping emitter areas with 35 Ω/sheet and 121 Ω/sheet. The main solution containing different concentrations of phosphorus doped-SOD source was synthesized in this work. The sheet-resistance dependence of n-Si emitter layers on the concentration of phosphorus acid in the SOD solution was studied in term of the volume ration of TEOS: H_3PO_4, as well as the thermal diffusion temperature. The suitable condition for forming n^+ Si selective emitters in one process step is 1000°C diffusion temperature for 30 minutes with the complementary SOD volume ratio of 4:1 and 2:1. SOD solution can be patterned by a screen printing or an inkjet printing.

1 Introduction

In last 40 years several types of crystalline silicon (c-Si) solar cells were increasingly developed for high efficiency laboratory cells where the processing relies on all available high-quality materials and techniques to reduce losses. Silicon solar cell efficiency of energy conversion to reach the fundamental efficiency limits at as high as 24% has achieved on laboratory [1]. On the other hand, the typical solar cell efficiency of industrial c-Si cells is in the range of 15-17%. However, the efficiency from the industrial process can approach the efficiency as high as laboratory cells with attenuating the technological factors. The main loss influencing cell efficiency is surface recombination losses. The major contributor is to look at an available and simple implementation to improve passivated surfaces of c-Si solar cells [2]. There are two different regions of surfaces in a c-Si solar cells to be concerned in order to reduce surface recombination velocity. Non-metalized region at interfaces of passivating and n-emitter layers is concerned in order to avoid the formation of a dead layer. Metal-heavy dope (n^+) emitter interfaces need to form for increasing blue light responsibility and reducing contact resistance. Therefore, an effective method as n^+ selective emitter layer on p-type Si wafer leads to these requirements. The most commercial selective emitter silicon (SE Si) solar cell involves lightly doping n^- emitter layer on front p-Si surface substrate followed by the standard antireflection process and heavy doping n^+ emitter region underneath silver contacts completed by screen printing after firing [3]. This common method to achieve n^+ type Si region is compatible with mass product technique to form front contact using silver paste with quickly firing. However, over firing may occur to result in silver migration through the junction that cause the leakage current. Recently various commercialized methods have been proposed to optimize the lightly-doped and heavily-doped region for applying SE technology to Si solar cell in mass production scale such as phosphorus ink [4], doped Si ink [5], and laser doped-SE [6]. Many SE techniques have involved the additional diffusion process steps which effect on low fabrication cost [7].

In this work, the preparation of different sheet resistances was presented to form lightly doped- and heavily-doped emitters in one diffusion process step. This method is an alternative to prepare lightly n-type emitters and heavily n^+ type emitters regions [8]. The different resistivity of emitter regions would completely feature with using minimum process sequences without using an acid-resistant mask to make shallow trenches. Firstly, a typical thermal diffusion method was exploited to drive dopant atoms into p-type Si substrates. Resistivity value of emitter layer finally obtained under one condition of thermal diffusion can be simply adjusted by varying the phosphorus concentrations in SOD solution. The different phosphorus doped-spin-on doping (SOD) sources were synthesized from the main solution of Tetraethylorthosilicate (TEOS) containing different concentrations of phosphorus acid. The dependence of sheet resistance of the emitter layer on the concentration of phosphorus acid in the SOD solution as well as the thermal diffusion temperature was studied.

* Corresponding author: thipwan@g.sut.ac.th

The optimal ratios of TEOS: H_3PO_4 with the same appropriate time and temperature conditions were investigated.

2 Experimental Methods

2.1 Synthesis of P doped-SOD solution

Lightly doped- and heavily doped-emitter layers were prepared by using SOD solution as a precursor. SOD solutions with different phosphorus concentration are the different SOD sources for the thermal diffusion process. Firstly, the SOD solution was obtained from starting solutions of tetraethylorthosilicate (TEOS, 98%), ethanol absolute (EtOH, 99%) and H2O in a volume ratio of 10:20:3. Subsequently, the prepared solution was stirred for 5 minutes at room temperature to ensure homogeneity. Phosphoric acid (H_3PO_4, 85%) was then added by different volume ratios of TEOS: H_3PO_4 such as at 5:1, 4:1, 3:1 and 2:1 and then stirred for 60 minutes. In general, the hydrolysis reaction of the TEOS mixture solution was catalysed at 70°C. While the SOD solution was stirred at room temperature for 30 minutes, it was cooled down at the same time. The schematic diagram of the process step for preparation of SOD solution is depicted in Fig.1.

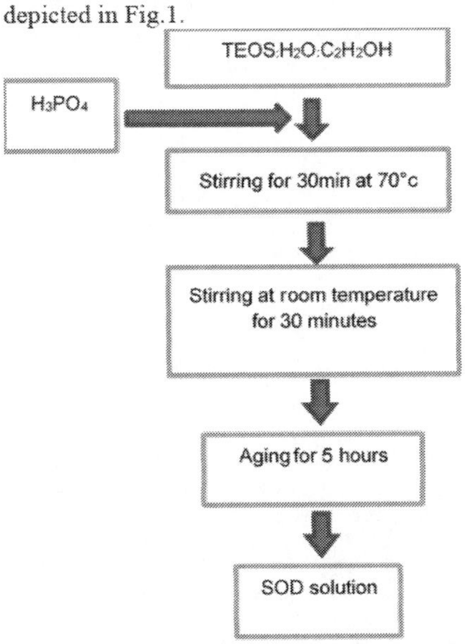

Fig. 1. Schematic diagram of process sequence for preparing phosphorus doped-SOD solution

2.2 Preparation of n-emitters

The P doped-SOD solution was coated on p-type Si substrates by spin-coating method. P dopant concentrations were controlled by a different amount of phosphoric acid in the volume ratios of TEOS: H_3PO_4. A thickness profile of SOD region covered on crystalline silicon (100) wafers (1 – 20 Ω • cm) used as substrates were cleaned with standard cleaning of Radio Corporation of America (RCA) and then dipped in dilute 5% hydrofluoric solution for 20 seconds and rinsed in deionized water in order to remove the native oxide on the Si surface. After aging of the SOD solutions for 5 days except for 2:1 and 3:1 SOD solutions without aging condition due to featuring higher viscosity than others, each solution was spun on the Si substrates at 2-speed rates automatically at 500 rpm/min for 5 seconds. and then 3000 rpm/min for 30 seconds. The p-type Si substrates coated with P doped-SOD films containing different phosphorus were dried at 200°C for 30 minutes in air and then thermally diffused in a high temperature furnace for the thermal diffusion process in a N_2 atmosphere. The dopant diffusion was operated at the temperature of 1000°C with various diffusion time from 30 minutes to 90 minutes. P atoms effectively forced by the high temperature were driven into the p-type Si wafer to form the n-emitter layer with different resistivity values obtained. Four-point probes measurement was employed to measure sheet resistance. It is possible to achieve a variety of doping profiles in the SOD coated regions.

Different SOD solution was used to form the n-emitter layer with an n^+ selective region by a spin-on-dopant technique and subsequently, thermal diffusion. Metal contacts on selective doping Si wafers were provided by screen printing and fired under rapid temperature annealing (RTA) process with temperature 915°C for 7 seconds. The selective emitter solar cell shows the structure in Figure 2.

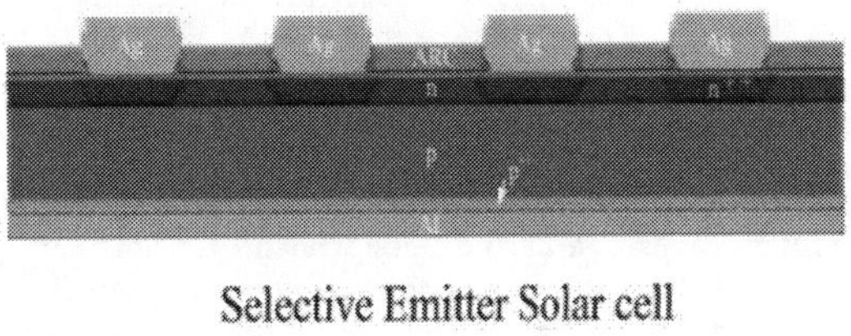

Fig. 2. Structure of selective emitter solar cell in this work by using different doped-SOD solution

3 Results and discussions

Figure 3 shows the thickness profile of the coated SOD region with 4: 1 ratio (TEOS: H_3PO_4) by verifying the optical profiler (VEECO) measurement. The SOD and PSG film thicknesses with the different volume ratios of TEOS: H_3PO_4 were plotted in Fig.4. The range of SOD film thickness between 64 nm and 609 nm range after drying results in the presences of PGS film after thermal diffusion process. The range of PSG film thickness is between 103 nm and 297 nm. It was presented that SOD thickness was not significantly related to PSG thickness. Typically, about 100 nm the thickness of the silicon oxide layer is chosen for designing antireflection coating (ARC) layer for Si solar cells in order to minimize reflectance for a wavelength of 600 nm. With feathering results, it seems that PSG film from a starting material of SOD solution in the ratio of 4: 1 TEOS: H_3PO_4 could be used for ARC to have an agreement with Yujie Tang's work [9] that involved dual tasks for preparing n^+ emitter and ARC layers.

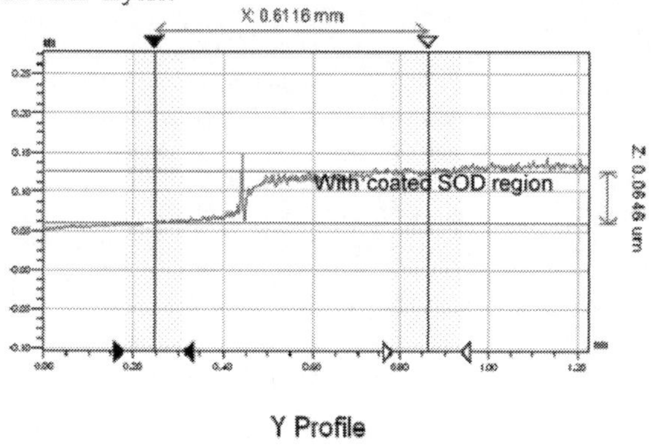

Fig. 3. A thickness profile of the SOD region layer

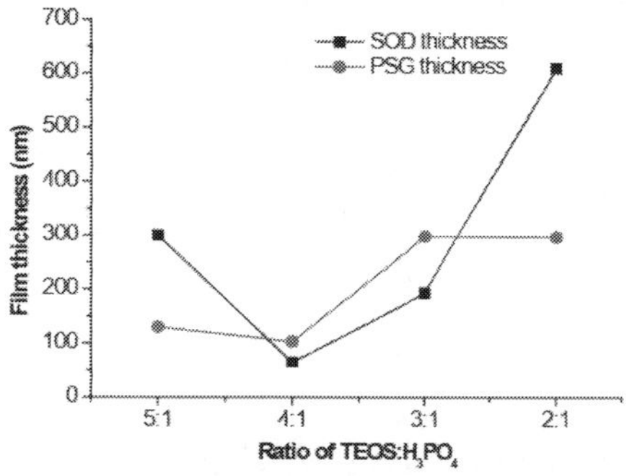

Fig. 4. SOD and PSG film thicknesses

Dopant density in the coating regions and in the field emitter can be adjusted independently from varying TEOS: H_3PO_4 ratio. Dependence of sheet resistance of n-type emitter layer on various phosphorus concentrations in the term of the SOD film and diffusion time was depicted in Fig. 5.

The results show that the period of time in the diffusion process effects to the decrease of sheet resistance of the n-emitter layer in all ratio conditions. The volume ratio of TEOS: H_3PO_4 = 5: 1 obtains sheet resistance in the range of 183 - 446 Ω/ sheet with decreasing time in diffusion from 90 min to 30 min. The volume ration at 4: 1 and 3: 1 obtained the sheet resistance in the minimum range of 12-35 Ω/ sheet and 16-36 Ω/ sheet, respectively. It is noticed that too large amount of H_3PO_4 for the volume ratio at 2: 1 resulted in the sheet resistance became increasing. It is mostly due to the presence of more defects from heavy dopants. Therefore, at the same condition of a diffusion process, different n^+ Si selective emitter regions can feature with complementary SOD solutions to form lightly doping n-type (<50 Ω/sheet) and heavily doping n^+ Si type (80-100 Ω/ sheet) emitters. The coating alignments such as screen printing needs in this technique. This result suggests that the complementary volume ratios of 4: 1 and 2: 1 or 3: 1 and 2: 1 can be chosen for diffusion process at 1000°C for 30 min to form the n^+ selective emitter underneath silver contacts and at the illuminated area. In addition, PSG film occurs inevitably after process can be used as an ARC.

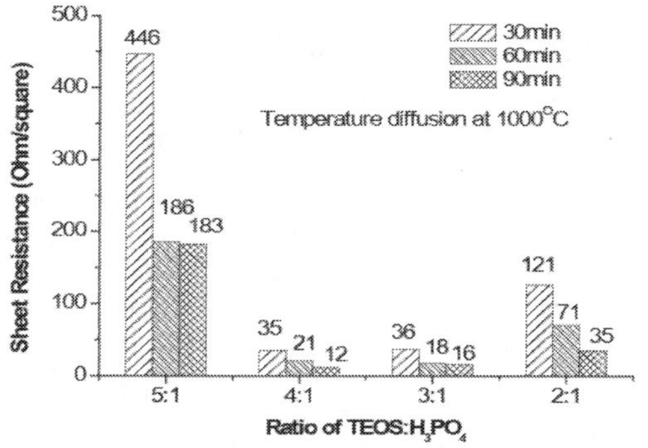

Fig. 5. Sheet resistance of N-type emitter layers as a function of TEOS:H3PO4 volume ratio with the different time in diffusion conditions

Figure 6 shows the comparison of J-V curve between the p/n Si solar cell and the selective solar cell with the n^+ Si region, respectively. It is noted that the SE solar cell generates the gain current density and the higher open voltage by 12% and 3.7%, respectively compared with a typical p/n Si solar cell. It also directly leads to gain its power conversion efficiency by 1.74% due to the enhancement of blue light absorption in lightly doping emitter.

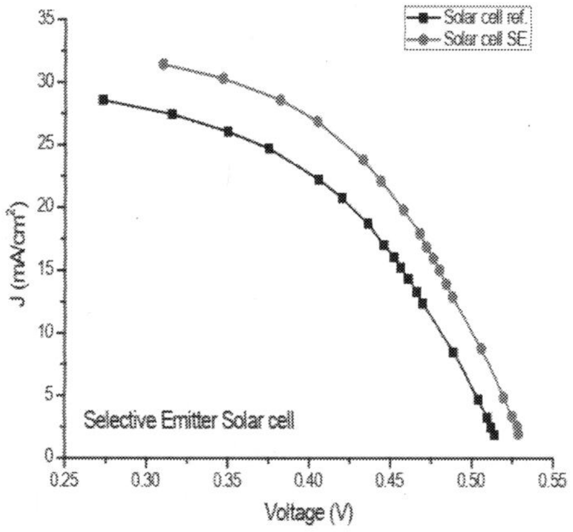

Fig. 6. J-V curve of the SE solar cell

4 Conclusions

In this work, the SOD solution was developed by a using mixture of TEOS and H_3PO_4 as a phosphorus source for using in the n^+ selective emitter purpose. After the thermal diffusion process, PSG residue obtained the suitable thickness for ARC with the minimal reflectance at 600 nm in the light wavelength. Complementary SOD solutions with 4: 1 and 2: 1 or 3: 1 and 2: 1 were coated in the silver contact region and illuminated region to form lightly (121 Ω/□) and heavily (35 Ω/□) doping emitters as an n^+ selective emitter layer after the thermal diffusion process at 1000°C for 30 minutes.

This alternative technique can be a simple implementation to merge several steps of n^+ Si selective emitter processing into one step without degrading the performance of solar cells. Both important factors of the SE solar cell including current density and open voltage increase by 12% and 3.7%, respectively compared with a typical p/n Si solar cell. These enhancement values directly result in the gain of its power conversion efficiency by 1.74%

Acknowledgement

This work was supported by Suranaree University of Technology, Nakhon Ratchasima Thailand.

References

1. J. Zhao et al., "24.5% effieciency silicon PERT cells on MCZ substrates and 24.7% efficiency PERL cells and FZ substrates", Progressive in Photovoltaics: Research and Applications. 7, pp.471-474 (1999).
2. A. G. Aberle, "Surface passivation of crystalline silicon solar cells: A review", Progressive in Photovoltaics: Research and Applications. 8, pp.473-487(2000).
3. A. Rohatgi et al., "Self-aligned self-doping selective for screen-printed solar cells", Proceeding of the 17th European Photovoltaic Solar Energy Conference, pp.1307-1310 (2001).
4. Sihua Zhong et al., "Mass production of high efficiency selective emitter crystalline silicon solar cells employing phosphorus ink technology", Solar Energy Materials & Solar Cells. 117, pp. 483–488(2013)
5. H. Antoniadis et al., "All screen printed mass produced silicon Ink selective emitter solar cells", Proceedings of the 35th IEEE Photovoltaic Specialists Conference, pp. 1193–1196 (2010).
6. Z.Hameiri et al., "18.7% efficient laser-doped solar cell on p-type Czochralski silicon", Applied Physics Letters. 97, pp.222111-1–222111-3 (2010).
7. Y. P. Chen et al., "Single-crystalline silicon solar cell with selective emitter formed by screen printing and chemical etching method: A feasibility study", Hindawi Publishing Corporation International Journal of Photoenergy. ID 510242, 9 pages (2013).
8. N. Phiwpha and T. Fangsuwannarak, "Surface Passivation of Point-Contacted Solar Cells by Inkjet Printing", Integrated Ferroelectrics, 149, pp. 102–106 (2013)
9. Y. Tang et al., "Preparation of N+ emitter onp-type silicon wafer using the spin-on doping method", Materials Science in Semiconductor Processing. 15, Issue 4, pp.359–363 (2012).

E3S Web of Conferences **122**, 02007 (2019) https://doi.org/10.1051/e3sconf/201912202007
REEE 2019

A Long-term Period Performance Assessment of a Building Integrated Photovoltaic System

Mustafa Araz[1,*], *Emrah* Biyikt[1], and *Arif* Hepbasli[1],

[1]Department of Energy Systems Engineering, Faculty of Engineering, Yasar University, 35100 Bornova, İzmir, Turkey

Abstract. Building Integrated Photovoltaic (BIPV) systems can be defined as PV modules, which can be integrated in building's envelope by replacing conventional building materials such as windows, tiles etc. and have an impact on the functionality of the buildings. Considering the huge share (40%) of buildings in total energy consumption and nearly zero-energy building target of the European Union (EU), BIPV systems present a sustainable solution and have gained increased interest in last years. In this study, the performance of a BIPV system, which was installed on Feb. 8, 2016 on the façade of a campus building at Yasar University, İzmir, Turkey within the framework a EU/FP7 project and has a capacity of 7.44 kW_p, is evaluated for a three-year period using first and second laws of thermodynamics. Within this context, real (experimental) monthly and yearly electricity productions are determined and compared with the results obtained from the simulations. Energy and exergy efficiencies and performance ratios of the system are also calculated based on the cell and total areas.

1 Introduction

Energy conservation in a building is an important issue from the point of view of global warming and depletion of fossil fuels. Nowadays, there is a strong need to design buildings, in which renewable sources of energy are utilized, in order to use minimum fossil fuel based conventional power [1]. At present, the share of buildings in total energy consumption is about 40% in the EU and there are some legislations such as the Building Energy Directive that targets nearly zero-energy buildings [2]. Considering these, solar energy seems to be a competitive option for buildings. On the other due to the limited roof area of buildings, integrating photovoltaic (PV) systems into the building façades and roofs represents a significant solution. In fact, the average compounded growth rate of the BIPV technology between 2014 and 2020 is estimated to be 39 and the anticipated annual global BIPV installation in 2020 is 11 GW, which was only 1.5 GW in 2014 [3]. BIPVs have also gained the attention of academicians from all over the world and therefore there are numerous studies in the literature related to them. Biyik et al. [4], conducted a comprehensive review where they examined studies related BIPVs and BIPV/Ts. Readers can refer to this review study for more information.

In this study, the performance of a BIPV system, which was installed on Feb. 8, 2016 on the façade of a campus building at Yasar University, İzmir, Turkey within the framework a EU/FP7 project and has a capacity of 7.44 kW_p, is evaluated for a three-year period using first and second laws of thermodynamics. Within this context, real (experimental) monthly and yearly electricity productions are determined and compared with the results obtained from the simulations.

* Corresponding author: mustafa.araz@yasar.edu.tr

2 System description

The BIPV system was installed to a south-east facing façade of a building at Yasar University and it consists of 48 mono-crystalin PV panels, each with a capacity of 155 Wp, making a total capacity of 7.44 kWp. These panels are installed in four rows as can be seen from Fig. 1 and there is a 150 mm gap between the wall and panels. 24 panels above and below the windows are connected in series and these two strings are connected parallelly to the 7 kW inverter, which has two independent MPPT inputs. The total area of the system is 57.6 m^2, while the cell area is only 42.08 m^2 due to the 30% transparency rate.

Various measurement instruments have been installed on the system and are illustrated in Fig. 2.

https://doi.org/10.1051/e3sconf/201912202007

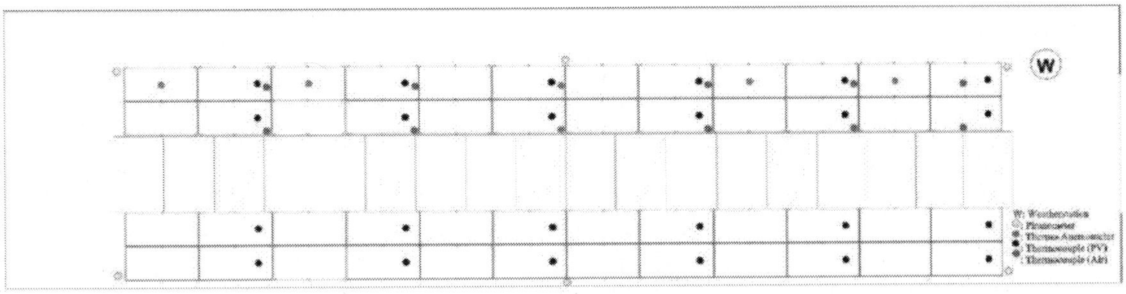

Fig. 2. P&ID of the system [5].

As it can be seen from this figure, the surface temperatures of the PV modules are measured on 24 points, while the air temperatures between the wall and the modules are measured at 12 different location. The air velocity behind the modules is also measured on two points. For the irradiation measurements, 6 piranometers are mounted on the system (4 in the corners and 2 at midpoints) while weather-station is located just next to the upper string. Wind velocity and direction, air temperature and humidity are measured at this point. All these measurements are continuously recorded on a 60-channel internet-connected data logger. The inverter on the other hand serves also as a measuring instrument and all electrical data are recorded to a cloud service by 15 minute intervals.

3 Analysis

Energy efficiency (η) and exergy efficiency (φ) of the system can be calculated by dividing the total power generation for a period of time (day, month, year) to the total irradiation received by the BIPV system at the same period [6].

$$\eta = \frac{\sum \dot{W}}{\sum AG} \tag{1}$$

$$\psi = \frac{\sum \dot{W}}{\sum \dot{Ex}_{sol}} \tag{2}$$

Here the exergy of solar irradiation can be found by using Petela's equation [7]:

$$\dot{Ex}_{sol} = AG\left[1 - \frac{4}{3}\left(\frac{T_0}{T_{sun}}\right) + \frac{1}{3}\left(\frac{T_0}{T_{sun}}\right)^4\right] \tag{3}$$

Another performance indicator for PV systems is the performance ratio (PR) and it can be obtained from Eq. 4. This can be calculated monthly or annualy for different time steps (τ).

$$PR = \frac{\frac{\tau \sum P_{AC}}{P_{STC}}}{\frac{\tau \sum I}{I_{STC}}} \tag{4}$$

4 Results and Discussion

In the first part of the study, monthly electricity generations for 2016, 2017 and 2018 are estimated and compared to the results obtained from TRNSYS [8] simulation. This comparison is illustrated in Fig. 3, below.

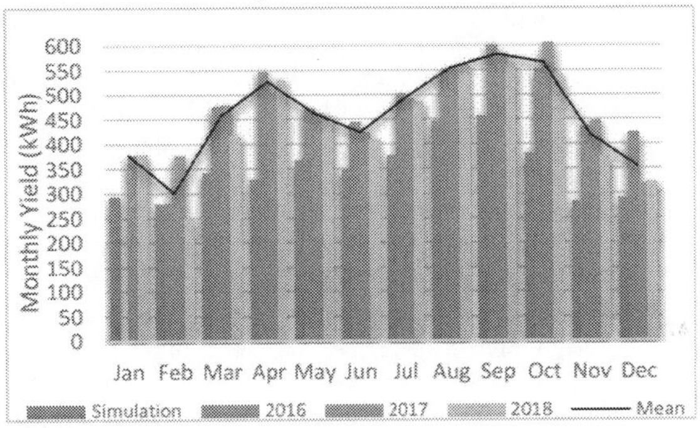

Fig. 3. Monthly yield comparison.

It can be clearly seen from this graph that real (experimental) electricity generation values were higher for almost all of the months and years. The yearly yields were 5278.02 kWh, 5616.24 kWh and 5283.4 for 2016, 2017 and 2018, respectively. The low electricity yield for 2016 is due to the installation date (Feb 8,2016) while the low yield of 2018 might be explained with weather conditions. When we compare the months among each other we see that the maximum yield was achieved during fall months such as September and October, while the minimum yield was in February.

In the next part of the study, monthly yields have been divided into 6 parts in Fig. 4 to see the distributions of each month. When we look at this figure we can see that during summer months the yield is medium, while the maximum yield was achieved during spring and fall months. This is pretty unconventional compared to a standard rooftop or stand-alone PV system and the reason for that is mainly the tilt angle being 90° for a façade. Because, when the tilt angle is 90°, the angle of incidence for a façade-type BIPV system becomes less in summer months and high in winter months and for this reason the peak does not happen in summer months although daily sunshine duration is higher.

 https://doi.org/10.1051/e3sconf/201912202007

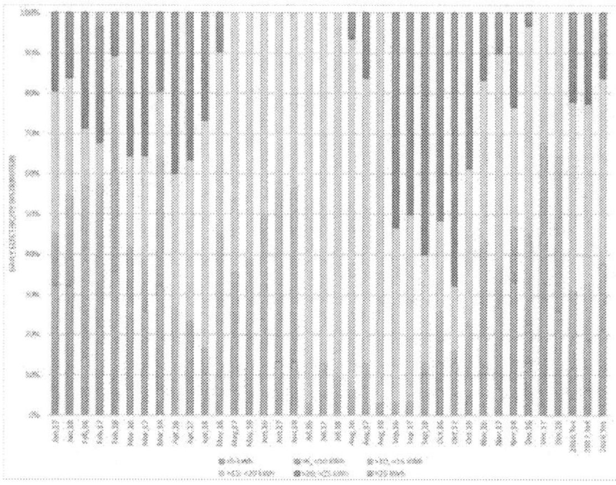

Fig. 4. Monthly average of daily electricity generation distribution.

To further investigate the unconventional profile of the BIPV system, one sunny day of each month has been selected for 2018 to be able to understand power distribution during the day. In this regard, peak powers, the time we reach to the peak power, sunhours and average air temperatures are illustrated in Fig. 5. It can be observed from this figure that although we have more sunny hours during summer, the air temperature is also higher which leads to a decrease in the efficiency of the system and the peak powers occurred during fall and winter months. In fact, the minimum peak power is found to be in June, which is mainly related to the low incidence angle. The peak times were also before midday, changing between 10 to 11 A.M.

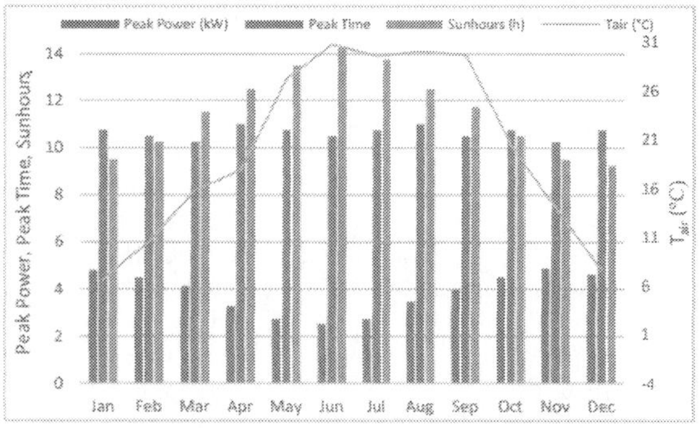

Fig. 5. Peak power, peak time, sunhours and air temperatures during the year.

Next, the power distributions during the day for each season are calculated using the average values of selected days and is shown in Fig. 6. One will notice from this figure that during spring and summer we have less power generation during the day but for longer periods, while the situation is the opposite for winter and fall.

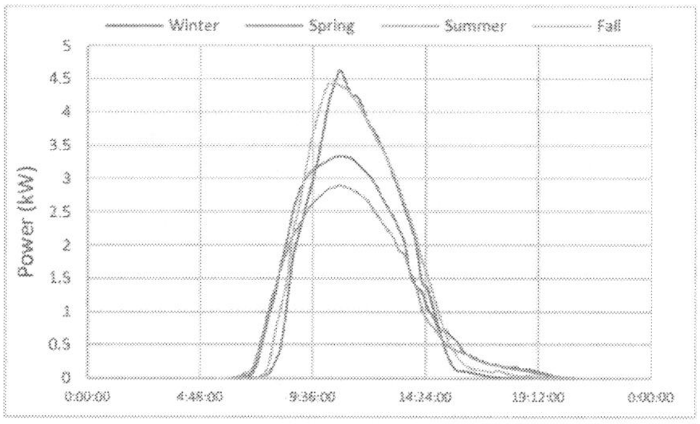

Fig. 6. Seasonal average power distributions during the day.

In the next part of the study, the energy efficiency of the system is estimated for each month of the year and compared with the results obtained from the simulation. This comparison is given in Fig. 7, where we can see that experimental efficiencies are higher for each month of the year. The higher experimental efficiencies are mainly related to the lower experimental surface temperatures. The difference between the experimental and simulated surface temperatures found to be around %16 for the whole year.

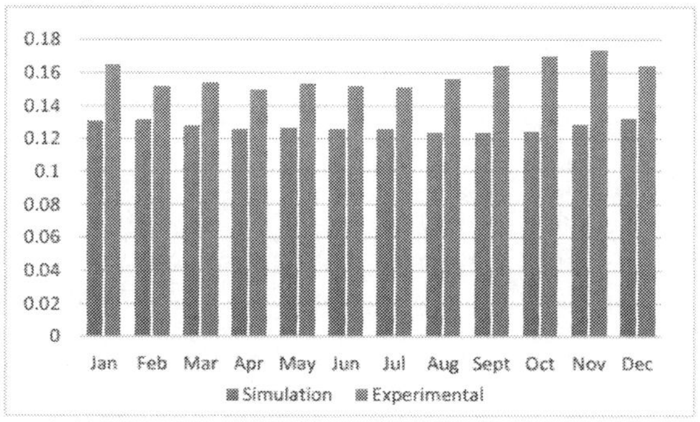

Fig. 7. Energy efficiency comparison.

Lastly, energetic and exergetic efficiencies and also PR values of the BIPV system are calculated based on the cell area and total area. The efficiency results are illustrated along with the irradiation and module surface temperature values in Fig. 8, while PR results are given in Fig. 9.

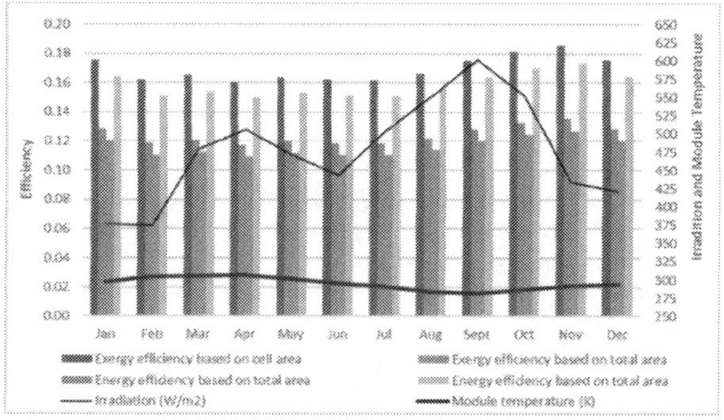

Fig. 8. Efficiency comparison.

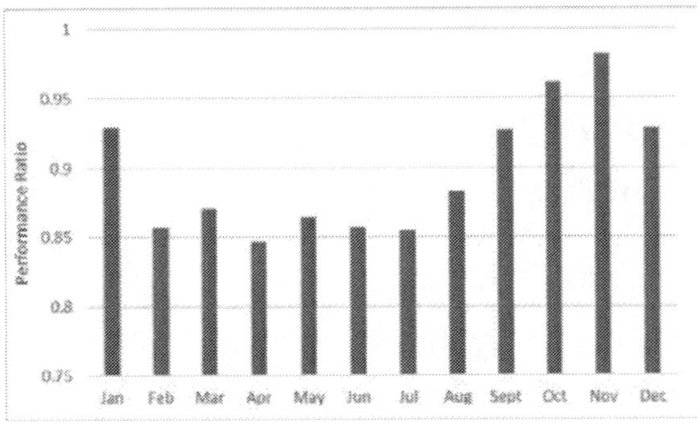

Fig. 9. Performance ratio values of each month.

When we look at Fig. 8, we see that efficiencies based on the cell area are always higher as expected and this is also the case for exergetic efficiencies being always higher than energy efficiencies. The difference between efficiencies based on total area and cell area were found to be around 4%, while the difference between exergy and energy efficiencies is approximately 1.5%. On the other hand we can also observe that efficiencies in hot months is lower than those of cold months.

Lastly, when we look at the obtained PR values, we see a similar trend like efficiency and the values ranged between 0.846 and 0.982. The highest PR is obtained in November while the lowest was in April.

5 Conclusions

In this study, a 7.44 kW_p BIPV system installed at Yasar University, İzmir, Turkey is experimentally evaluated for a three-year period, using some performance indicators such as energy efficiency, exergy efficiency and PR. The main conclusion that can be drawn from this study can be listed as follows:

(i) BIPV is a promising technology in achieving the targets of EU towards nearly zero energy buildings.

(ii) The experimental monthly electricity generation values were almost always higher than the values obtained from simulation.

(iii) The maximum monthly yield was on October, 2017 with 604 kWh, while the minimum yield has occurred on Feb, 2018 (250 kWh).

(iv) Daily electricity yields were between 10 to 20 kWh for summer days, while the maximum yields were during season changes.

(v) Experimental efficiency values were found to be around 3% higher than those of simulation's. Lower experimental surface temperatures can be addressed for this situation.

(vi) Energy efficiencies based on total area ranged between 11.1% to 12.7%, while exergy efficiencies based on total area were in the range of 11.8-13.6%.

(vii) Performance ratios followed a similar trend to efficiencies having a maximum of 0.982 in November and a minimum of 0.846 in April.

(viii) For further work, weather corrected performance ratios can be calculated to reduce the fluctuation. Also, one of the methods covered in IEC 60891 can be applied to the system to convert the measured values to standart test conditons and understand the degrading rate of the system.

Acknowledgements

The presented work was developed within the framework of project "REELCOOP - Research Cooperation in Renewable Energy Technologies for Electricity Generation", co-funded by the European Commission (FP7 ENERGY.2013.2.9.1, Grant agreement no: 608466).

References

1. G.N. Tiwari, H. Saini, A. Tiwari, A. Deo, N. Gupta, and P.S. Saini, *Periodic Theory of Building Integrated Photovoltaic Thermal (BIPVT) System*, Sol. En., **125**, 373-380 (2016).
2. EU Buildings, https://ec.europa.eu/energy/en/topics/energy-efficiency/buildings (Access Date: 19 January 2017).
3. M. Tabakovic, H. Fechner, W.V. Sark, A. Louwen, G. Georghiou, G. Makrides, E. Loucaidou, M. Ioannidou, I. Weiss, S. Arancon, *Status and outlook for building integrated photovoltaics (BIPV) in relation to educational needs in the BIPV sector*, Energy Procedia, **111**, 993–999 (2017).
4. E. Biyik, M. Araz, A. Hepbasli, M. Shahrestani, R. Yao, L. Shao, E. Essah, A.C. Oliveira, T. del Caño, E. Rico, J.L. Lechón, L. Andrade, A. Mendes, Y. Baver Atlı, *A key review of building integrated photovoltaic (BIPV) systems*, Eng. Sci. and Tech., an Int. J., **20**, 833-858 (2017).
5. REELCOOP, Research Cooperation in Renewable Energy Technologies for Electricity Generation report. Seventh Framework Programme of the European Union (2015).
6. R. Rawat, R. Lamba, S.C. Kaushik, *Thermodynamic* study *of solar photovoltaic energy conversion: An* overview, Ren. and Sust. En. Rev., **71**, 630-638 (2017).
7. R. Petela, *Exergy of undiluted thermal radiation*, Sol. En., **74**, 469-488 (2003).
8. S.A. Klein et al., TRNSYS 17: A transient system simulation program. 2009, Wisconsin: Solar Energy Laboratory, University of Wisconsin.

The Impacts of Power Infrastructure Development in the Socio-Economic Situations in Sub-Sahara Africa

Ngcobo Ntebo, *Kasenge* Mathe and *Emmanuel* Oikelomen Ayorinde*

Department of Civil Engineering Technology, University of Johannesburg, South Africa

Abstract. Infrastructural development is fundamental in nation's economic growth and development. Power infrastructure is imperative for nation building and helps create employment and improved the well-being of its citizens. This research paper identifies the impact of power infrastructure development on the economic growth in sub-Saharan Africa (SSA). The study adopted a quantitative research approach with data gathered from the respondents within power infrastructure development in the region. Information gathered were analysed using mean item score, standard deviation and factor analysis. The findings revealed the impact of power development on the economic benefit in SSA to be wealth creation, boost in citizens' income, health care improvement, improved educational systems were seen as the direct impact of infrastructure development on the economic situation in sub-Sahara Africa. Since the lack of infrastructural facilities of an economy can lead to various setbacks in the nation's economic development, it thereby requires adequate participations by stakeholders to deliver sustainable power infrastructure development in the society. The study power infrastructure development can contribute to regional and national growth, urbanization challenges, and improvement in the environmental through the provision of clean energy which foster social and economic inclusion.

1 Introduction

The expenditures in infrastructure development is the highest in emerging and developing countries, especially the ones with the highest growth, such as India and China in particular. These two nations, together with Brazil, South Africa, and Russia, make up the so-called BRICS, which in 2012 produced one-quarter of the world's gross domestic product (GDP) (1). Therefore, emerging nations require huge capital investment as a way of developing infrastructure for the purpose of economic growth and development.

Also, according to the Organization for Economic and Cooperation Development (OECD), the ever-increasing size in globalization and urbanization by the year 2030 for telecommunication, water, transportation and power generation, transmission and distribution will amount to $71 trillion. Therefore, the government cannot meet this target alone through annual budgets carried out by the awarding of direct contracts. There is a need for collaboration with the private sector to invest in infrastructure projects, to receive aids and bring about industrial growth and socio-economic development (2).

(3) Stated that urgent adequate unprecedented infrastructure is needed to tackle the increase in the population for a better standard of living through the provision of infrastructural" amenities such as power facilities. Sub-Saharan Africa (SSA) is in a profound crisis of low power output, which has constrained its economic growth rapidly and restrained development in the region. It is a known fact that the combined 48 countries of sub-Saharan Africa have about 80 gigawatts (GW), while a single country like Spain with a far smaller population generates more. Capital growth in the sub-Saharan region has been stagnant for the last three decades, with growth rates barely half of those in other developing countries. It has widened the gap between sub-Saharan Africa countries and the rest of the world, even compared to other nation groups in the same income bracket (4).

Therefore, it is fundamental for government and private investors to make huge investments in large, medium and small irrigation facilities. The large facilities will be responsible for the availability of water provision, irrigation and hydroelectric power (5). Investments in the energy sectors have resulted in improvements in the standard of living of citizens, thereby eradicating poverty and boosting the economic growth in the world (5).

Developed nations in the world have striven to remain above the subsistence economic growth through the secured access to electricity delivered to its citizens (6). Secured access to energy services has improved education, improved incomes for the citizens, and improved the healthcare sector, improved gender equality as well as several other socio-economic growths (7; 8). The development of human resources and the economic growth of a country requires secure energy services (9). It is known that adequate provision of secured energy services in SSA will increase the socio-economic growth of the region, and help in the

* Corresponding author: engrkulz@gmail.com

actualization of the Millennium Development Goals (MDGs) (10). The study of (11) agrees with that of (12) that a secured power generation transmission and generation will boost industrialization and economic growth in SSA.

Effective power infrastructure development is a prerequisite for economic growth, improved competition, industrialization and earned recognition and induction into the world economy (13). The effective growth in the energy sector in India for about 20 years between 1972 and 1992 accounted for about half of the Indian industrialization growth (11).

2 Methodology

The method used in this study was quantitative research approach with the motive of achieving the aim of the study, which is to identify the impacts of power development on the socio-economic situations in SSA. Quantitative methods relate to positivism and factual data (14). The questionnaire was developed from a wide review of the literature and is not part of any existing survey instrument. Practicing power infrastructure professionals in the power sector in SSA were engaged in the collection of the primary data on the impact of power development on the economic benefit in SSA.

The Likert scale (strongly agree = 5, agree = 4, neutral = 3, disagree = 4, strongly disagree = 5) Mean item score (MIS) was used to present the research findings from the Likert scale in a decreasing order.

Exploratory factor analysis (EFA) is one of the two types of factor analysis (FA), and is often deployed during the initial stage of research by researchers in order to collate information about the interrelationships within a set of variables (15). The EFA of the results were obtained to comfirm the validity and reliabilty of the impacts of power development on the socio-economic situations in SSA, with the highest likelihood with an eigen value of more than one, together with the varimax rotation EFA was used specifically for this study. SPSS software version 21.0 was used to conduct the EFA for this research. The descriptive results show the rankings of all the factors from the first to last according to the variables, with the table representing the individual variables' mean score as well as the standard deviation of the variables.

2.1 Data analysis

Two descriptive statistics were carried out, which are in the form of mean item score and factor analysis. The ranking of the variables was done with mean item score, likewise factor analysis was carried out to outline the variables measuring same underlying effects (16).

2.2 Mean item score

The mean ranking of the variables presented depicts the individual views reached on by the respondents. The result for the test is shown in the table below. The mean table represented below also include the standard deviation of the variables.

Table 1. Socio-economic impact of power development

The Impact of Power Infrastructure Development	Mean	Standard Deviation	Rank
Creation of wealth	4.89	0.333	1
Creates jobs	4.86	0.412	2
Reduces cost of productions	4.77	0.472	3
Ensures a functioning economy	4.77	0.421	4
Ensures a viable economic growth in the country	4.76	0.430	5
Increases incomes	4.75	0.452	6
Improves the standard of living of the people	4.75	0.435	7
Ensures a socio-economic improvement	4.73	0.460	8
Increase productions	4.73	0.443	9
Increases the profitability of organizations	4.68	0.484	10
Increased competition to boost economic growth	4.66	0.536	11
Improves the confidence of investors in the economy	4.65	0.509	12
It creates convenience, i.e. makes life easier for everyone	4.63	0.500	13
Improves health care and education quality of the country	4.45	0.584	14
It creates room for innovation	4.44	0.542	15
It creates connectivity between organizations and international investors	4.22	0.543	16

3 Results from factor analysis

The EFA results of the impacts of power infrastructure development in economic situation are depicted in tables 1, 2, 3, 4, 5, 6 as well as fig 1. With the sixteen variables outlined, two (2) of the variables were missing which are 'increases production' (EG11), and 'it creates convenience i.e. makes life easier for everyone' (EG13). The following were the fourteen variables identified with the potential of roles of power infrastructure development in economic growth.

Table 2. Definition of identified impacts of power infrastructure development on economic situation

Variable	Variable name	Definition
EG1	Wealth creation	Increases the wealth of citizens.
EG2	Creates jobs	Room for more capital projects.
EG3	Reduces cost of production	Decreases the high cost of production.
EG4	Increases profitability of organization	Increases the profits establishments.
EG5	Increases income	Better standard of living.
EG6	Ensures a functioning economy	Viable economy.
EG7	Increases competition to boost economic growth	Healthy competition between competitors.
EG8	Increases the confidence of investors in the economy	The high rate of investments.
EG9	Ensures a socio-economic growth of the country	Boost the gross domestic product (GDP).
EG10	Improves health care and education of the country	Better health care and learning.
EG12	Ensures a viable economic growth of the country	Improved economy.
EG14	Improves the standard of living of the people	Improves citizens incomes.
EG15	Creates room for innovation	It ensures creativity and manpower production.
EG16	Increases connectivity between organizations and international investors	Smooth communication for transactions.

Factor analysis is vital in breaking down numbers of large variables and breaking them into more simple clusters for better interpretations (17). Table 3-6 and fig. 1 below shows (table 3) Kaizer-Meyer-Olkin (KMO), the measure of sampling adequacy attained a high score of 0.802 (Table 4) The Bartlett test of sphericity also was important, this suggest that the matrix of population is not an identical matrix. In addition, the Cronbach alpha that measures internal consistency is 0.802, thus suggest that the reliability of the instrument used by the researcher in the research is quite good.

Table 3. KMO and Bartlett's test

Kaiser-Meyer measure of sampling adequacy		.802
Bartlett's test of sphericity	Approx. chi-square	924.479
	df	91
	Sig.	.000

The data was regulated with principal component analysis (with varimax rotation). The Eigen values has a high value of more than 1. As represented in table 5 and also see figure 1, the factor loading extracted was eight components with the eigenvalue more than 1 and 0.5 (see fig. 1 scree plot). For the total variance (see table 6), as explained by each component extracted; component 1 (41.135%), component 2 (13.893%), component 3 (9.082%). Therefore, the result from the principal component analysis (PCA) and the factors extracted amounted to 64.11% of the total cumulative variance.

Table 4. Correlation matrix of factor analysis

C. matrix	EG1	EG2	EG3	EG4	EG5	EG6	EG7	EG8	EG9	EG10	EG12	EG14	EG15	EG16
EG1	**1.000**	.722	.526	.358	.228	.263	.437	.366	.313	.093	.246	.185	.176	-.081
EG2	.722	**1.000**	.537	.496	.338	.338	.398	.414	.401	.147	.318	.224	.285	-.062
EG3	.537	.537	**1.000**	.517	.268	.315	.234	.240	.424	.128	.328	.093	.155	.047
EG4	.358	.496	.517	**1.000**	.541	.542	.197	.538	.544	.408	.434	.381	.333	.268
EG5	.228	.338	.268	.541	**1.000**	.381	.339	.315	.523	.376	.510	.496	.327	.288
EG6	.263	.338	.315	.542	.381	**1.000**	.331	.483	.633	.486	.453	.480	.274	.254
EG7	.437	.398	.234	.197	.339	.331	**1.000**	.596	.343	.450	.334	.188	.519	.259
EG8	.366	.414	.240	.538	.315	.483	.596	**1.000**	.450	.434	.413	.328	.503	.224
EG9	.313	.401	.424	.544	.523	.633	.343	.450	**1.000**	.452	.483	.544	.348	.204
EG10	.093	.147	.128	.408	.376	.486	.450	.434	.452	**1.000**	.411	.301	.521	.597
EG12	.246	.318	.328	.434	.510	.453	.334	.413	.483	.411	**1.000**	.449	.264	.262
EG14	.185	.224	.093	.381	.496	.480	.188	.328	.544	.301	.449	**1.000**	.405	.331
EG15	.176	.285	.155	.333	.327	.274	.519	.503	.348	.521	.264	.405	**1.000**	.473
EG16	-.081	-.062	.047	.268	.288	.254	.256	.224	.204	.597	.262	.331	.473	1.000

Table 5. Rotated factor matrix

Factor Matrix	**Factors**		
	1	2	3
Ensures socio-economic improvement	**.747**		
Improves the standard of living of the people	**.724**		
Increases income	**.699**		
Ensures a functioning economy	**.692**		
Increases profitability of organizations	**.684**		
Ensures a viable economic growth of the country	**.657**		
Creates room for innovation		**.784**	
Increased competition to boost economic growth		**.749**	
Improves health care and education of the country		**.698**	
Increases connectivity between organizations and international investors		**.632**	
Improves the confidence of investors on the economy		**.600**	
Wealth creation			**.857**
Creates job			**.833**
Reduces cost of production			**.673**
Extraction method: Principal component analysis. Rotation method: Varimax with Kaiser Normalization.[a] a. Rotation converged in 7 iterations			

Table 6. Total variance explained

Factors	**Initial eigenvalues**			**Extraction sums of squared loadings**			**Rotated sums of squared loadings**		
	Total	**% of Variance**	**Cumulative %**	**Total**	**% of Variance**	**Cumulative %**	**Total**	**% of Variance**	**Cumulative %**
1	**5.759**	41.135	41.135	5.759	41.135	41.135	3.580	25.575	25.575
2	**1.945**	13.893	55.028	1.945	13.893	55.028	2.698	19.273	44.848
3	**1.271**	9.082	64.110	1.271	9.028	64.110	2.697	19.262	64.110
4	.822	5.873	69.983						

https://doi.org/10.1051/e3sconf/201912203001

5	.741	5.296	75.279						
6	.660	4.714	79.993						
7	.571	4.077	84.070						
8	.499	3.561	87.631						
9	.458	3.274	90.905						
10	.374	2.669	93.574						
11	.305	2.182	95.756						
12	.245	1.748	97.504						
13	.190	1.360	98.864						
14	.159	1.136	100.000						
Extraction method: Principal axis factoring									

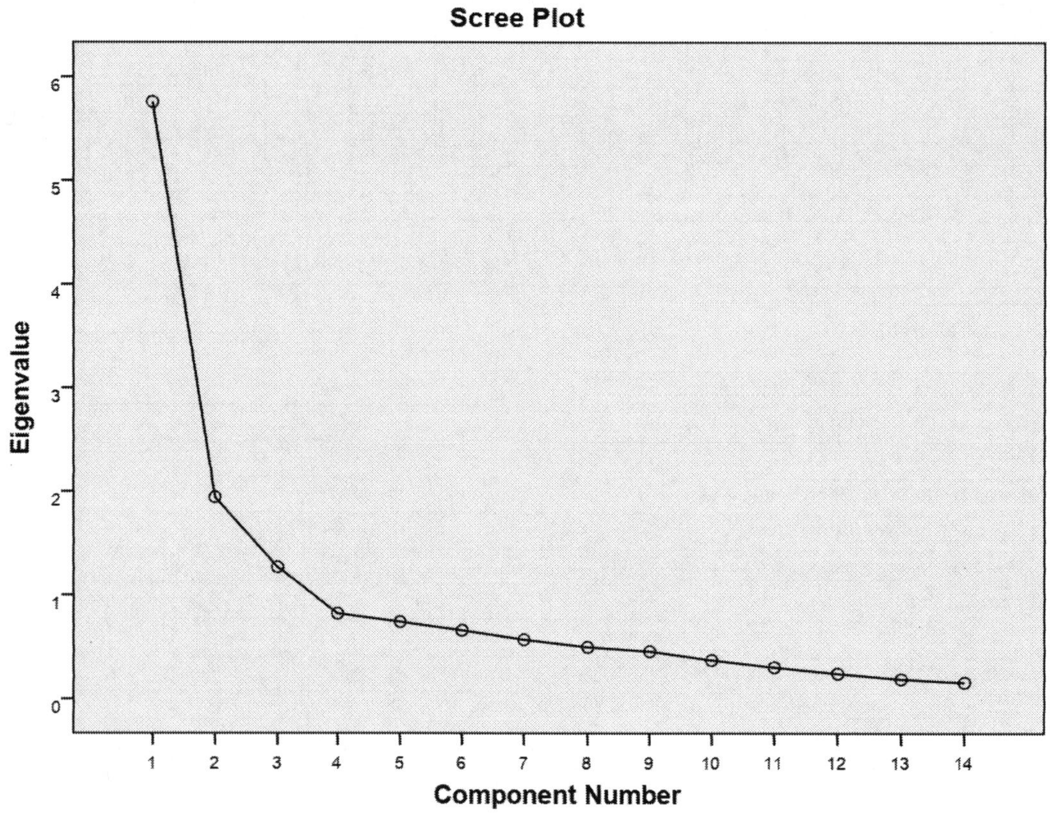

Fig. 1. Scree Plot for Factor Analysis

The principal axis factoring used showed that three (3) were present with eigen value greater than 1 as represented in the table 6. above. Due to the careful observation of the inherent connections among each of the variables under each factor, the following assessments were made. Factor 1 was described as **economic and social development**; factor 2 was described as **environmental and welfare development** and factor 3 was termed **income and industrial improvement.** The term used in describing these factors was gotten as a result of closely observing the variable within each of the factors. The three factors extracted and its constituents indicators are explained below, together with a comprehensive descripytion on how it was described the three factor sections.

4 Discussion of results

Factor 1: economic and social development

As shown in table 5 above, the three (3) was extracted as the impact of power infrastructure development on economic growth. For factor 1 were 'ensures socio-economic improvement (74.7%), improves the standard of living of the people (72.4%), increases incomes (69.9%), ensures a functioning economy (69.2%), increases profitability of organisations (68.4%), ensures a viable economic growth in the country (65.7%). The number in the parenthesis showcase the individual factor loadings, also the definition of these variables are explained in table 6 this cluster acounted for 41.135% of the variance.

Factor 2: environmental and welfare development

As shown in table 6 above, three (3) was extracted as the impact of power infrastructure development on economic growth. For factor 2 were it creates room for innovation (78.4%), increased competiton to boost economic growth (74.9%), improves health care an education of the country (69.8%), it increases connectivity between organisations and international investors (63.2%) and improves the confidence of investors on the economy (60.0%). The number in the parenthesis showcase the individual loadings, also the definition of these variables are explained in table 6 above. This cluster accounted for 13.893% of the variance.

Factor 3: income and industrial improvement

As shown in table 5 above, three (3) was extracted as the impact of power infrastructure development on economic growth. For factor 3 were 'wealth creation' (85.7%), 'creates jobs' (83.3%), and 'reduces cost of productions' (67.3%). The numbers in the parentheses show the individual loadings. The definitions of these variables are explained in table 6 above. This cluster accounted for 9.028 per cent of the variance.

5 Implication of study

From the findings, the theoretical reviews conform to the empirical findings. It can be concluded from the study that wealth creation is the most important impact power infrastructure development has in the economic growth in SSA. It creates jobs, reduces cost of production, ensures a functioning economy, increases productions, and increases income. To be able to achieve these benefits the development of power infrastructure development in SSA is essential. Preference must be given to the power sector to be able to achieve these benefits for an industrialised and functioning economic in SSA. Finally, the role power infrastructure development has in the economic growth in SSA and other nations cannot be over-emphasized. SSA, therefore, requires urgent implementation of this study to achieve the required growth.

6 Conclusion

Results from the literature review established that the following are the impact of power infrastructure development in economic growth of SSA. It creates employment, alleviates poverty, provides better health care, and improves the country's gross domestic product (GDP). The literature has further shown that other benefits power infrastructure development has on the economic growth are an increase in the income of the citizens, and more investment in the economy by investors.

Results from the findings of the secondary data i.e. the questionnaire survey indicate that there are ten main impact of power infrastructure development on the economic growth in SSA are: wealth creation, job creation, reduction in the cost of productions, a functioning economy, viable economic growth of the country, increased income, improvement in the standard of living of the people, a socio-economic improvement, improved health care and education of the country and increased competition to boost economic growth. It can be conclusively be said that the research objectives for this study has been answered.

References

1. Wilson, D. & Purushothaman, R. (2003). Dreaming with BRICs: The path to 2050. *Global Economics Paper,* (99):1.
2. Emenike, D. (2016). A qualitative case study of Nigeria electric power outage and its economic consequence.
3. Estache, A. (2008). Infrastructure and development: A survey of recent and upcoming issues. Rethinking infrastructure for development, 247.
4. Youm, I., Sarr, J., Sall, M., Ndiaye, A. & Kane, M. (2005). Analysis of wind data and wind energy potential along the northern coast of Senegal. Review.energy.renewable, 8(2005):95-108.

5. Tortajada, C. (2016). Policy dimensions of development and financing of water infrastructure: The cases of china and India. *Environmental Science & Policy*, 64177-187.
6. World Bank. (2000a). *Fuel for thought. An environmental strategy for the energy sector.*
7. International energy agency (IEA). (2003). the power to choose: Demand response in liberalized electricity markets.
8. Cabraal, R.A., Barnes, D.F. & Agarwal, S.G. (2005). Productive uses of energy for rural development. *Annual Review of Environment and Resources*, pp.30117-144.
9. Modi, V., McDade, S., Lallement, D. & Saghir, J. (2005). Energy services for the millennium development goals. Energy services for the millennium development goals.
10. Tunde, A.O., 2005, July. Small hydro schemes-taking Nigeria's energy generation to the next level. In *Power Engineering Society Inaugural Conference and Exposition in Africa, 2005 IEEE* (pp. 112-119). IEEE.
11. Hulten, C.R., Bennathan, E. & Srinivasan, S. (2006). Infrastructure, externalities, and economic development: A study of the Indian manufacturing industry. *The World Bank Economic Review*, 20(2):291-308.
12. Yadoo, A. & Cruickshank, H. (2012). The role for low carbon electrification technologies in poverty reduction and climate change strategies: A focus on renewable energy mini-grids with case studies in Nepal, Peru and Kenya. *Energy Policy*, 42591-602.
13. Oshikoya, T.W. (2008). Nigeria in the global economy. *Business Economics*, 43(1):31-43.
14. Burns, N. & Grove, S. K. (2005). *The practice of nursing research conduct critique and utilization* (5th edition). Pennsylvania: Elsevier.
15. Pallant, J. (2011). *Multivariate analysis of variance. SPSS survival manual.* Allen & Unwin. Pallant, J. (2010). *SPSS survival manual: A step-by-step guide to data analysis using SPSS.* Maidenhead. *England: McGraw-Hill Education.*
16. Ledwaba, M.J. (2012). *Informal settlements and organisations in post-apartheid South Africa: The case of Bethlehem, Tshwane* (Mini dissertation). Available online at: http://hdl.handle.net/10210/532.[Accessed 27 February 2017].
17. Ahadzie D, Proverbs D, Olomolaiye P. Critical success criteria for mass house building projects in developing countries. *International Journal of Project Management*. Pp-26(6):675-87, Aug 31, 2008.

A hybrid model for China's power grid investment demand forecasting based on variational mode decomposition, regularized extreme learning machine and support vector machine

Qianqian Wu[1], *Shaowen* Zhu[1], *Jinchao* Li[1,2,*], *Wenjun* Chen[1,2], *Yunna* Wu[1,2]

[1]School of Economics and Management, North China Electric Power University, Beijing, China
[2]Beijing Key Laboratory of New Energy and Low-Carbon Development, North China Electric Power University, Beijing, China

Abstract. With the continuous maturity of China's power grid as well as the advancement of electricity market reform in China, accurate and efficient investment decision has become an inevitable requirement of power grid enterprises. However, China's Power grid investment demand has complicated nonlinear and non-stationary characteristics due to it's complex causes of formation, thus make it hard to be forecasted. Aiming at this problem, this paper puts forward a novel hybrid VMD-RELMLOO-PSOSVM forecasting model based on variational mode decomposition (VMD), leave-one-out cross validation error based optimal regularized extreme learning machine (RELM-LOO) and support vector machines optimized by particle swarm optimization algorithm (PSO-SVM). Firstly, the VMD method is employed to decompose the original power grid investment data sequence into several modes which have specific sparsity properties while producing main signal. Then, according to the different characteristics of each subsequence, the RELM-LOO and PSO-SVM model will be used to forecast different modes, respectively; Next, the prediction results of all modes are aggregated to obtain the final prediction results of China's power grid investment demand. Finally, this paper predicts China's power grid investment demand from 2018 to 2020 under 5 different scenarios based on the proposed VMD-RELMLOO-PSOSVM hybrid forecasting model.

1 Introduction

In recent years, the advancement of electricity market reform in China has greatly improved marketization degree of power industry. Against this background, power grid enterprises have to continuously improve their decision-making quality and efficiency to enhance enterprise's operational efficiency and market competitiveness. For power grid enterprises, power grid investment demand decision is critical for enterprises' development due to it's huge investment amount along with long payback period [1]. However, China's power grid investment demand is influenced by many factors such as economic development, price level, national policy and other uncertainties, complex causes of formation bring China's power grid investment demand with complicated nonlinear and non-stationary characteristics. So, in this paper, we devote to put forward an accurate and effective forecasting model for China's power grid investment, thus to improve the investment decision-making quality and efficiency of power grid enterprises.

The study on power grid investment forecasting model is still less. ZHAO Huiru, etc. [2] proposes a co-integration theory and error correction model to forecast the power grid investment demand, this model finds out the logarithmic linear relation between the power grid investment and influencing factors of the power grid investment; However, the power grid investment has complicated nonlinear nonstationary characteristics, the proposed forecasting model in this paper hardly excavate the internal laws and characteristics of the power grid investment. Shuyu Dai,etc. [1] proposes a hybrid support vector machine optimized by differential evolution algorithm and grey wolf optimization algorithm(DE-GWO-SVM) to forecast China's power grid investment based on the historical data from 1990 to 2016; The support vector machine can effectively process the complex nonlinear data and brilliant parameters optimization ability of the intelligent algorithm, the DE-GWO-SVM model has an outstanding prediction performance for China's power grid investment. Jinchao Li, etc. [3] utilize the empirical mode decomposition(EMD) method to decompose the original power grid investment data series into several intrinsic mode functions and a residual component which contain different characteristics of the original power grid investment data series, respectively; These subsequences decomposed have simpler frequency feature and stronger correlations which are easier to build the prediction model compared to the original data series, thus further improve the prediction accuracy prominently.

* Corresponding author: lijc@ncepu.edu.cn

 https://doi.org/10.1051/e3sconf/201912203002

In this paper, a new hybrid intelligent method based on variational mode decomposition (VMD) is employed to forecast China's power grid investment based on the history data series from 1990 to 2017. First, the VMD method is employed to decompose China's power grid investment into several modes which have specific sparsity properties. Then, the time series of each decomposed mode will be forecasted by appropriate methods. Finally, the forecasting results of all decomposed modes will be aggregated to obtain the final forecasting result of China's power grid investment. In view of the lack of the sample numbers of China's power grid investment and enhancing the generalization ability of the prediction model to further improve the accuracy and effectiveness of the VMD-based power grid investment forecasting model, the support vector machine optimized by particle swarm optimization (PSO-SVM) algorithm and the leave-one-out cross validation error based optimal regularized extreme learning machine (RELM-LOO) algorithm will be employed to forecasting the decomposed modes.

The structure diagram of the VMD-RELM-PSOSVM hybrid forecasting model is shown in Figure 1.

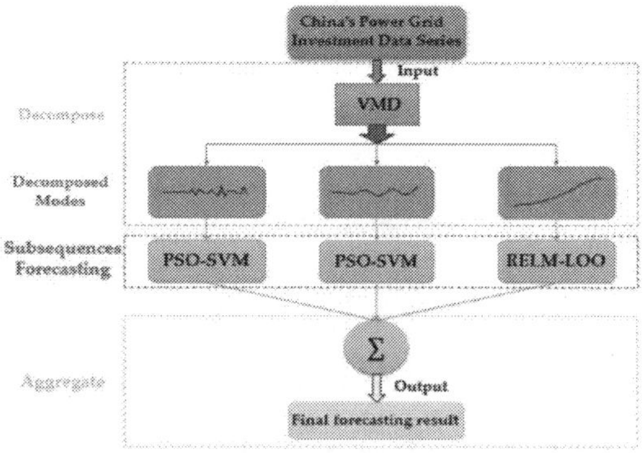

Fig. 1. Structure diagram of the VMD-RELMLOO-PSOSVM hybrid forecasting model.

The rest of this article is arranged as follows: Section 2 introduces the methodologies; Section 3 carries out empirical analysis to verify the validity of the proposed model for the power grid investment prediction in China; Section 4 forecasts the power grid investment in China from 2018 to 2020 under 5 different scenarios; Section 5 summarizes the whole paper.

2 Methodology

2.1 Variational mode decomposition

Variational mode decomposition (VMD) was proposed by Dragomiretskiy and Zosso [4] in 2014, it is a newly developed multiresolution technique for adaptive and non-recursive signal decomposition. The purpose of the VMD is to decompose an input signal y into k discrete number of sub-signals (modes), where each mode has limited bandwidth in spectral domain. And, each mode y_k is required to be mostly compact around a center pulsation ω_k determined along with the decomposition.

2.2 Regularized extreme learning machine optimized by leave-one-out cross validation (RELM-LOO)

A simple and efficient learning algorithm for single-hidden layer feedforward neural networks (SLFNs) called extreme learning machine (ELM) which was proposed by Huang et al [5]. However, since ELM doesn't consider heteroskedasticity in real applications, its performance will be affected seriously when outliers exist in the dataset. In order to address these drawbacks, regularized extreme learning machine based on structural risk minimization principle and weighted least square was proposed by Deng et al [6]. For the regularized ELM described in [6], the search for the optimal regularization parameter is important, as different regularization parameter has different generalization performance. Aiming at this problem, Cao et al [7] proposed a LOO error based optimal regularized ELM for classification problem, this method can achieve better classification performance by employing the leave-one-out cross validation (LOO) approach to find out the best regularization parameter of the regularization extreme learning machine.

2.3 support vector machine optimized by particle swarm optimization (PSO-SVM)

The support vector machines (SVMs) were proposed by Vapnik. It is a statistical learning algorithm based on VC (Vapnik-Chervonenkis) dimension theory and structural risk minimization (SRM) principle.SVM has a strong capacity for processing nonlinear data [8]. The basic principle of support vector machines is to find a nonlinear mapping function $\phi(x)$ to make the linear inseparable data x in the low-dimensional feature space projected into the high-dimensional feature space F to make it a linear separable problem.

Particle swarm optimization (PSO) was first proposed by Kennedy and Eberhart in 1995. It is an emerging population-based meta-heuristic that simulates social behavior such as birds flocking to a promising position to achieve precise objectives in a multidimensional space. Like evolutionary algorithms, PSO searches for optimal solution in solution space using population of particles. It can be used for parameter determination and feature selection in the SVM. The operation steps of PSO-SVM are described in detail in [9].

3 Case study

In this section, we establish the hybrid VMD-RELMLOO-PSOSVM model for China's power grid investment prediction. To further validate the effectiveness and accuracy of the proposed VMD-RELM-PSOSVM hybrid forecasting model. The VMD-PSOSVM model, VMD-RELMLOO model, EMD-based hybrid forecasting models and the single regularized extreme learning machine model as well as PSO-SVM model are set as benchmark models of the proposed

VMD-RELM-PSOSVM model. The mean absolute percentage error (MAPE) and the root mean square error (RMSE) will be the evaluation criteria for quantitative analysis of different forecasting models.

3.1 VMD-RELMLOO-PSOSVM hybrid forecasting model for China's power grid investment

Since the VMD requires a predetermined number of decomposition k, the latter is varied depending on the number of IMFs obtained by EMD [10]. If the number of IMFs is d, then the parameter k used to determine the number of decomposition components to obtain by EMD is $d-3, d-2, d-1, d, d+1, d+2, d+3$.

In this paper, we determine the value of k according to the MAPE and RMSE of different VMD-RELMLOO-PSOSVM model which have different decomposition value k. Table 1 shows the calculation results under different models with different k value.

Table 1. MAPE and RMSE of different VMD-RELMLOO-PSOSVM model with different *k* value.

Indicator	k=3	k=4	k=5	k=6
MAPE	3.3%	3.7%	3.9%	4.8%
RMSE	120.7	126.1	139.3	172.2

According to the calculation results, we can conclude that when k=3, the proposed VMD-RELMLOO-PSOSVM has the best forecasting performance, so in this paper, we decompose China's power grid investment into 3 modes. Figure 2 gives the decomposed result of China's power grid investment by VMD method.

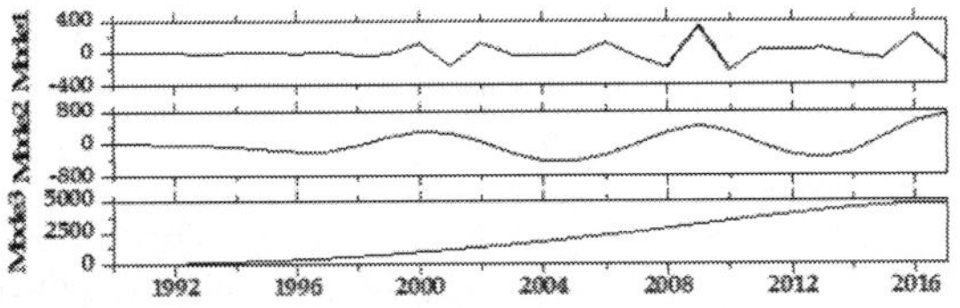

Fig. 2. VMD decomposition diagram.

As we can see from Figure 2, mode 1 has a significant stochastic characteristic while mode 2 has a significant periodic characteristic. Time series prediction method is suitable for processing stationary data series which have stochastic or periodic characteristics. Traditional time series prediction method like ARIMA has a major defect for forecasting nonlinear data series while neural network algorithm and support vector machine algorithm have great properties to process nonlinear data series. However, China's power grid investment data series is relative short, the data samples of mode1 and mode 2 are too small for the training of the neural network while support vector machine has a great property to process problems with small number of samples. So in this paper, the support vector machine optimized by particle swarm optimization (PSO-SVM) algorithm will be used to forecast mode1 and mode 2 which have significant nonlinear and high frequency characteristics.

As we can see from Figure 2, mode 3 is a smooth increasing curve which has a significant trend characteristic. And we use the regularized extreme learning machine (RELM) combined with leave one out cross validation to forecast mode 3. Mode 3 reveals the basic development trend of China's power grid investment, it is directly affected by the development level of China's economic and social as well as electricity production and consumption, so in this paper, we select GDP growth rate, the share of secondary industry, urbanization rate, total electricity installed capacity, total electricity consumption, share of non-fossil energy power generation installed capacity and population as inputs of the RELM-LOO model for the forecasting of mode 3.

3.2 Results and discussion

In this section, we will analyze the effectiveness and accuracy of the proposed VMD-RELMLOO-PSOSVM model based on the forecasting results presented in Table 2. And three conclusions can be found based on these forecasting results.

First, we compare the forecasting results of the VMD-RELMLOO and VMD-PSOSVM model with the proposed VMD-RELMLOO-PSOSVM model. We can find out that the proposed VMD-RELMLOO-PSOSVM model has better APE performance than the other two VMD-based forecasting models, these results confirm the effectiveness and accuracy of the proposed VMD-RELMLOO-PSOSVM model compared to other VMD-based models.

Second, the forecasting results of EMD-based models will be compared with VMD-based models. All the VMD-based models have better forecasting performance compared with their corresponding EMD-based model. This confirms the superiority of VMD method compared with EMD method for China's power grid investment.

Finally, the forecasting results of RELMLOO and PSOSVM model will be compared with VMD-RELMLOO-PSOSVM model. Both of them have bad forecasting accuracy in extremes while the proposed VMD-RELMLOO-PSOSVM model has primely solved this problem. This further confirms the robustness and effectiveness of the proposed VMD-RELMLOO-PSOSVM model.

Table 2. Calculation results of MAPE and RMSE of different forecasting models.

Year	VRS	VR	VS	ERS	ER	ES	R	S
MAPE	0.033	0.052	0.043	0.068	0.082	0.071	0.067	0.069
RMSE	120.7	183.1	126.9	241.6	291.3	222.9	264.4	349.1

4 Scenario Analysis

In this section, China's power grid investment demand from 2018 to 2020 under five different scenarios will be forecasted based on the proposed VMD hybrid forecasting model. First, the PSO-SVM model will be employed to forecast the high frequency model and mode 2 components separately in advance for three years. Then, the mode 3 component which has a trend characteristic will be forecast by RELM-LOO model under different scenarios. Finally, the forecasting results of the power grid investment demand in the next three years will be obtained by aggregating the forecasting results of all the modes.

According to the 13th Five-Year plan (2016-2020) which was published by Chinese government in 2016, the target of the GDP growth was set to more than 6.5% and the urbanization rate will reach 60% by the end of 2020, so in this paper the GDP growth rate under different scenarios will be set to 6.5%, 6.6%, 6.7%, 6.8% and 6.9% respectively in the next three years. And the urbanization rate in the next three year will be set to a fixed value under all scenarios, reaching 59%, 59.5% and 60%, respectively. And the share of secondary industry will be set to 40%, 39%, and 38% in the next three years, respectively.

According to the "Electric power development planning in 13th Five-Year" (2016-2020), the total installed capacity of power generation will increase at an annual growth rate of 5.5% while the target share of installed capacity of non-fossil energy was set to 39% by the end of 2020, so in this paper the growth rate of total installed capacity of power generation will set to 5.5% from 2018 to 2020 under different scenarios, and the share of installed capacity of non-fossil energy will be set to 37%, 38% and 39% in the next three years, respectively. Meanwhile, the growth rate of total electricity consumption was set to 3.6%-4.8% in the planning, so the annual growth rate of total electricity consumption under different scenarios will be set to 3.6%, 3.9%, 4.2%, 4.5% and 4.8% in the next three years, respectively. Refer to the 13th Five-Year national family planning development plan, we set the growth rate of population to 0.6% from 2018 to 2020.

The forecasting results of China's power grid investment under different scenarios are shown in Figure 3.

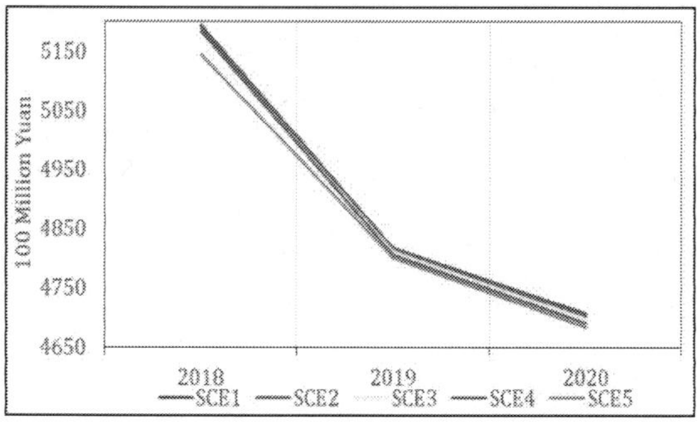

Fig. 3. Forecasting results of China's power grid investment from 2018 to 2020 under different scenarios.

Compared with the rapid growth of the power grid investment from 2012 to 2016, the forecasting results indicate that the power grid investment will not substantially increase in the next three years. In contrast, the downward trend of the power grid investment demand in the next three years would be happened.

5 Conclusion

In order to improve power grid enterprises' investment decision-making quality and efficiency, this paper puts forward a newly hybrid VMD-RELM-PSOSVM forecasting model for China's power grid investment demand. Firstly, the VMD method is employed to decompose the original power grid investment into three modes. Then model and mode 2 with high frequency characteristics are forecasted by PSO-SVM models while mode 3 which has a trend characteristic will be forecasted by RELM-LOO model. Then, the prediction results of different modes are summing up to get the final power forecasting results of China's power grid investment demand. Next, the VMD-PSOSVM model, VMD-RELMLOO model, the previously popular EMD-based forecasting models and single RELM-LOO model as well as PSO-SVM model are set as benchmark models of the proposed VMD-RELM-PSOSVM hybrid forecasting model; The RMSE and MAPE calculation results of different forecasting models suggest that the VMD-RELM-PSOSVM hybrid model has a better prediction performance than all the benchmark models. Finally, we forecast China's power grid investment from 2018-2020 under five different scenarios; The forecasting results indicate that China's power grid investment will decline in the next three years.

Acknowledgments

This work has been supported by the Fundamental Research Funds for the Central Universities (no. 2018ZD14) and the Science and Technology Project of SGCC "Research on lean investment optimization method based on benefit analysis of the whole process of power grid".

References

1. S.Y. Dai, D.X. Niu, Y.R. Han. *Applied Sciences*, **8**, 636 (2018)
2. H.R. Zhao, L. Yang, C.J. Li, X. Ma. *Power System Technology*, **35**, 193-198 (2011)
3. J.C. Li, S.W. Zhu, Q.Q. Wu, P.F. Zhang. *MATHEMATICAL PROBLEMS IN ENGINEERING* (2018)
4. K. Dragomiretskiy, D. Zosso. *IEEE Trans Signal Process*, **62**, 531-44 (2014)
5. G.B. Huang, Q.Y. Zhu, C.K. Siew. *Neurocomputing*, **70**, 489-501 (2006)
6. W. Deng, Q. Zheng, L. Chen. *2009 IEEE Symposium on Computational Intelligence and Data Mining*, Nashville, TN, 389-395 (2009).
7. J.W. Cao, K. Zhang, M.X. Luo, C. Yin, X.P. Lai. *Neural Networks*, **81**, 91-102 (2016).
8. V. Vapnik, S. Golowich, A. Smola. *Adv. Neural Inf. Process. Syst*, **9**, 281-287 (1996).
9. A. Selakov, D. Cvijetinović, L. Milović, S. Mellon, D. Bekut. *Applied Soft Computing*, **16**, 80-88 (2014).
10. R. Zhang, Z.Y. Dong, Y. Xu, K. Meng, K.P. Wong. *IET GENERATION TRANSMISSION & DISTRIBUTION*, **7**, 391-397 (2013).

A Review of Power Infrastructure Financing: The Brazil and India Approach

Kasenge Mathe[1], *Ngcobo* Ntebo[2] and *Emmanuel* Oikelomen Ayorinde[3,*]

[1,2,3] Department of Civil Engineering Technology, University of Johannesburg, South Africa

Abstract. Infrastructure development promotes human capital development. Access to quality education and health care are some of the incentives of Infrastructure development. The development of power infrastructure in Brazil and India has greatly facilitated it gross domestic product (GDP) and social-economic development. This study assesses the approach of development of power infrastructure in Brazil and India as a prerequisite for other emerging nation's economy to follow. A confirmatory literature involving power infrastructure development in India and Brazil was explored for the purpose of this study. The paper highlights the challenges, policies effects, legislations effects, ways to finance, philosophical approach and the economic benefits involves in power infrastructure development in Brazil and India. Power infrastructure will strive when the challenges involves in its development is addressed. The adoption of this approach of implementing the developmental challenges, better policies, innovative financing scheme and effective legislations that has aided the development of power infrastructure development and should be replicated in emerging nations i.e. sub-Sahara Africa (SSA). In addition, direct implementation of this approach will bring forth the benefits of infrastructure development. This will contribute to the knowledge of power infrastructure development in emerging nations across the globe.

1 Introduction (Brazil and the state of power infrastructutre development)

Brazil occupy's an area of 8.5 million km^2, with a growing population of 191 million people, with average income of US$ 10,414 (Purchasing Power Parity- PPP, 2010), with GDP of 2.017 trillion, Brazil is the most developed country in the Latin America, according to the (1). Brazil power infrastructure facilities development can be traced back over 40 years down the line, after the fall in oil in the year 1970s, where there was decrease in the imports of oil and there was increased in the domestic production with mixed directives which brought about powerful growth of sugarcane ethanol, placing Brazil independent in the transportation sector European Journal on Power Electronics and Electrical Drives (2). Since 1990 the countries demand electricity has doubled, closing on 300 mtoe by 2013 and has exceeded the increased in the gross domestic products (GDP) to the average increase in growth which amounts to 3.1 per cent and that of GDP of 2.8 percent respectively. For the more than ten years in Brazil for the supply of energy; Natural gas, oil and it's subsequent has accounted for about 80 percent of the increase in energy supply for the past ten years in Brazil (3). According to (3), renewal energy mechanisms remains a solid aspect in the local fuel mix, which represents 41 percent in 2013, as against 60 percent in 1970, making Brazil with 1.55 Tco2/toe (CO2 intensity) emission element is the least carbon-intensive economy in the world as compared to the world average which is 2.37 tco2/toe.

* Corresponding author: engrkulz@gmail.com

Laws that help to improve the power infrastructure aspects of Brazil was made in the year 1988 during the military regime, which settle the ownership of both natural resources and hydraulic power potentials under the government and stated guild lines for negotiations and requirements to facilitates them. This process was not implemented until 1996 and later reviewed in 2002, and this government control over natural resources act has allowed Brazil as a nation to determine its power infrastructure development. Uncertainties have been the case of the players in the power infrastructure markets and investors, because of changes in legislations as relating to power infrastructure development in Brazil (4). Biomass, energy efficiency and hydropower have allowed Brazil to be among the least carbon-intensive economy in the world. The destabilization in the oil industry in the year 1975, led to the use bioethanol, with the compulsory use of 25 percent (E25), these bioethanol of "gasoline-bioethanol" mix, is Major elements in the increase in oil prices and economic development as a result of this over 90 percent of cars manufactured in Brazil are "flex-fuel" which either contains E25 or E100. The composition of bio-ethanol and bioenergy from sugar cane source amounts up to 17 percent of the primary power supply in the country (3).

The increase in the growth in the non-renewable provision of energy over the year 2000s, specifically that

 https://doi.org/10.1051/e3sconf/201912203003

of natural gas, is because of unfavorable weather conditions challenges to hydropower expansion. The effect of drought caused excessive blackouts in 2001; these have led to the building of gas plants to contain such excesses in the future (5). Also, about 70 percent of Brazil's hydropower capacity is constructed in the Amazon area (6). In the year 2002 a motivational program was created called the PROINFA, these in every case increased the competitions between the small hydropower, biopower, photovoltaic power (PV) and wind power plants. The Wind will be the next big thing from Brazil renewal energy drive as a result of drought, even if the wind performance is seasonal, few wind farms have recorded exceptional rise in capacity factors in the surplus of 50 percent (5). Wind farm development capacity in Brazil increased by 84 percent in 2013 (7).

1.1 Challenges and Policies Affecting Power Infrastructure in Brazil

Brazil with India, China, and Russia is the world top four emerging nations in terms of infrastructure development and an increase in GDP, these makes Brazil one of the best places for economic growth, due to the excessive increase in the power infrastructure facilities (8). For the past twenty years Brazil's growth has exceeded other nations in the world, these gave them the opportunity to host the FIFA World Cup in 2014 and Olympics in 2016, and has the opportunity to be among the G5 economy in the world, but these have since declined due to bad policies and inconsistency (8 and 9). (10 and 11) stated that due to the unfavorable economic policies and lack of preparedness to drought faced by hydroelectric facilities, which is the provider of about 71.58% of electricity in 2011, which declined to 61.96% due to drought as represented in fig.1 the map of Brazil. This problem faced by the Brazil power sector has opened the door to other means of generating electricity, to meet the need of the growing population; these other sources are thermal power facilities which amount to 25.65% for 2011, this makes Brazil 100% sufficient in terms of power infrastructure development, and later the thermal power facilities increased to 26.83% (10). These, however, demonstrated that Brazil needed to import from other nations to be sufficient in providing electricity for the citizens. Though Brazil boosts an increase in electricity consumption with about 3.3 increase between 2000 to 2011, this is still lacking when compared to the world's two emerging nations, China and India with an increase in the rate of consumption of 4.3% and 5.4% (12). The problem of drought has made Brazil locate their hydroelectric facilities in the Amazon region; this is because of reduced amount of rainfall (12).

Brazil has the potential to produce about 260GW of electricity from all the sources of generating electricity in Brazil which includes; hydroelectric power facilities, which is the major provider of electricity in Brazil, thermal power plant which produces most of the energy capacity after hydroelectric facilities and wind farm which supplement the other sources of electricity to meet 100% sufficiency (10).

A map of Brazil showing drought in different cities

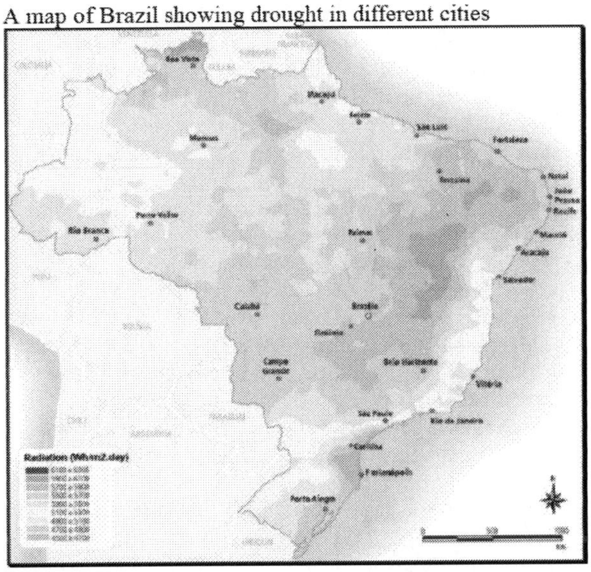

Fig.1. Source: (13)

1.2 Present and Future Policies Situation

The two important organizational deeds that have helped Brazil in the energy policymaking are both the long-term (14) and yearly-improved medium-term energy policy. The PNE 2030 was produced in 2007 and is expected to be updated for the year 2050 (16). While that of the medium-term policies PDE, expand to 2022 and it will include the following; affordable tariff prices, energy stability, reduction in the emissions of gasses, all-around access to power (16). For several years now, there have been different ways of financing infrastructure in Brazil, but not always work in funding for power infrastructure facilities. The decision taken to finance energy infrastructure is directed by a useful policy framework, which ensures funding gotten is cost effective (17). General infrastructure development investment in Brazil is spearhead by the Brazil Development Bank (BNDES). The bank decreases the circulation for infrastructure services the institution also prolong the agreement of some its grants ranging from US$ 3.4 billion in 2003 to an increase of US$ 17.5 billion in 2008 and US$ 25 billion in 2009 for infrastructure development (16 ;18). Brazil policy since 1970, placed the "national energy planning and investment" on a yearly energy balance, made known by the Ministry of Mines and Energy. For the last decade now it is published by the Planning and Research sector "Empressa de Pesquisa Energetica" (19).

1.3 Ways of Financing Power Infrastructure in Brazil

The following according to (16) are the sources of power infrastructure financing in Brazil:

- Budgeting: it is gotten from electrobas and petrobas. This is the annual budgetting assigned for the purpose of developing power infrastructure projects in the country.

This has however not been able to meet the demand and financial capacity of the brazil power sector, which has led to sourcing for other alternative sources in the financing of power infrastructure in Brazil (16).

• Global energy fund: these types of funding is obtained from the "global reversion reserve" and this is the expansion funding given for the purpose of improving and expanding power development global and also it supports renewable energy development (16).

• International finance: this is a source of financing power infrastructure projects mandated through Brazil electricity regulatory agency (ANNEL) to all power utilities company, asking them to spend 0.5% of their working revenue on power cost-effectiveness measures. This has however raked in US$ 378m (16).

• National development banks (BNDES): the national development banks of Brazil has been the main means of financing power infrastructure projects in Brazil. And also, responsible for the granting of loans to the private investors investing in the power infrastructure sector in Brazil and approve finances on behalf of the Brazil government for the purpose of developing power infrastructure. By 2011 the BNDES has approved a total of US$ 16.5m as an credit efficiency loan to an power distribution company, for the purpose of improving and expansion of power infrastructure development (20).

• National climate finance: this is also known as Fundo Nacional Sobre Mudanga do Clima. This kind of financing mechanism is gotten from the ministry of Environment and jointly controlled by the National Development Bank (BNDES). This focuses on reduction and adaptation of the following seven factors; urban flexibility, sustainable municipality, productive equipment, low carbon energy sources, concrete waste and briquette. A total of US$ 252m was allocated through the budget, and partly gotten from petroleum resources for this purposes (20).

• Social Fund: this mechanism was enacted in 2010, and receives finances from petroleum resources. They are also used in the area of reduction and adaptation and other specific areas with the purpose of boosting development in power infrastructure (21).

Clean development mechanism (CDM): this has been used to finance quite a number of sustainable energy in Brazil. The Brazil government through the BNDES has created numbers of credit facilities for this purpose, this will however Improvement power development in the region (16).

2 India and the State of Power Infrastructure Development

Since the 1970s in India power infrastructure remained a natural monopoly whereby it was operated and financed by the traditional financiers, which is the government (4). From the early stage of the 1970s, the traditional operators and financiers of this model i.e. the central government was unable to meet the demands growing population of the citizens, this is what led to the general reforms and deregulation of the sector, which ushered concessions in executing power infrastructure development facilities as a means to meet the demand of the growing population in India (22). The power sector in India has improved significantly both in size and capacity over these years, from total existing generation ability of 1,362MW in 1947 to 147,966 in 2009 Ministry of power (23). The main reasons for the reform is t to attain improvements in the areas of distribution and generation. The case of India power sector, policies was made to encourage the involvements of the private sector in investing in the power sector, as a means of meeting the demands of the populations in the area of energy stability (24). Despite the fact that there was a reform in 2004, that was set out to achieve low tariffs and also attracts private investments in the financing of power infrastructure developments; there has been little gains in the involvements of private investments in the power developmental projects. The power is still dominated by the public-owned utilities with very low involvements of private investments, examples are; in 2009 overall existing capacities was 1, 47,715MW where 48,970MW (34 percent) was from the central government, 76,115MW (53 percent) was state-owned, and 22,625MW (13 percent) was private investment (25). The failure to boost private sector involvement in power infrastructure development has led to inefficiencies in the supply of electricity (26). In distribution, the citizens have not realized an increase in tariffs as the total losses continue to increase commercially, the average AT&C loss in India is against that of developing countries which is 30 percent in India from 2007-08 and developing countries which is 10 to 12 percent respectively. It is also understandable that increased in competitions will boost efficiency and increase in the private investments in the power sector Priority Flow Control (27).

2.1 Past and Present Situations of India power development

In India to tackle the history problems of the provision of enough energy security for the entire citizen and improved economic development, this has led to the improving the capacity of renewable energy to the power sector systems for efficiency (28) the India government has undertaken a giant plan in building of 20,000MW of solar power by 2020 to boost the supply of electricity and directly improving the economy of the nation. India is known to be the fifth biggest wind farm turbines around the world, with total installations of 16GW by 2011. This ambition of the India government is been put to the test due to the finances needed in the actualization of such giant stride, this is because the government alone cannot on its own undertake such projects, so therefore there is need for private investment in the power infrastructure development sector to actualize this objective in India (29).

According to (30) and (31) which stated that there will be the need of an around US$ 1 trillion to be able to increase from 8 percent GDP of 2011-12 to 10 percent increase in GDP for 2016-17 for a five-year plan, therefore there is a gap of US$ 300 billion in the infrastructure investment sector, and is as a result of non-availability of long-term capital. From the 1991 Act, the development financial institutions of India was the major financial of long-term capital for infrastructure development in which power infrastructure development is included, there has been decline in the role of the Development Financial Institutions (DFIs) due to the deregulations of financial markets, which has relegated the DFIs in competing with the commercial banks in the financial markets. This coupled with the naive corporate bond has a deepened the gap in the financing power infrastructure facilities (29). Historically, public owned utilities have always been the drivers of power infrastructure facilities in India and coupled with the decline in services given by the central government. There is a forecast of 94 percent of the 11th plan of the 2007-12 target is aimed at getting from 230MW in 1994-95 to 16,078 at 2011 (32).

2.2 Financing Power Infrastructure Facilities in India

According to the Global trends in renewable (33), there are several investors involved in the financing of India power infrastructure developments, they are banks, legitimate companies, institutional financiers who can be public owned utilities or owned by two or three government of different countries. They are also financed by private finance and public finance, venture finance, and private equity finance. The returns of the investment by these bodies depend greatly on the where the capital is gotten from and the risk involved in a particular project, these usually depends on the duration of the power infrastructure project. The report of Bloomberg (28) showed that in the period between 2006 09, India's power infrastructure facilities investments was in between US$ 4 billion and US$ 5 billion, there has been great increase in the investment from that period, from US$ 4.2 billion in 2009 to a rise of US$ 12.3 billion in the year 2011.

Table 1. Financing Infrastructure In Brazil

Variable names	Definition	References
Commercial banks	these involves the provision of investment capital for power infrastructure development in India, they give investment capital to public sector banks, private sector banks, and foreign banks with the aim of aiding power development and a boost in the economy of India and its GDP	29
Equity investors	This is one the mechanisms of financing power infrastructure developments in India, by providing finance in the form of private equity and venture capital for the purpose to aid power facilities development, for the purpose of energy sufficiency and growth in the economy of India.	34
Institutional investors	This mechanism of financing power infrastructure facilities, provides private finance as insurance finance to investors for the growth and development of the sector and the economy	34
Development Banks	The development banks provide capital to the development financial institutions in India though has declined in their capacity in the delivery of capital investment for the development of infrastructure in India still has a significant role in financing power infrastructure delivery in the India and economic development	29
Public-Private Partnership	Though most of the shareholders of a public-private partnership are the private sector, and a fewer representative from the public sector usually a representative of either the federal government, state or local government agencies (35). Example of a PPP project in India through the Independent power project (IPP) was the Dabhol power project (DPP), with government of India and ministry of power collaborated with the Enron Development Corporation (EDC) in delivery this massive project of 2015MW which will be attached to the Maharashtra state electricity board (MSEB) network via 440KV transmission, the Dabhol Power Company (DPC), a Special Purpose Vehicle (SPV) served as medium in bringing together private investors and government organizations in achieving this purpose.	36

2.3 India Electricity Policies effects

Proper and well-structured government policies on energy infrastructure development can pave the way for new way and mechanisms in financing power infrastructure facilities (37). A steady regulatory will be needed even with the various changes that come with political parties. The national electricity Act of India 2003 transformed the power industry, which about changes such as deregulation of the power generation, transmission, and distribution, these gave the state electricity regulatory commission (SERCs) the authority to set the procurement methods of the power sector (39 Ministry of Power, 2003). According to the study of (28), there is a mandate of improving the shares of the power industry in the energy supply chain. Presently in India, the government effective policies and incentives that support the different means of power infrastructure facilities (37). In India states, electricity regulatory commission (SERCs) has adopted some measures towards power development facilities and renewable purchase obligation (RPO). And despite these favorable policies in place by the state and central government, there has been a broad difference in the capacity of each state for instance, states like Tamil Nadu, Maharashtra, Gujarat and Karnataka wind capabilities is more than 1000MW while states like West Bengal and Kerala produces less than 50MW (29).

2.4 Effects of Legislation on power infrastructure in India

According to the report of (38), the following was identified as the policies that can affect the rate of the cash flow from investors in the financing of power infrastructure development.

2.4.1 Time of revenue support

This entails the length of rates or tariff, purchasing power rapport, or another source of financing support i.e. the more time given allows debts to be paid of gradually; with an efficient result, that reduces the financial burden all through the time of the project (38).

2.4.2 Revenue certainty

It is important to have a stable source of financing; this kind of support will offer more assurance of financing in the project (38).

2.4.3 Risk perceptions

There is high financial risk associated with power infrastructure development, due to the length of the projects. This is done to know the cost of the project with risky perceptions (38).

2.4.4 Completion certainty

this is done due to the cost of delays associated with power project completion rates brought by regulations delaying the completion of power infrastructure projects with the setback in realizing the financial benefits (38).

2.4.5 Cost certainty

Regulations can affect the cost of completion of a power infrastructure project, by adding extra requirements in the process of construction (38).

2.5 Challenges of Power Development in India

The following according to Reserve Bank of India, (39) are the challenges facing power infrastructure financing in the India context;

- Construction risk; the lower the construction risk in the financing of power infrastructure projects in India by an investor, the lower the pricing in the allocation of loans from the banks (39).
- Limited time for payment: In the India context, the time given for loan refunding on the power infrastructure projects is a period of five to ten years. This however is a challenge faced by investors in the India power industry due to the difficulty involved by foreign investors in assessing India power sector in 5 to 10 years (39).
- Currency risk: this is also a factor affecting the financing power infrastructure development in India; this is due to currency fluctuations, which can bring about great loss to investors. This is because the main aim of investors, investing in the sector is to make profit and currency risk can hinder their investments (39).
- Policy framework: favorable policy frame work needs to be enacted to guide the nature of power infrastructure financing in India, with the purpose of boosting development in the power industry of India (39).
- Political risk: due uncertainties surrounding politics globally, is a challenge in financing of power infrastructure projects. No private investors will be willing to invest in a political unstable atmosphere, as these might harm the investments (39).

2.6 Economic Benefits of Power Development in India

India located in Asia, a continent that house about 4.3 billion of the world and a home to four of the world biggest economy: China, India, Japan and Republic of Korea. Together these great countries account for about 30% of the world GDP (40). The development of power infrastructure in India has increased the economic growth in the country, undermining the fact that there are gaps in service delivering between the rural areas and urban centers with the rural centers having poor service delivery to the urban centers (41). India level of power infrastructure development has been credited with the improved economic growth in the country (30).

The government of India has made huge investments in large, medium and small irrigation facilities. The large facilities have been responsible for the availability of water provision, irrigation and hydroelectric power (42).

Investments in the energy sectors have resulted to improvements in the life and standard of citizens, thereby eradicating poverty and boosting the economic growth of India (42).

3 Research Methodology

This study is carried out with respect to existing literature, from published researchers with the aim of reviewing the level approach of power infrastructure development in Brazil and India as a model to developing nations economy, inorder to develop their power infrastructure. This study mostly focuses on the power/energy infrastructure development in Brazil and India. In addition, this study reviewed power infrastructure literature on policies effects, challenges, ways of financing and effective legislation in Brazil and India, quantitative research methodology was used in carrying out this study. Also, the study reviewed literature relating to sustainble infrastructure growth in India, Brazil and emerging nations. This study reviewed a wide range of existing literature from the year 1997 to 2017 for a more robust study.

4 Discussions of Findings from Literature

This section discussed the understanding and mode of power infrastructure development in Brazil and India power industry. To understand power infrastructure development in the context of the country under review, a brief overview of the history were explored. The like terms between these countries (BRAZIL and INDIA) is that both are among the five emerging nations in the world (BRICS), and has put in place favorable policies that will transform the power sector.

Brazil power sector has emerged as the top renewal energy in the South American region. As one of the top five emerging nations in the world (BRICS), Brazil with 1.55 Tco2/toe (CO2 intensity) emission element is the least carbon-intensive economy in the world as compared to the world average which is 2.37 tco2/toe is one of the top low carbon economy in the world due to effective renewable policies. In the year 2002 PROINFA, was created these in every case increased the competitions between the small hydropower, biopower, photovoltaic power (PV) and wind power plants. Because of drought that has affected the main source of electricity in Brazil, ensure that about 70 percent of hydroelectric power in Brazil are located in the Amazon area. This has made Wind farm the next big thing from Brazil renewal energy drive even if the wind performance is seasonal, few wind farms have recorded exceptional rise in capacity factors in the surplus of 50 percent. Wind farm development capacity in Brazil increased by 84 percent in 2013. Brazil economic growth because of power infrastructure facilities in the past 20 years gave them the opportunity of hosting the FIFA World Cup and Olympics in the year 2014 and 2016 respectively. In addition, these gave the opportunity to be among G5 economy in the world, it has long taken a nosedive due to recent unfavorable legislative regulation as regards to infrastructure development (Power). General infrastructure development investment in Brazil is sponsored by the Brazil Development Bank (BNDES). The bank decreases the circulation for infrastructure services the institution also prolong the agreement of some its grants ranging from US$ 3.4 billion in 2003 to an increase of US$ 17.5 billion in 2008 and US$ 25 billion in 2009 for infrastructure development (43).

In India, power infrastructure was developed by the government as a natural monopoly from the 1970s. As the population grew over the years it became imperative for a concession in the financing of power infrastructure development, because central government no longer possess the capacity to finance the infrastructure solely. This led the government to collaborate with other partners and shareholder which has greatly improve the stock of India power sector significantly, both in size and capacity over these years, from total existing generation ability of 1,362MW in 1947 to 147,966 in 2009. In 2004, a reform was undertaken to achieve a low tariffs system and improve the participation of private investment. However, this has amounted to little gain, because there has been little benefits to that effect; the power development is still dominated by government owned utilities. In 2009 overall existing capacities was 1, 47,715MW where 48,970MW (34 percent) was from the central government, 76,115MW (53 percent) was state-owned, and 22,625MW (13 percent) was private investment. India is known to be the 5th biggest wind farm turbines around the world, with total installations of 16GW by 2011, this ambition of the India government is been put to the test due to the finances needed in the actualization of such giant stride. Indian power development is financed by both public and private investors. These consist of development banks, institutional investors, private sector and government collaboration in the form public-private partnership, commercial banks, equity investors etc.

5 Conclusion

This article explored literature on power infrastructure development in the Brazil and India. The explored literature showed that there are several approaches adopted by the Brazil and India government in the development of power infrastructture that has positively affected the horizon in terms of quality education, provision of quality health care, employment, wealth creation. This has confirmed the literature that stated that secured energy would aid nation's development in terms of GDP and also boost regional growth, reduce social related crisis, eradicate poverty and unemployment.

References

1. Unit, E.I., 2009. E-readiness rankings 2009: The usage imperative. The Economist. A report from the Economist Intelligence Unit written in cooperation with the IBM Institute for Business Value.

2. EPE (2014b), Ten Year Energy Expansion Plan 2023, Rio de Janeiro.
3. E.P.E (2014). Brazilian energy balance 2014 year 2013, Rio de Janeiro, a.
4. Horner, N., Horner, N., de Paula Oliveira, Antonio Geraldo, de Paula Oliveira, Antonio Geraldo, Silberglitt, R., Silberglitt, R., Khaled Poppe, M., Khaled Poppe, M., Bressan Rocha, B. & Bressan Rocha, B. (2016). Energy foresight, scenarios and sustainable energy policy in Brazil. Foresight, 18(5):535-550.
5. Organisation for Economic Co-operation, Development, and International Energy Agency (2013), "Brazil energy outlook", World Energy Outlook, OECD/IEA, Paris, pp. 301-417.
6. EPE (2007), Brazilian Energy Plan 2023, Rio de Janeiro.
7. Spatuzza, A. (2014). Brazil wind capacity factor at 50%. Recharge, 21.
8. Victor,D.G. & Heller,T.C. (2007). The political economy of power sector reform: the experiences of five major developing countries Cambridge University Press.
9. Oliveira, C., Zulanas, C. & Kashiwagi, D. (2016). A long-term solution to overcome the problems caused by droughts in the Brazilian power systems. Procedia engineering, 145948-955.
10. EIA, (2014, June 17). Hydropower supplies more than three-quarter of Brazil's electric power. U.S. Energy Information Administration. Retrieved May 23, 2015.
11. Petroleum, B., 2015. BP Statistical Review of world energy 2015. Statistical review of world energy, 2016, p.65.
12. Liming, H. (2009). Financing rural renewable energy: A comparison between china and india. Renewable and sustainable energy reviews, 13(5):1096-1103.
13. Amaral, S. and D'Alge, J.C.L., 2009. Spatial data availability and its implications for sustainable development of the Brazilian Amazon. Earth science informatics, 2(4), pp.193-203.
14. De Pesquisa Energética, E. (2007). Plano nacional de energia 2030–PNE 2030. Rio de janeiro: EPE/MME.
15. IHU, Instituto Humanitas Unisinos. 2014. 'MME prevê necessidade de novas usinas nucleares'. 12 February. [LINK].
16. Luomi, M. (2014). Sustainable energy in Brazil–Reversing past achievements or realizing potential.
17. Runhaar, H., Driessen, P. & Vermeulen, W. (2005). Policy competences of environmental sustainability professionals. Greener management international, (49):25.
18. Martins, F.R. & Pereira, E.B. (2011). Enhancing information for solar and wind energy technology deployment in Brazil. Energy policy, 39(7):4378-4390.
19. Vichi, F.M. & Mansor, M.T.C. (2009). Energia, meio ambiente e economia: O brasil no contexto mundial. Química nova, 32(3):757-767.
20. Meltzer, J.P. (2016). Financing low carbon, climate resilient infrastructure: The role of climate finance and green financial systems.
21. PBMC, Painel Brasileiro de Mudancas Climáticas. 2013. Sumário Executivo. Mitigacão das Mudancas Climáticas. Contribuicão do Grupo de Trabalho 3 ao Primeiro Relatório de Availiacão Nacional do Painel Brasileiro de Mudancas Climáticas. Brasília, DF.
22. Yi-Chong, X., 2004. Electricity Reform in China, India and Russia. Books.
23. Siddhartha Potluri, S. & Rajan A, T. (2010). Merchant power plants in India: Risk analysis-using simulation. International journal of energy sector management, 4(1):87-112.
24. Ranganathan,V. (1997). International Energy Markets, Competition and Policy, 18th USAEE/IAEE North American Conference, 1997. Conference proceedings of the Xth conference held in Y. International Association for Energy Economics.
25. CEA Data from available at: www.cea.nic.in (accessed 20 April 2017).Chaudhary, A., Krishna, C. & Sagar, A. (2015). Policy making for renewable energy in india: Lessons from wind and solar power sectors. Climate policy, 15(1):58-87.
26. Koppenjan, J.F. and Enserink, B., 2009. Public–private partnerships in urban infrastructures: reconciling private sector participation and sustainability. Public Administration Review, 69(2), pp.284-296.
27. PFC (2009), Report on State Power Utilities, Power Finance Corporation, New Delhi, available at hhtp://pfc.gov.in/report_spu.pdf, 2008.
28. Nelson, D., Shrimali, G., Goel, S., Konda, C. & Kumar, R. (2012). Meeting India's renewable energy targets: The financing challenge. CPI-ISB report, climate policy initiative (www.climatepolicyinitiative.org),
29. Shrimali, G., Nelson, D., Goel, S., Konda, C. & Kumar, R. (2013). Renewable deployment in india: Financing costs and implications for policy. Energy policy, 6228-43.
30. Planning Commission, Government of India, 2008a Eleventh Five-Year Plan (2007-12). Volume 1: inclusive growth. Oxford University Press, Delhi.
31. Garg, P. (2012). Energy scenario and vision 2020 in India. Journal of sustainable energy & environment, 3(1).
32. BP, 2012. BP statistical review of World energy, 2012. Available at: (http://www.bp.com/sectionbodycopy.do?categoryId=7500&contentId=7068481

33. McCrone, A., Usher, E., Sonntag-O'Brien, V., Moslener, U. & Grüning, C. (2012). Global trends in renewable energy investment 2012. Frankfurt school UNEP collaborating centre for climate and sustainable energy finance.
34. Initiative, C.P., Nelson, D., Shrimali, G., Goel, S., Konda, C. & Kumar, R. (2012). Meeting India's renewable energy targets: The financing challenge. Climate policy initiative, San Francisco, USA.
35. Gupta, J.P. & Sravat, A.K. (1998). Development and project financing of private power projects in developing countries: A case study of India. International journal of project management, 16(2):99-105.
36. Chowdhury, A.N. & Charoenngam, C. (2009). Factors influencing finance on IPP projects in asia: A legal framework to reach the goal. International journal of project management, 27(1):51-58.
37. Chowdhury, S., Sumita, U., Islam, A. and Bedja, I., 2014. Importance of policy for energy system transformation: Diffusion of PV technology in Japan and Germany. Energy policy, 68, pp.285-293.
38. Singh, R. & Sood, Y.R. (2011). Current status and analysis of renewable promotional policies in Indian restructured power sector—a review. Renewable and sustainable energy reviews, 15(1):657-664.
39. Reserve Bank of India (RBI) (2014), "Risk management systems in banks", available at: hhtp://rbidocs.rbi.org.in/ (accessed 1 June 2017).
40. Bhattacharyay, B. (2010). Estimating demand for infrastructure in energy, transport, telecommunications, water, and sanitation in Asia and the pacific: 2010-2020.
41. Bhattacharyay, B.N. (2012). Estimating demand for infrastructure, 2010–2020. Infrastructure for Asian connectivity, 19.
42. Tortajada, C. (2016). Policy dimensions of development and financing of water infrastructure: The cases of china and India. Environmental science & policy, 64177-187.
43. Heller, T.C., Tjiong, H.I. & Victor, D.G. (2003). Electricity restructuring and the social contract. Program on energy and sustainable development, estados unidos, stanford university, working paper, (15).

Power energy system sustainability performance evaluation and improving path of two regional economic circle in China

Xiangyu Yi[1], *Jinchao* Li[1,2,*], *Yunna* Wu[1,2], *Wenjun* Chen[1,2]

[1] School of Economics and Management, North China Electric Power University, Beijing, China
[2] Beijing Key Laboratory of New Energy and Low-Carbon Development, North China Electric Power University, Beijing, China

Abstract. The power energy system is the foundation of the sustainable development of the human society and has an important impact on the economic, environmental and security of human society development. This paper first establishes the input and output evaluation index system of sustainable development capability of power energy system from the economic, social, environmental and security dimensions, and then uses the CCR-DEA model considering the undesired output and Malmquist index to dynamically measure the sustainable development efficiency of the power system in the Beijing-Tianjin-Hebei and Yangtze River Delta from 2005 to 2016 and decompose the corresponding items. The future development situation was analyzed through scenario analysis. The results show that: (1) Beijing and Shanghai is better than that in other provinces and cities in the same region, showing an upward trend; (2) The changes in the efficiency of sustainable development of power energy systems in Beijing, Shanghai, Zhejiang and Anhui provinces are mainly caused by technical efficiency changes. That in Tianjin, Hebei and Jiangsu are from two aspects: technical efficiency and scale efficiency; (3) The sustainable development capacity of the power system in the Beijing-Tianjin-Hebei economic circle is higher than that of the Yangtze river delta power system, and this gap has the tendency of further widening according to the strategic development plan of the two regional power energy systems; (4) In the future, China's regional electric energy system development plans should draw on the energy development strategic plan of the Beijing-Tianjin-Hebei region, especially the Xiong'an National New District, so as to achieve coordinated and sustainable development of power energy systems in various regions.

1 Introduction

With the deepening of urbanization and industrialization, there is an ever-growing demand for energy in human society. As the world's fastest growing economies, China has become the world's largest energy consumer now. The total energy consumption amount of China has reach 4.486 billion tons standard coal in 2017, about one third of total global energy consumption. Among them, China's urban energy consumption presents obvious agglomeration characteristics in spatial distribution. Energy consumption is concentrated in developed areas along the east coast and several regional central provinces and cities. Especially in Yangtze River delta, the Pearl River delta and the Beijing-Tianjin-Hebei regions, large urban agglomerations are highly concentrated in energy consumption. In 2017, the three regions' energy consumption accounted for about 15%, 20% and 10% of China's total energy consumption respectively. However, due to the differences in power supply structure, resource endowment and management level in different regions, there are gaps in the aspects of economy, safety and environmental protection in the power system, that is, the sustainable development ability of the power energy system in different regions varies. This paper takes the Beijing-Tianjin-Hebei region and the Yangtze River delta region as the research objects to carry out relevant research on the assessment of the sustainable development ability of electric power and the improvement path.

Data Envelopment Analysis (DEA) can decompose the total production productivity with consideration of the multi-input and output indexes, which is an excellent method to calculate directional distance function. Among current study of total factor productivity model based on DEA method, a large number of literature have combined Total Factor Index (TFP) and Malmquist function[1] to measure the spatial difference and spatial aggregation effect of total factor productivity in different regions of China under the constraint of environmental resources, and used panel model to decompose the influencing factors of TFP[2-5]. Among the existing studies, there are few studies using total factor production efficiency to evaluate and compare the energy sustainable development ability of urban agglomerations.

In this paper, the energy sustainable development capacity of major cities in the Yangtze River delta (Shanghai, Jiangsu, Zhejiang and Anhui) and the Beijing-Tianjin-Hebei urban agglomeration is compared

* Corresponding author: lijc@ncepu.edu.cn

under the framework of DEA full-efficiency energy efficiency model. This paper is structured as follows: the second part constructs the DEA full efficiency model to measure the energy consumption efficiency of the Yangtze River delta region and the Beijing-Tianjin-Hebei region. The third part analyzes and compares the energy sustainable performance of the two major urban agglomerations by the TFP index of energy consumption and its decomposition results.

2 Sustainable development evaluation model of power energy system

The use of energy produces both desirable and undesirable outputs. Here, every city of the Yangtze River delta region and the Beijing-Tianjin-Hebei region is a DMU (Decision-making unit).

Data envelopment analysis (DEA) uses mathematical programming and statistical data to identify the relatively efficient production frontier. The decision-making unit is projected onto the production frontier and then their relative validity is evaluated by comparing the extent to which decision-making units deviate from the production frontier. The DEA method has its unique advantages. It is suitable for evaluating the validity of multiple-input and multiple-output. Besides, there is no need for non-dimensional data processing when applying this method. It also eliminates many subjective factors without any weight assumption. Because the output indexes include desirable and undesirable, here use the Unified efficiency DEA model to evaluate the sustainable development ability of power energy system. The formulas of the unified efficiency DEA model are as follows:

Fare [6] proposed the following directional distance function:

$$Max\{\theta \mid (G+\beta\zeta_g, B-\beta\zeta_b)\in P(X)\} \quad (1)$$

Here, P(X) = {(G, B): X can produce (G, B)}. The P(X) indicates a production possibility set, which has a column vector of inputs(X) that can produce not only a column vector of desirable outputs (G) but also a column vector of undesirable outputs (B). $\zeta=(\zeta_g,-\zeta_b)$ is suggested as $(1, 1, \ldots 1, -1, -1, \ldots, -1)^T$ which contains $s+h$ components.

Mandal and Madheswaran [7] assumed that if the firm's objective is to simultaneously expand the desirable output and reduce the undesirable one by same proportion without increasing the inputs, the directional technology distance function becomes:

$$\vec{D}_T(x,y,b;0,y,-b)=\sup\left[\beta:\left[(1+\beta)y,(1-\beta)b\right]\in P(x)\right] \quad (2)$$

The value β represents technical inefficiency. The direction vector $g=(g_x,g_y,-g_b)=(0,y,-b)$ determines the direction in which efficiency is measured. Given the technology and direction vector, the directional distance function measures the maximum feasible expansion of desirable output and directional distance function β is zero. The directional distance function β is obtained by solving the maximization problem in model (3).

$$\begin{aligned} &Max\beta \\ s.t.\ &\sum_{j=1}^{n} x_{ij}\lambda_j \le x_{ik} \quad (i=1,\ldots \\ &\sum_{j=1}^{n} g_{rj}\lambda_j \ge grk+\beta g_{rk} \quad (r=1,\ldots \\ &\sum_{j=1}^{n} b_{fj}\lambda_j \le b_{fk}-\beta b_{fk} \quad (f=1,\ldots \\ &\sum_{j=1}^{n}\lambda_j=1 \\ &\beta\ge 0, \lambda_j \ge 0 \quad (j=1,\ldots \end{aligned} \quad (3)$$

Here, the outputs in the *jth* DMU are separated into desirable outputs (g_{rk}) and undesirable outputs (b_{fk}). This model measures the efficiency by $\theta=1-\beta$, where β is obtained from optimality of the Model (3).

In addition to Model (3), Zhou and Ang [8] proposed the following model to measure the unified efficiency of the energy firms:

$$\begin{aligned} &Min\theta \\ s.t.\ &\sum_{j=1}^{n} x_{ij}\lambda_j \le x_{ik} \quad (i=1,\ldots \\ &\sum_{j=1}^{n} e_{qj}\lambda_j \le \theta e_{qk} \quad (q=1,\ldots \\ &\sum_{j=1}^{n} g_{rj}\lambda_j \ge g_{rk} \quad (r=1,\ldots \\ &\sum_{j=1}^{n} b_{fj}\lambda_j = b_{fk} \quad (f=1,\ldots \\ &\theta \ge 0 and \lambda_j \ge 0 \quad (j=1,\ldots \end{aligned} \quad (4)$$

Here, the *jth* DMU input is divided into non-energy (x_{ij}:i=1) and energy-related input (e_{ij}: q=1,…). Model (4) can be considered as an extension of CCR (Charnes-Cooper-Rhods) and the production possibility set of Model (4) is shaped by constant RTS (returns to scale).

Then Malmquist index is used to decompose the sustainability performance of power energy system. Besides, the Malmquist productivity index (TFPCH) can be decomposed into the technical efficiency change index (EFFCH) and the technical change index (TECHCH). The transformation of the Malmquist index is as follows:

$$\begin{aligned} M(x^{t+1},y^{t+1},x^t,y^t) &= \frac{D^{t+1}(x^{t+1},y^{t+1})}{D^t(x^t,y^t)}\times\left[\frac{D^t(x^{t+1},y^{t+1})}{D^{t+1}(x^{t+1},y^{t+1})}\times\frac{D^t(x^t,y^t)}{D^{t+1}(x^t,y^t)}\right] \\ &= \text{EFFCH}\times\text{TECHCH} \end{aligned} \quad (5)$$

When the revenue scale changes, the EFFCH can be further decomposed into pure technical efficiency change (PECH) and scale efficiency change (SECH).

$$\begin{aligned} \frac{D^{t+1}(x^{t+1},y^{t+1})}{D^t(x^t,y^t)} &= \frac{D^{t+1}(x^{t+1},y^{t+1}/V)}{D^t(x^t,y^t/V)}\times\frac{S^{t+1}(x^{t+1},y^{t+1})}{S^t(x^t,y^t)} \\ &= \text{PECH}\times\text{SECH} \end{aligned} \quad (6)$$

The final transformation of the Malmquist index is as follows:

$$M(x^{t+1},y^{t+1},x^t,y^t) = \text{EFFCH}\times\text{TECHCH}$$

$$=PECH\times SECH\times TECHCH \quad (7)$$

$M(x^{t+1}, y^{t+1}, x^t, y^t)$ represents the variation of the TFP level. If EFFCH>1, it indicates that the relative technical efficiency of t and t+1 period is increased, whereas, in the contrary, it is the reverse. If TECHCH> 1, it indicates that t+1 period has technological progress compared to t period, whereas, in the contrary, it is the reverse. PECH indicates whether the technology is fully utilized. If it is larger than 1, it shows that the resource allocation is reasonable, whereas, in the contrary, it is the reverse. SECH expresses the index of the change of scale efficiency in two periods. If SECH is larger than 1, it means that the scale efficiency is optimized, whereas, in the contrary, it is the reverse.

3 Power energy system sustainability performance evaluation of urban agglomeration in Yangtze river delta region and the Beijing-Tianjin-Hebei region

Energy intensity means energy consumed per unit of GDP output, which is a common indicator to measure the energy efficiency of different economies, but it cannot fully reflect the input of labor, capital or other resources. Energy sustainable development performance is an ideal indicator to measure regional energy utilization efficiency and evaluate energy sustainability on the basis of resource input and energy cost in the Yangtze River delta region and the Beijing-Tianjin-Hebei region. The comprehensive consideration of multi-factor inputs by Total Factor Productivity is effective to judge the sustainability of regional energy sustainable development and can provide a scientific reference for the government's energy utilization policies.

3.1 Data processing

The evaluation index of the sustainable development capability of power energy system includes input and output. Input indexes represent the input factors of the labor force in the energy industry with the employment population of power, steam, hot water production and employment in supply industry respectively. Investment in the production and supply of electricity, steam and hot water represents delegate the financial input. The installed scale and transmission line length represent material input. In terms of output indicators, GDP represents the output of the power and energy system supporting economic development. Carbon dioxide emissions from power systems represent undesired outputs from an environmental perspective. The average outage time represents the undesired output from the safety perspective. The power supply population represents output in social services.

The study is based on the relevant data of the Yangtze River delta region (Shanghai, Jiangsu, Zhejiang and Anhui) and the Beijing-Tianjin-Hebei region (Beijing, Tianjin and Hebei) from 2005 to 2016. Data are from China statistical yearbook (2005-2016) and electric power statistical yearbook.In order to eliminate the influence of price level, the urban economic output value was processed based on 2005.

3.2 Measurement and comparison of sustainability performance of electric power system

The total factor energy consumption efficiency of model (5) is used to measure the energy sustainable performance of each provinces and cities in these two regions. The results of the energy sustainable development performance of seven provinces and cities in the two delta regions are shown in table 1 and figure 1.

Table 1. Beijing-Tianjin-Hebei and Yangtze River delta regions total factor energy consumption efficiency

Region	Provinces and Cities	Energy consumption TFP rate of change												Average value
		2005	2006	2007	2008	2009	2010	2011	2012	2013	2014	2015	2016	
Beijing-Tianjin-Hebei	Beijing	0.985	0.956	0.972	1.005	0.967	1.185	1.129	1.015	1.242	1.044	2.692	1.063	1.188
	Tianjin	1.159	0.998	1.003	0.913	0.953	1.023	0.961	0.983	1.061	1.018	0.936	0.974	0.999
	Hebei	1.057	1.031	1.028	0.999	1.015	1.203	0.961	0.896	0.942	0.848	0.765	0.702	0.954
	Average	1.067	0.995	1.001	0.972	0.978	1.137	1.017	0.965	1.082	0.970	1.464	0.913	1.047
Yangtze River delta	Shanghai	0.998	0.987	0.992	0.976	0.975	1.097	1.132	0.999	1.062	1.244	2.119	1.074	1.138
	Jiangsu	0.941	0.988	1.015	1.019	1.006	1.467	1.268	1.030	0.936	1.204	1.079	1.174	1.094
	Zhejiang	0.708	0.722	0.721	0.627	0.653	0.686	0.721	0.726	0.679	0.678	0.720	0.668	0.692
	Anhui	1.126	1.052	0.990	1.077	1.161	1.173	1.041	1.203	0.863	0.807	0.703	0.647	0.987
	Average	0.943	0.937	0.930	0.925	0.949	1.106	1.041	0.990	0.885	0.983	1.155	0.891	0.978

As can be seen from table 1:

(1) Overall, the performance of energy sustainable development in the Beijing-Tianjin-Hebei region from 2005 to 2016 was higher than that in the Yangtze River delta region, and the energy efficiency index remained above 0.9. The performance of Zhejiang province in the Yangtze River delta region is the lowest, between 0.6 and 0.8.

(2) The change rate of energy consumption in most provinces grew fastest from 2008 to 2010, and the index growth rate of Beijing and Shanghai reached a new peak in 2015.

(3) As to the change trend, the energy consumption efficiency of most provinces remains at the original level and fluctuates to some extent. The change rate of TFP of energy consumption in the capital Beijing has been greater than 1 since 2010, indicating that the energy performance keeps increasing. However, the total factor energy efficiency value of Anhui province gradually showed a downward trend from 2012, while that of Hebei province showed a downward trend from 2013. The poor performance of energy sustainability is worth our vigilance.

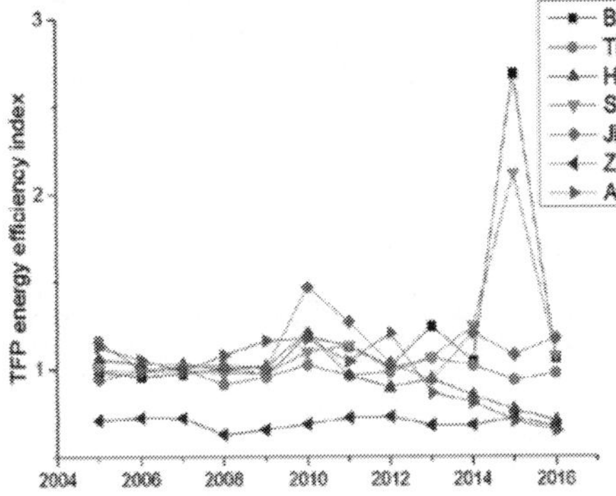

Fig. 1. Comparison of energy sustainable development performance among Yangtze River delta and Beijing-Tianjin-Hebei regions

3.3 Comparison of TFP index among provinces and cities with similar economy scale

Considering the different levels of economic development, it is more reasonable to compare the energy sustainability of provinces with similar economy scales. Therefore, this paper divides the Beijing-Tianjin-Hebei region and the Yangtze River delta into four groups according to the urban economic output value: Very high, high, medium and low. Then the cumulative TFP index is analyzed under upper four groups. The lower group include Tianjin and Anhui. The middle group includes Shanghai and Beijing. The high group include Hebei and Zhejiang. Because the economic output value of Jiangsu province is much higher than that of the other six provinces and cities, so Jiangsu province belongs to the very high group. Figure 2 shows the dynamic change trend of the cumulative TFP index of the four groups. The change rate of TFP index of energy consumption in the lower group fluctuated around 1 from 2005 to 2016. Due to its economic and political core position, the middle group is the focus of national energy consumption efficiency improvement. In recent years, energy consumption efficiency in Beijing and Shanghai has been on the rise. The energy sustainable performance of the higher group was lower and showed a downward trend. The sustainable energy performance of Jiangsu province, which has the highest economic output value, is also high and has an overall trend of growth.

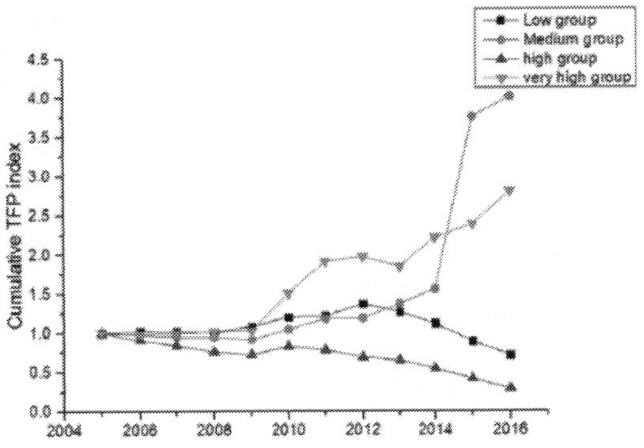

Fig. 2. The dynamic trend of the cumulative TFP index of the two urban agglomerations grouped by economic aggregate

3.4 Comparison and decomposition of urban agglomerations average annual energy consumption efficiency

From 2005 to 2016, the average annual total factor productivity of power energy sustainable performance of the four major provinces and provinces in the Yangtze River delta and the Beijing-Tianjin-Hebei region showed an upward trend. The decomposition results are showed in Figure 3 and Table 2.

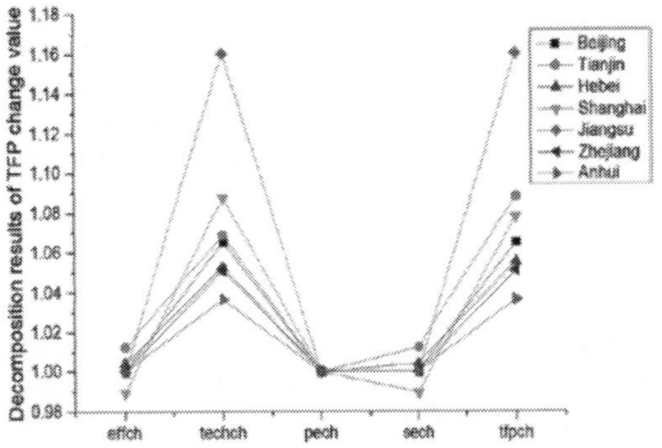

Fig. 3. Decomposition of urban agglomerations average annual TFP change value

Table 2. Average annual TFP change and decomposition of power energy system in the Yangtze River delta and Beijing-Tianjin-Hebei region

District	Effch	Techch	Pech	Sech	Tfpch
Beijing	1.00000	1.06527	1.00000	1.00000	1.06527
Tianjin	1.01236	1.06873	1.00000	1.01236	1.08818
Hebei	1.00418	1.05291	1.00000	1.00418	1.05555
Shanghai	0.98945	1.08755	1.00000	0.98945	1.07827
Jiangsu	1.00000	1.16036	1.00000	1.00000	1.16045
Zhejiang	1.00000	1.05118	1.00000	1.00000	1.05118
Anhui	1.00000	1.03655	1.00000	1.00000	1.03655

3.4.1 Yangtze River region

The improvement of urban energy sustainable development performance in the Yangtze River delta region in the past decade comes from technological progress. From 2006 to 2016, the average annual growth rate of technological progress in the Yangtze River delta region was 8.39%, while the technical efficiency decreased by 0.3%. Lower technical efficiency of which comes from the low scale efficiency, annual scale efficiency by 0.03%. It means the improvement of urban energy sustainable development performance in Yangtze River region depended on advanced technology and technical innovation in past ten years, but as a result of ignoring the improvement of the technical efficiency, the actual development model still needs to be improved, the performance growth comes from economies scale are becoming weakened gradually.

3.4.2 Beijing-Tianjin-Hebei region

The improvement of technical progress and technical efficiency in the Beijing-Tianjin-Hebei region jointly drives the improvement of total factor production efficiency, and the contribution of technical progress is far greater than that of technical efficiency, with the growth rate being 0.55% and 6.23% respectively. Among the increase of technical efficiency, the pure technical efficiency contributed 0%, and the annual increase of scale efficiency was 0.55%, indicating that the improvement of technical efficiency in the region comes from scale effect. The results show that the introduction of advanced technology has achieved certain effects in reducing energy consumption and improving the sustainability of energy use in the Beijing-Tianjin-Hebei region.

4 Conclusion

Based on the super-efficiency data envelopment analysis model and Malmquist index, the sustainable performance of power energy system in the Yangtze River delta, Beijing-Tianjin-Hebei regions were analyzed, calculated and decomposed the performance, and analyzed the main reasons and trends of performance changes. The specific conclusions are as follows:
(1) Beijing and Shanghai is better than that in other provinces and cities in the same region, showing an upward trend; (2) The changes in the efficiency of sustainable development of power energy systems in Beijing, Shanghai, Zhejiang and Anhui provinces are mainly caused by technical efficiency changes. That in Tianjin, Hebei and Jiangsu are from two aspects: technical efficiency and scale efficiency; (3) The sustainable development capacity of the power system in the Beijing-Tianjin-Hebei economic circle is higher than that of the Yangtze river delta power system, and this gap has the tendency of further widening according to the strategic development plan of the two regional power energy systems; (4) In the future, China's regional electric power system development plans should draw on the energy development strategic plan of the Beijing-Tianjin-Hebei region, especially the Xiong'an National New District, so as to achieve coordinated and sustainable development of power energy systems in various regions.

Acknowledgments

This work has been supported by the Fundamental Research Funds for the Central Universities (no. 2018ZD14) and the Science and Technology Project of SGCC "Research on lean investment optimization method based on benefit analysis of the whole process of power grid".

References

1. S. Honma, J.L. Hu, Energy Policy, **36**: 821-823 (2008)
2. G. Hao, Y. XU, X. ZHANG, et al., Chin. Pop. Res. Environ.， **28**: 75-83 (2018)
3. J.C. Li , D.X. Niu , J.Y. Li , J.X. Qi , Power Dem. Side Manage., **11**: 11-15 (2009)
4. Q.W. Wang, D.Q. Zhou , L.T. Zhang, Ind. Technol. Econ., **27**: 53-55 (2008)
5. Q.L. TAN, H. SUN, X.P. ZHANG, PROC. CHIN. SOC. ELECTRICAL ENG., **32**: 105-111 (2012)
6. R. Färe, S. Grosskopf, D.W. Noh, et al., J. ECONOMETRICS, **126**:469-492 (2005)
7. S.K. Mandal, S. Madheswaran, Energy Policy, **38**:1108-1118 (2010)
8. P. Zhou, B.W. Ang, Energy Policy, **36**:2911-2916 (2008)

 https://doi.org/10.1051/e3sconf/201912204001

A New Pitch Angle Control Method of Wind Turbine Generators Based on Feedforward Wind Speed Information

Mouayad Sahib [1,*] *and Thaker* Nayl [1]

[1]University of Information Technology and Communications, College of Engineering, Baghdad, Iraq

Abstract. In this work, a new strategy to control the pitch angle of wind turbine generator is proposed. The strategy is based on designing an intelligent control system capable of maintaining a stable minimum fluctuating power generation. This can be achieved by providing the wind speed information to the controller in advance and hence allowing the controller to take the optimum action in controlling the blade pitch angle. A model based optimizer uses Model Predictive Control (MPC) technique to predict the wind turbine generator future behaviour and select the optimal control actions assisted by the wind speed information while satisfying the power generation constraints. The simulation results show that a significant improvement can be made using the proposed control method.

1 Introduction

Nowadays, the world is testifying an increasing demand of using renewable energy sources. Among these various energy sources is wind. Wind turbine generator (WTG) is a system that converts the wind's kinetic energy into electrical energy. WTG become the most prevalent power generation resource for its clean and low maintenance factors [1]. One of the operational challenges for WTG is the oscillations in the output power caused by the varying speed of wind. As a result, this variability leads to a difficult conversion of constant energy and thus a control system is needed. One solution to control fluctuating power generation is through blade pitch angle control [2].

With respect to wind speed, WTG has three regions of operation; stop region, partial-load region, and full-load region [3]. Figure 1, shows the generated output power and the variation rang of the blade pitch angle beta as a function of the wind speed. When the wind is below the cut-in speed value the WTG rotational speed is zero and it does not generate any power. If the wind is above the cut-in and below the rated speed values the control system is utilized to capture maximum power from the wind. This control method is termed as maximum power point tracking (MPPT) [4]. When the wind speed exceeds the rated value the pitch angle control is activated to maintain the generated output power constant at its rated value [5].

In pitch angle control the angle of the WTG blades are changed as a control variable for power regulation above the rated wind speed. The proportional-integral (PI) is the conventional controller type for pitch angle control [6]. Despite its simplicity, PI controller may not achieve the intended optimum performance due to the nonlinear dynamics of the WTG [7]. To compensate for the unknown time-varying nonlinearities and disturbances various control structures has been used such as nonlinear PI (N-PI) [8], fractional PI [9], cascade control structure [10] and Fault-tolerant control (FTU) to Compensate for the pitch actuator fault [11]. Other strategies to enhance the pitch control include; fuzzy logic control (FLC) [12][13][14], observer-based blade-pitch controller [15] and finite-frequency identification PI controller [16].

Knowledge of future wind conditions assisting model predictive control (MPC) has also been proposed in literature [17]. MPC enables optimal solution while considering system constraints. In addition, it can predict behaviour in future using a plant's model. MPC showed significant improvement in the WTG control system [18]. However, from practical point of view, the main drawback of MPC is the requirement to solve a quadratic programming problem on line [18].

From the aforementioned research studies, the performance of pitch angle control can be improved using MPC if the predication computational complexity can be overcome. This can be achieved with future knowledge of wind behaviour. Instead of prediction, wind speed information can be passed from one WTG to another. Then, this information can be fed to the MPC to produce the optimal action in advance. In this paper, a new strategy to control the pitch angle proposed. The strategy is based on designing an MPC controller capable of maintaining a stable minimum fluctuating power generation. The concept of the new strategy is based on feeding forward the wind speed information in advance to the WTG. This will allow the controller to take the optimal control action in advance hence minimizing undesired power fluctuation.

This article is organized as follows. In Section 2, the mathematical model of the wind turbine power

* Corresponding author: mouayad.sahib@uoitc.edu.iq

https://doi.org/10.1051/e3sconf/201912204001

generation is introduced. The proposed control method is presented in section 3. In section 4, simulation results and discussion is presented. Finally, section 5 concludes the paper research.

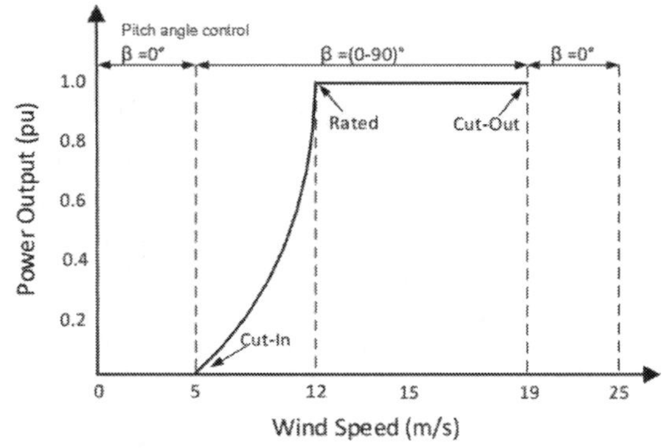

Fig. 1. Operation regions of WTG.

2 WTG Modelling

Wind turbine generators extract energy from the wind and convert it into mechanical energy, then into electrical power. The wind power equation is given by [13]:

$$P = \frac{1}{2}\rho A V^3 \tag{1}$$

where, ρ is the air density, $A = \pi R^2$ is the circular area swept by the blades of radius R, and V is the wind velocity.WTG converts the wind power into electrical power at ideal maximum efficiency of 59%. This power conversion limit is termed as Betz limit [19]. The generated power is expressed by the following nonlinear equation [9]:

$$P_g = \frac{1}{2}\rho A C_p(\lambda, \beta) V^3 \tag{2}$$

The amount of generated power P_g extracted from the total wind power is determined by the power coefficient C_p. The power coefficient is a nonlinear function of the blade pitch angle and the blade tip speed ratioλ(TSR). The following equation is used to approximate C_p in the full load region [5]:

$$C_p = 0.22\left(\frac{116}{x} - 0.4\beta - 5\right)e^{\frac{-12.5}{x}} \tag{3}$$

where,

$$\frac{1}{x} = \frac{1}{\lambda - 0.08\beta} - \frac{0.035}{1 + \beta^3} \tag{4}$$

The TSR λ is the ratio of the angular rotor speed of the wind turbine to the linear wind speed at the tip of the blades given by:

$$\lambda = \frac{\omega R}{v} \tag{5}$$

where ω is the angular velocity of the turbine rotor (rad/s) and v is the wind velocity (m/s). using equations (3) and (4) the relation of the power factor C_p with the TSR λ for different values of the pitch angle β is shown in figure 2.

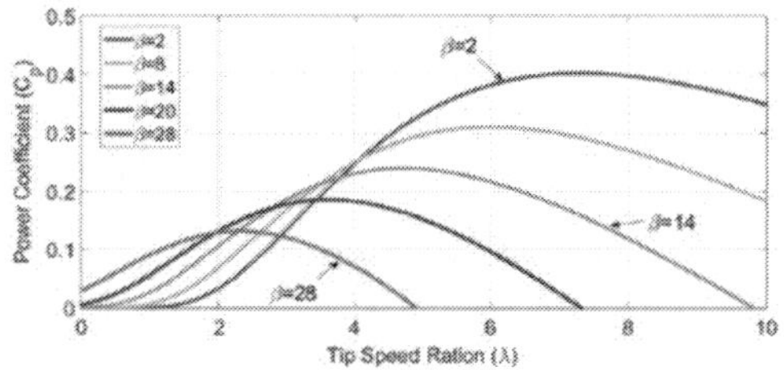

Fig. 2. C_p curves vs TSR for different β.

In each C_p curve in figure 2, there is a unique optimal TSR (λ_{opt}) value at witch the power coefficient is maximum. In partial-load region (region II), the control objective is to maintain the TSR at its optimal value λ_{opt} with variable wind speed to extract maximum power. However, in full-load region (region III) where the wind speed exceeds the rated value the control objective is to limit power generation. The optimal rotor speed is the rated constant speed ω*. In equation (5), λ, is inversely proportional to the variation of the wind velocity v. To maintain the power coefficient C_p constant, the variation of λ is compensated by varying the pitch angle β in equations (3) and (4). Therefore, a control method is required to adjust the pitch angle based on the deviation of the output power compared to a set point value equal to the rated power. One common controller used for pitch angle control is the Proportional Integral (PI) controller explained in the following section.

3 Proposed Control Method

The conventional pitch angle control system is shown in figure 3.

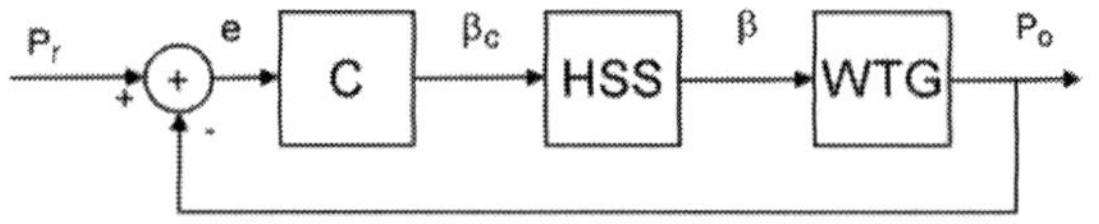

Fig. 3. Pitch angle control system block diagram.

The controller C is used to command the hydraulic servo system (HSS) with β_c based on the error signal e between the set point power P_r and the output power P_o. The HSS is an actuator model used to turn the blades along their longitudinal axis. The HSS model describes the dynamic behaviour between a pitch command β_c from the pitch controller and an actuating pitch angle β to the WTG. The dynamics of the blades are non-linear with saturation limits on both pitch angle and pitch rate [19]. Due to these limitations the controller performance drops and cannot cope with high wind speed variations affecting the output power. Linear model predictive control (LMPC) can be used together with predictions of future wind speed to improve controller performance in above rated wind conditions [17]. However, the controller performance drops due to wind estimation errors.

The proposed control method in this paper is based on providing the MPC with future wind velocity information. Wind information can be transmitted from near WTG tower. In designing the layout of wind farm, a minimum spacing distance (d) between turbines should be taken into consideration. With a wind of velocity v, the time it takes to travel from turbine i to turbine j is:

$$t = \frac{d}{v} \tag{6}$$

The delay time t represent the maximum prediction horizon used by the MPC controller. When the prediction horizon is well matched to the lag between future and current wind velocities the controller can achieve improved performance.

4 Simulation Results

In order to evaluate the performance of the proposed algorithm this section presents and discuss simulation results. The proposed MPS is compared with the conventional PI controller. Table 1 lists the WTG simulation parameters [5].

Table 1. Simulation parameters

Wind Turbine Parameter	Value
Rotor radius (R)	38 m
Rated rotor speed (ω_r)	2.273 rad/s
Air density (ρ)	1.205 kg/m^3
Cut in wind speed (v_{ci})	5 m/s
Rated wind speed (v_r)	12 m/s
Cut out wind speed (v_{co})	19 m/s
Rated power (P_r)	2 MW
Pitch angle limit (β)	0- 90°
Pitch rate limit ($\dot{\beta}$)	± 10°/s

In the simulated scenario, the constraints of the proposed MPC controller are imposed on the control pitch angle. The parameters of the MPC controller are tuned according to the change of the wind speed and the information provided in advance. Moreover, the prediction horizon and the control horizon selected $N_p = 10$, and $N_c = 2$ respectively, with a control interval of 1.0 sec. The aim of the MPC controller is to calculate a proper control action in advance. The overall objective of the control design is to stabilize the power output at its rated value which is 2 MW. Two types of wind speed profile signals are used in simulation for comparison; sine and square waves. In the sine wave type the wind varies between 12 to 14 m/s with frequency equal to (π/6) rad/sec. However, for the square wave type the wind varies between 12.5 to 13.5 m/s with frequency equals to (π/5) rad/sec. These artificial wind profiles are selected to evaluate the tracking performance of the proposed controller. Figures 4-7 show the performance of the PI and proposed MPC controllers for the two types of wind signal profiles.

It can be observed from figures 4-7 that the proposed MPC controller outperforms the conventional PI controller for both types of wind profiles. The WTG output power has smoother behaviour with the new controller compared to PI.

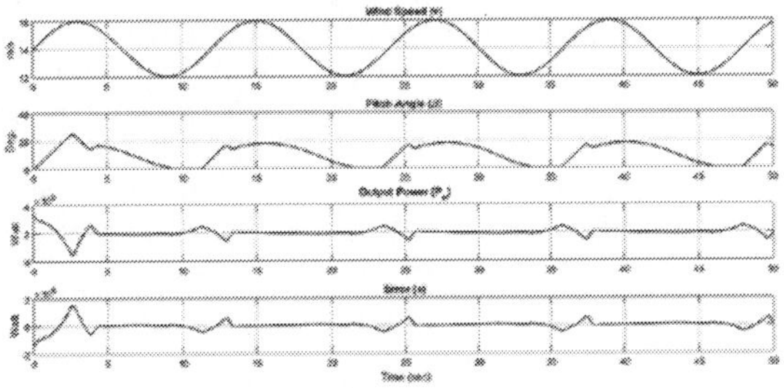

Figure 4, PI controller with sine wave wind profile.

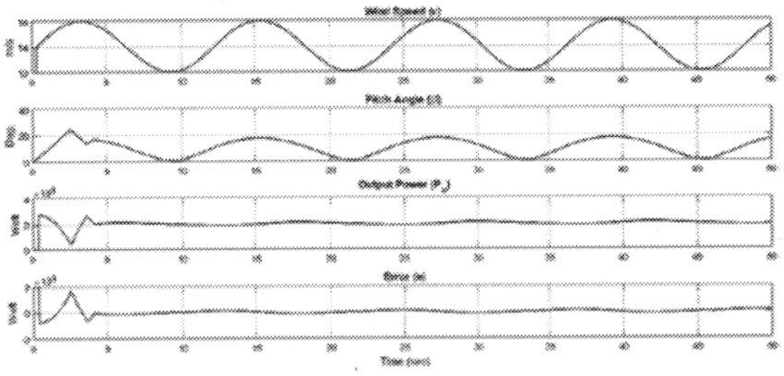

Figure 5, MPC controller with sine wave wind profile.

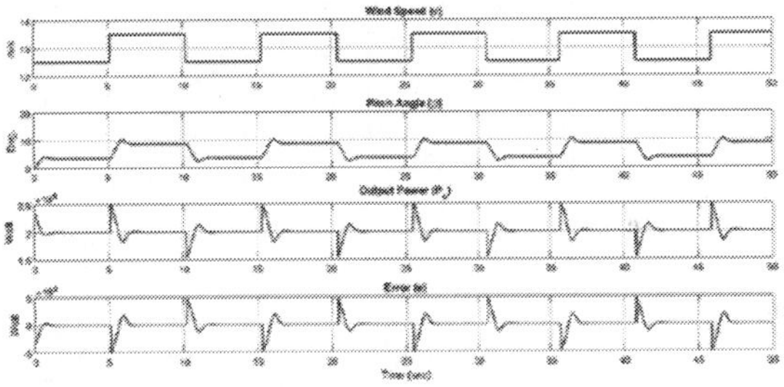

Figure 6, PI controller with square wave wind profile.

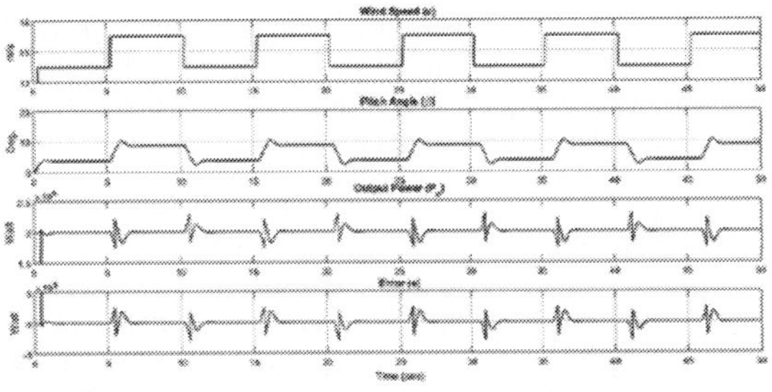

Figure 7, MPC controller with square wave wind profile.

The improved control performance of the proposed MPC is due to the advance action the controller takes to compensate for wind variations. This action is based on the feedforward information of the wind speed provided to the controller. Figure 8, shows a comparison between the PI and the proposed MPC error input signals. As can

https://doi.org/10.1051/e3sconf/201912204001

be seen from figure 8, the proposed MPC possess less peaks amplitudes and fluctuations compared to the PI controller for both wind types. Table 2 lists a comparative result between the proposed MPC and PI controllers. The numerical values represent the maximum overshoot and minimum undershoot percentages of the error signals. It is evident that the proposed MPC outperforms the conventional PI in having less output deviation from the rated power.

It is worth noting that, the proposed control strategy can be generalized in a distributed control wind farm system. Each WTG unit acts as an agent to share its current control status and wind speed to advice adjacent control units in calculating the appropriate control action in advance in order to maintain smooth power output.

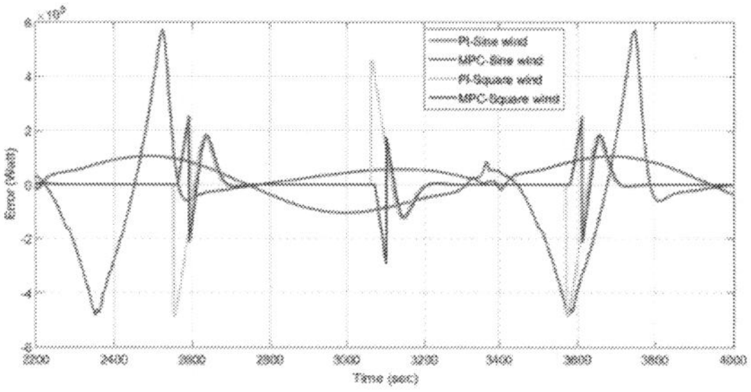

Figure 8, Error Comparison.

Table 2, Performance Comparison.

Controller Type	Wind Type	Max. Overshoot (%)	Min. Undershoot (%)
PI	Sine	28.54	-24.06
MPC		**05.33**	**-05.21**
PI	Square	23.11	-24.48
MPC		**12.66**	**-14.73**

5 Conclusion

In this paper a new pitch angle control method is used to smooth wind turbine generator output power. The control strategy is based on providing future wind speed information to a model predictive controller. The wind information is provided (feed forward) from an upstream wind turbine. With this method, the proposed MPC can take the appropriate control action in advance and hence maintain a smooth power generation.

References

1. R. Sitharthan, C. Sundarabalan, K. Devabalaji, T. Yuvaraj, and A. Mohamed Imran, "Automated power management strategy for wind power generation system using pitch angle controller," Meas. Control, vol. **52**, no. 3–4, pp. 169–182, Mar. (2019).
2. D. Ochoa and S. Martinez, "Frequency dependent strategy for mitigating wind power fluctuations of a doubly-fed induction generator wind turbine based on virtual inertia control and blade pitch angle regulation," Renew. Energy, vol. **128**, pp. 108–124, Dec. (2018).
3. L. Cavanini, M. L. Corradini, G. Ippoliti, and G. Orlando, "A Control Strategy for Variable-Speed Variable-Pitch Wind Turbines within the regions of partial-and full-load operation without wind speed feedback," in 2018 European Control Conference (ECC), (2018), pp. 410–415.
4. A. Dahbi, N. Nait-Said, and M.-S. Nait-Said, "A novel combined MPPT-pitch angle control for wide range variable speed wind turbine based on neural network," Int. J. Hydrogen Energy, vol. **41**, no. 22, pp. 9427–9442, Jun. (2016).
5. H. Matayoshi, A. M. Howlader, M. Datta, and T. Senjyu, "Control strategy of PMSG based wind energy conversion system under strong wind conditions," Energy Sustain. Dev., vol. **45**, pp. 211–218, Aug. (2018).
6. M. Ben Smida and A. Sakly, "Smoothing wind power fluctuations by particle swarm optimization-based pitch angle controller," Trans. Inst. Meas. Control, vol. **41**, no. 3, pp. 647–656, Feb. (2019).
7. Y.-J. Joo and J.-H. Back, "Power Regulation of Variable Speed Wind Turbines using Pitch Control based on Disturbance Observer," J. Electr. Eng. Technol., vol. **7**, no. 2, pp. 273–280, Mar. (2012).
8. Y. Ren, L. Li, J. Brindley, and L. Jiang, "Nonlinear PI control for variable pitch wind turbine," Control Eng. Pract., vol. **50**, pp. 84–94, May (2016).
9. M. Oussama, A. Choucha, and L. Chaib, "Fractional Order PI Controller Design for Control of Wind Energy Conversion System Using Bat Algorithm," in Lecture Notes in Networks and Systems, vol. **62**, Springer, Cham, (2019), pp. 270–278.
10. D. Song et al., "A novel wind speed estimator-integrated pitch control method for wind turbines with global-power regulation," Energy, vol. **138**, pp. 816–830, Nov. 2017.
11. J. Lan, R. J. Patton, and X. Zhu, "Fault-tolerant wind turbine pitch control using adaptive sliding mode estimation," Renew. Energy, vol. **116**, pp. 219–231, Feb. (2018).
12. T. L. Van, N. K. Dang, X. N. Doan, T. H. Truong, and H. N. Minh, "Adaptive Fuzzy Logic Control to Enhance Pitch Angle Controller for Variable-Speed Wind Turbines," in 2018 10th International Conference on Knowledge and Systems Engineering (KSE), (2018), pp. 225–229.
13. Z. Civelek, M. Lüy, E. Çam, and H. Mamur, "A new fuzzy logic proportional controller approach applied to individual pitch angle for wind turbine load mitigation," Renew. Energy, vol. **111**, pp. 708–717, Oct. (2017).
14. K. A. Naik and C. P. Gupta, "Fuzzy logic based pitch angle controller/or SCIG based wind energy system," in 2017 Recent Developments in Control, Automation & Power Engineering (RDCAPE), (2017), pp. 60–65.

15. M. L. Corradini, G. Ippoliti, and G. Orlando, "An observer-based blade-pitch controller of wind turbines in high wind speeds," Control Eng. Pract., vol. **58**, pp. 186–192, Jan.(2017).
16. A. G. Alexandrov, V. N. Chestnov, and V. A. Alexandrov, "Identification Based Control For Wind Turbine," IFAC-PapersOnLine, vol. **50**, no. 1, pp. 2272–2277, Jul. (2017).
17. M. D. Spencer, K. A. Stol, C. P. Unsworth, J. E. Cater, and S. E. Norris, "Model predictive control of a wind turbine using short-term wind field predictions," Wind Energy, vol. **16**, no. 3, pp. 417–434, Apr. (2013).
18. J. Pahasa and I. Ngamroo, "Model predictive control-based wind turbine blade pitch angle control for alleviation of frequency fluctuation in a smart grid," in 2014 International Electrical Engineering Congress (iEECON), (2014), pp. 1–4.
19. A. Hwas and R. Katebi, "Wind Turbine Control Using PI Pitch Angle Controller," IFAC Proc. Vol., vol. **45**, no. 3, pp. 241–246, Jan. (2012).

Interior Point Algorithm Applied to the Optimization of the Power Supplied by a Wind Farm with a BESS

Pablo Durán[1], *Daniel* Villanueva[1*], *Andrés* E. Feijóo[1], *and Neeraj* D. Bokde[2]

[1]Electrical Engineering Department, University of Vigo, 36310-Vigo, Spain
[2]National Institute of Technology, Nagpur, India

Abstract. Wind power constitutes a variable energy source that introduces unbalance in electrical network management because it cannot be programmed. Then, the possibility of storing wind energy becomes very important. The lack of control is a drawback that disappears when the combination of a wind farm (WF) and a battery energy storage system (BESS) is considered. In that case, the goal is to adjust the power plant output and the load requirements of electrical network, i.e., to contribute to system adequacy as much as possible. Considering the features of the problem, it can be defined as an optimization problem. Two algorithms are proposed to solve it: the primal dual algorithm and the Mehrotra predictor-corrector one. In both cases, the best solution of the proposed problem is reached in an efficient manner. The primal dual algorithm performs better in terms of time and the Mehrotra predictor-corrector one needs fewer iterations.

1 Introduction

The lack of control in electrical network management due to wind power grows with the installed wind power. Therefore, in order to improve the balance between generation and consumption in the electrical network, storage systems need to be installed altogether with wind farms. The main objective of having the capability to store wind power is to inject the power into the electrical network when needed, combining periods of decreasing wind farm production and increasing ones. The power surplus has to be stored somewhere and is given back when wind farm production is not enough to supply the required power.

Among others, a BESS is a profitable mean to store wind power due to its efficiency, adaptability and no need of specific installation conditions.

When dealing with a system formed by a WF and a BESS, the first issue to consider is the battery size [1,2], which is critical. However, it is not dealt with here and the value used in the case study is based on a real power plant. The equations and constraints that define the behavior of the system were obtained from [3-11].

Another issue is the application time interval of the BESS, that can be in an hourly, daily, weekly or monthly term [12], but in this paper the objective is particularized to a daily term. Besides, the prediction is not considered here, it is assumed that the wind speed is known with enough anticipation to apply the algorithm, which is realistic due to the daily term analyzed.

It is also important to consider the model used for the BESS. Several models appear in the related literature [7,13,14] and they have been simplified here, considering certain efficiency value during the charging and discharging processes.

Finally, it is also important to decide the optimal allocation of the BESS in the network and its configuration [15-17]. In this paper, the BESS is located adjacently to the WF and the chosen configuration is dual inverter [17].

Then, in order to carry out a strategy based on the adaptation of the wind power produced by a power plant made up by a WF and a BESS, to the power requested along a period, an optimization problem arises. The objective should be that the power plant generates more power when it is more valuable. However, the power plant will have some technical constraints that limit its operating capacity, most of them related with the BESS but also with the whole installation, turning the optimization problem into a non-linear one.

Linear programming [18,19] is a part of mathematical optimization, which deals with the kind of problems where a linear objective function must be optimized, being subject to a set of linear constraints. Due to the specific circumstances of this type of problem, its optimal solution has to be located on the border of the set of feasible solutions. In those cases, the simplex algorithm is a good option to obtain the optimal solution. However, this algorithm goes over the border, trying to find the optimal one and, when dealing with a lot of variables and restrictions computation, the time consumption results very high. As an alternative to this algorithm there is the possibility of designing an algorithm that goes through the interior of the set of feasible solutions until the optimal one is reached. This algorithm is known as interior point algorithm [20].

* Corresponding author: dvillanueva@uvigo.es

In 1979, Leonid Kachiyan [21] designed an interior point algorithm, the ellipsoid algorithm, and proved that a linear programming problem can be solved efficiently in proper time by means of it. Later on, in 1984, Narendra Karmarkar introduced the interior point algorithm of projective transformation with great success [22]. In Karmarkar's work, primal-dual algorithm [23] and Mehrotra predictor-corrector one [24] are used to solve this type of optimization problem.

The organization of this paper is the following: section 2 includes the concepts, expressions and constraints to be considered in a hybrid WF-BESS system, section 3 explains the primal-dual algorithm, section 4 is an outline of the Mehrotra predictor-corrector algorithm, in section 5 the application of both algorithms to the problem proposed is given and the conclusions are stated in section 6.

2 Power Supplied by a Wind Farm with a BESS

A power plant consisting of a WF and a BESS is considered in this section. The equations and constraints that describe the behavior of this type of power plants are presented.

Generally speaking, the power plant output in a time interval is the sum of the power supplied by both the WF and the BESS in the same interval, as in (1).

$$P_{g_i} = P_{b_i} + P_{w_i} \qquad i = 1, \dots, n_i \tag{1}$$

where n_i is the number of intervals assessed, P_{g_i} is the power provided by the power plant in the interval i, P_{b_i} is the power provided by the BESS to the electrical network in the interval i and the power supplied by the WF in the interval i is named P_{w_i}. Notice that, in case that P_{b_i} is negative, it means that the total power provided to the electrical network is lower than the power supplied by the WF, but during this time interval, the BESS receives the rest of the power.

On the other hand, in eq. (2) the charging process of the BESS is expressed. The energy stored in the BESS in an interval is obtained as the sum of the energy previously stored in the BESS and the energy received during the interval. Notice that, the energy received is the product of the power provided by the BESS and the length of the interval with negative sign as expressed in (2). The reason for the negative sign is that P_{b_i} is defined as positive when supplied.

$$E_{b_{i+1}} = E_{b_i} - \Delta T \cdot P_{b_i} \qquad i = 1, \dots, n_i \tag{2}$$

where E_{b_i} is the energy stored in the BESS in the interval i and ΔT is the interval length. Therefore, the possibilities are: $E_{b_{i+1}} > E_{b_i}$ during the charging process, $E_{b_{i+1}} < E_{b_i}$ if the BESS is discharging or $E_{b_{i+1}} = E_{b_i}$ in case there is no power from or to the BESS during the interval. There are losses during the charging and discharging processes and some kind of efficiency has to be considered. However, those losses are not taken into account here due to the complexity of the problem. Once the problem is solved, those losses can be assessed.

The constraints that must be considered to apply equations (1) and (2) are the following:

i) BESS minimum and maximum charge levels (3): The reason for establishing a minimum charge level is to extend its useful life. The maximum charge level of the BESS is given by its maximum energy storage capacity. So, on each interval, constraints shown on (3) have to be taken into account.

$$E_{b_{min}} \le E_{b_i} \le E_{b_{max}} \qquad i = 1, \dots, n_i \tag{3}$$

where $E_{b_{min}}$ is the minimum charge level and $E_{b_{max}}$ is the maximum charge level of the BESS.

ii) Maximum power transferred to or from the BESS (4): These limits are due to the heating processes that appear when charging or discharging the BESS. Therefore, there are constraints in both directions and the minimum one is negative.

$$P_{b_{min}} \le P_{b_i} \le P_{b_{max}} \qquad i = 1, \dots, n_i \tag{4}$$

where $P_{b_{min}}$ is the maximum power that can be transferred to the BESS in absolute value. $P_{b_{max}}$ is the maximum value of power that can be provided by the BESS.

iii) Maximum and minimum level of power that can be supplied by the power plant (5): The reason for those levels is the installation configuration, because the limits depend on the whole power plant: it cannot consume power and the output power has to be between zero and its rated power.

$$0 \le P_{g_i} \le P_{g_r} \qquad i = 1, \dots, n_i \tag{5}$$

where the rated power of the power plant is P_{g_r}.

Therefore, the optimization problem to be solved can be summarized as in (6), considering the relationships in (1) and (2).

$$\max f = \textstyle\sum_i \left(\omega_i^2 \cdot P_{g_i}\right)$$

$$\text{s.t.} \begin{cases} E_{b_{min}} \le E_{b_i} \le E_{b_{max}} \\ P_{b_{min}} \le P_{b_i} \le P_{b_{max}} \\ 0 \le P_{g_i} \le P_{g_r} \end{cases} \tag{6}$$

where ω_i^2 are the weights of each time interval i and define comparatively the need to supply power.

They are squared in order to adapt the solution to the characteristics of the problem.

3 Primal Dual Algorithm

In this section, a full explanation of the primal dual algorithm is given in order to let the reader know the process to obtain the solution of the proposed optimization problem.

3.1 Main idea

The primal-dual algorithm is based on the use of a logarithmic barrier. The idea is to replace the border of the feasible set by an element in the objective function that penalizes its value as it is approached. For example, it is supposed a standard linear programming problem, such as the one defined in (7):

$$\begin{cases} \min_{x_j} f = c^t x \\ Ax = b \\ x_j \geq 0 \end{cases} \quad (7)$$

where f is the objective function, x is the vector of variables of the primal problem and $Ax = b$ is the constraint of the minimization problem.

The equivalent problem given by (8) can be obtained.

$$\begin{cases} \min_{x_j} f = c^t x - \mu \sum_{j=1}^{n} \log(x_j) \\ Ax = b \\ \mu \geq 0 \end{cases} \quad (8)$$

where the parameter μ is called the barrier parameter.

Considering the standard format of a linear programming problem and its dual, (9), [25,26]

$$\begin{matrix} \min_{x_j} f = c^t x & \min_{x_j} f = -b^t y \\ Ax = b & A^t y + s = c \\ x_j \geq 0 & s \geq 0 \\ & y \text{ not constrained} \end{matrix} \quad (9)$$

where y is the vector of variables of the dual problem and s is the slack vector.

The following assumptions are made for the primal-dual algorithm:

i) The set of solutions of the primal problem, named S, is not empty.

$$S = \{ x \in R^n : Ax = b, x > 0 \} \neq \emptyset$$

ii) The set of solutions of the dual problem, named T, is not empty.

$$T = \{ y, s \in R^m, R^n : A^t y + s = c, s > 0 \} \neq \emptyset$$

iii) The range of A is maximum.

Considering these constraints, it can be stated that both problems, primal and dual, have optimal coincident solutions. Besides, the feasible regions of both problems are bounded.

The logarithmic barrier function is applied to the sign constraints in both problems, following the process explained above as in (10).

$$\begin{matrix} \min_x c^t x - \mu \sum_{j=1}^{n} \log(x_j) \\ Ax = b \\ \max_y b^t y + \mu \sum_{j=1}^{n} \log(s_j) \\ A^t y + s = c \\ s \geq 0 \\ y \text{ not constrained} \end{matrix} \quad (10)$$

As $\mu \to 0$, it can be expected that the optimum solution of the problem P_μ (Eq. 10-top) converges to the optimum solution of the original primal problem. In order to prove it, it is considered that the objective function is strictly convex and additionally using the duality theorems (Karush-Kuhn-Tucker conditions, KKT [18]), the system (11) is obtained.

$$\begin{cases} Ax = b \quad x > 0 & \text{(Primal Feasibility)} \\ A^t y + s = c \quad s > 0 & \text{(Dual Feasibility)} \\ XSe - \mu e = 0 & \text{(Th. Compl. Slackness)} \end{cases} \quad (11)$$

where A and S are diagonal matrices with the values of the variables x and s, and e is a vector of ones.

From the assumptions i, ii and supposing that the feasible region of the primal problem is bounded, it can be derived that the problem P_μ is feasible and it has just one minimum in $x(\mu)$ for each $\mu > 0$. Therefore, the system in (11) has just one solution.

The system in (11) also provides the necessary and sufficient conditions (KKT) for $y(\mu)$ and $s(\mu)$ to be the maximal of the problem D_μ (Eq. 10-bottom).

When the condition iii) is true, variable x determines uniquely the variable y from (11). Besides, the duality gap can be expressed as in (12).

$$c^t x(\mu) - b^t y(\mu) = \left(c^t - y(\mu)^t A\right) x(\mu) = n\mu \quad (12)$$

Therefore, as $\mu \to 0$ the duality gap tends to zero and it implies that they are the optimal solutions of the primal and dual problem.

So the set of optimum solutions of the types of problem P_μ and D_μ form a path to the optimum solution of the original problem.

Therefore, this algorithm starts from an initial point (x^0, y^0, s^0), after which a sequence of points is generated, $(x^k, y^k, s^k) \in S, T$, choosing a movement direction and right step size, β_k.

3.2 Movement direction and step size

This is one of the fundamental stages of the algorithm and Newton method is used on it.

Newton method consists of a procedure to treat the solution of a system of nonlinear equations by means of the consecutive approximation of the system by linear equations. For example, dealing with a nonlinear function, $F(z)$, defined from R^n to R^n, treating to find

 https://doi.org/10.1051/e3sconf/201912204002

a vector z that nullifies the function. Using the multivariable series of Taylor, a linear approximation can be obtained, as in (13)

$$F(z^k + \Delta z) \approx F(z^k) + J(z^k)\Delta z \tag{13}$$

where J is the Jacobian matrix of function F( ), with the rows formed by the partial derivatives of the respective functions of F( ). Each element can be expressed as in (14).

$$\left[\frac{\partial F_i(z)}{\partial z_j}\right] \tag{14}$$

As the intention is to assess the root $F(z) = 0$, the system in (15) is obtained.

$$J(z^k)\Delta z = -F(z^k) \tag{15}$$

The solution to this system permits to iterate from z^k to z^{k+1} with a unit step. This method converges quadratically when the starting point z^0 is located close to the solution, but this good convergence is just a local behaviour. In a nonlinear generic case, if the starting point is located far from the solution, the Newton method can fluctuate permanently without reaching the solution

When applying the Newton method to the system in (11), in order to move from a given point (x^k, y^k, s^k) to another $(x^{k+1}, y^{k+1}, s^{k+1})$, so that x and y are no negatives, the system in (16) is obtained, providing the Newton directions corresponding to each variable.

$$\begin{bmatrix} A & 0 & 0 \\ 0 & A^t & I \\ S_k & 0 & X_k \end{bmatrix}\begin{bmatrix} d_x^k \\ d_y^k \\ d_s^k \end{bmatrix} = -\begin{bmatrix} Ax^k - b \\ A^t y^k + s^k - c \\ X_k S_k e - \mu^k e \end{bmatrix} \tag{16}$$

Developing the system (16), the relationships in (17) are obtained.

$$\begin{aligned} Ad_x^k &= r_1 \\ A^t d_y^k + d_s^k &= r_2 \\ S_k d_x^k + X_k d_s^k &= r_3 \end{aligned} \tag{17}$$

where:

$$\begin{bmatrix} r_1 \\ r_2 \\ r_3 \end{bmatrix} = \begin{bmatrix} b - Ax^k \\ c - A^t y^k - s^k \\ \mu^k e - X_k S_k e \end{bmatrix} \tag{18}$$

In case that points x, y, s are feasible, the rests r_1 and r_2 will be null.

3.3 Practical implementation

The practical version of the algorithm does not need that the initial point satisfies all the constraints. As long as the iterations grow, the sequence of points enters the feasible region.

Solving the system, the Newton direction for x shown in (19) is obtained.

$$d_x^k = \mu^k D_k P_k D_k X_k^{-1} e - D_k P_k D_k c + CA^t (AD_k^2 A^t)^{-1} r_1 \tag{19}$$

where $D_k^2 = X_k S_k^{-1}$ and $P_k = I - D_k A^t (AD_k^2 A^t)^{-1} AD_k$, that is the projection matrix in the kernel of matrix AD_k.

The elements in (19) have an interesting geometric interpretation: The first one is the centering direction, because it is the projection of the push vector $(1/x^k)$, and it helps keep the algorithm far from the border of the primal feasible region.

The second element is the direction of reduction of the objective function, because it is the projection of the negative gradient of the objective function of the primal problem, and it helps reduce quickly its value.

The third one is the feasibility direction, because r_1 is a measurement of the feasibility of the problem.

It can be pointed out that the centering direction and the reduction direction of the objective function belong to the kernel of A. Therefore, the third direction is the only one that joins in the feasibility of the primal problem.

At the beginning, one of the most important directions is the feasibility one, because the initial point chosen does no need to belong to the feasibility region. As long as the iterations move along, these elements are losing its importance with respect to the other two.

Once the Newton directions are obtained when solving the system, the variables have to be updated in order to move to the next iteration. The values are obtained in eq. (20).

$$\begin{aligned} x^{k+1} &= x^k + \beta_P d_x^k \\ y^{k+1} &= y^k + \beta_D d_y^k \\ z^{k+1} &= z^k + \beta_D d_z^k \end{aligned} \tag{20}$$

In order to obtain the β coefficients that determine the step size used on each iteration, equations (21) and (22) are applied.

$$\beta_P = \frac{1}{\max\{1, -d_{xi}^k / \alpha x_i^k\}} \tag{21}$$

$$\beta_D = \frac{1}{\max\{1, -d_{si}^k / \alpha s_i^k\}} \tag{22}$$

where $\alpha < 1$. This criterion is used in order to avoid updating the variables to negative values or zero.

The direction of movement in an iteration (k) is determined by the value of the barrier parameter. In fact, this translation should be made several times for a fixed value of μ^k, so that the Newton steps converge to the central path corresponding to that μ^k. However, this could demand a large number of calculations to the algorithm, therefore, on each iteration the Newton steps are calculated just once and the value of the barrier parameter is reduced.

In order to reduce the barrier parameter (23) is used.

$$\mu^k = \frac{1}{10}\frac{s^t x}{n} \tag{23}$$

As stopping criteria for the algorithm, elements in (24) are used.

$$\begin{aligned} \mu^k &\leq \epsilon_1 \\ \text{norm}(r_1) &\leq \epsilon_1 \\ \text{norm}(r_2) &\leq \epsilon_2 \end{aligned} \tag{24}$$

where the function norm( ) provides the absolute value of the vector. The first criteria in (24) ensures that the duality interval is small enough to verify that the optimum is close, the other two criteria ensure that the problem is feasible.

4 Mehrotra Predictor-Corrector Algorithm

As an alternative to the primal dual algorithm, the Mehrotra predictor-corrector algorithm is explained pursuing the same objective as the presented in previous section. This algorithm is a refined version of the primal-dual algorithm. It consists of two stages, the predictor stage and the corrector one.

Applying the step of the Newton directions in the former algorithm to the system, the system in (25) is obtained.

$$\begin{aligned} A d_x^k &= 0 \\ A^t d_y^k + d_s^k &= 0 \\ S_k d_x^k + X_k d_s^k + d_x^k d_s^k &= \sigma \mu^k e - X_k S_k e \end{aligned} \tag{25}$$

where σ is the parameter that indicates the rate of reduction of the barrier parameter.

Here, the term $d_x^k d_s^k$ has not been neglected. The general idea is to make a first estimation of d_x^k and d_s^k, and then to replace the product $d_x^k d_s^k$ in (25) and to solve the system.

In this algorithm it is very important the Cholesky decomposition, because it can be used in both stages and it speeds up the resoluteness of the system.

4.1 Predictor stage

In this stage, the system in (26) is going to be solved.

$$\begin{bmatrix} A & 0 & 0 \\ 0 & A^t & I \\ S_k & 0 & X_k \end{bmatrix} \begin{bmatrix} d_x^{kaff} \\ d_y^{kaff} \\ d_s^{kaff} \end{bmatrix} = - \begin{bmatrix} Ax^k - b \\ A^t y^k + s^k - c \\ X_k S_k e \end{bmatrix} \tag{26}$$

The solution of the system provides the affined movement directions, used to estimate the product $d_x^k d_s^k$ of the system in (25). Additionally, these values provide a specific σ, using the expression in (27).

$$\sigma = \left(\frac{\mu_{aff}}{\mu}\right)^3 \tag{27}$$

where the values of μ_{aff} and μ are obtained using (28) and (29).

$$\mu_{aff} = \frac{\left(x + \beta_P^p d_x^{kaff}\right)^t \left(s + \beta_D^p d_s^{kaff}\right)}{n} \tag{28}$$

$$\mu = \frac{s^t x}{n} \tag{29}$$

The β parameters are calculated using the same criteria as in the primal-dual algorithm.

4.2 Corrector stage

Once the values of d_x^{kaff} and d_s^{kaff} are calculated, the system in (30) is solved in order to obtain the corrected movement directions.

$$\begin{bmatrix} A & 0 & 0 \\ 0 & A^t & I \\ S_k & 0 & X_k \end{bmatrix} \begin{bmatrix} d_x^k \\ d_y^k \\ d_s^k \end{bmatrix} = - \begin{bmatrix} Ax^k - b \\ A^t y^k + s^k - c \\ X_k S_k e + D_x^{kaff} D_s^{kaff} + \sigma \mu e \end{bmatrix} \tag{30}$$

where D_x^{kaff} means a diagonal matrix formed by the elements of d_x^{kaff}. The step sizes are obtained again after the final movement directions, and the points are updated to move to the next iteration. The stopping criteria are the same as in the primal-dual algorithm, i.e., those given in (24).

5 Case Study

5.1 Algorithms assessment.

As explained above, two algorithms have been used to solve the proposed problem. Primal-Dual and Mehrotra Predictor-Corrector algorithms have been tested with the aim of comparing results for the same cases.

Initially, ten cases are considered consisting of randomly generating ten different wind power vectors, corresponding to a period of 24 hours. Besides, an interval of 10 minutes has been taken, resulting in a total amount of 144 variables.

The results of the time performance tests for both algorithms can be seen in table 1, where the number of iterations and the objective function have been included for the ten cases.

Table 1: Results for time performance tests.

 https://doi.org/10.1051/e3sconf/201912204002

	Primal-dual			Predictor-corrector		
n	Secs.	Iter.	Obj. Func.	Secs.	Iter.	Obj. Func.
1	2.039	19	-238.214	1.887	11	-238.214
2	2.089	20	-168.959	2.356	14	-168.959
3	2.446	24	-550.613	2.52	15	-550.613
4	2.141	21	-517.382	2.201	13	-517.382
5	1.981	19	-150.012	3.007	18	-150.012
6	1.968	19	-194.019	2.51	15	-194.019
7	2.125	21	-437.988	2.181	13	-437.988
8	2.146	21	-587.902	2.517	15	-587.901
9	2.146	21	-266.105	2.16	13	-266.105
10	1.981	19	-141.292	2.16	13	-141.292

As can be seen, no differences have been detected in terms of the optimal value of the objective function. In order to compare the elapsed time and number of iterations, a graphical comparison is given in figure 1, where the mean values of both algorithms are compared.

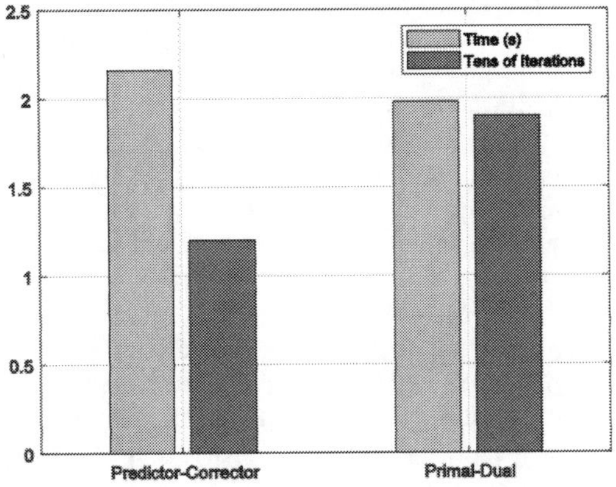

Fig 1. Comparison of elapsed time and number of iterations using both algorithms.

In figure 1, it can be seen that the primal-dual algorithm performs better in terms of time and the predictor–corrector one does the same when referring to the number of iterations. Using this result as a reference, it can be inferred that the primal-dual algorithm is the best choice in order to solve the problem described in this paper. However, the predictor–corrector algorithm should not be discarded completely due to the low number of iterations, i.e., it may be of application when the operations involved in an iteration are higher than in the case of considering a period of one week.

5.2 Estimation of the optimal capacity of a wind farm with BESS

In this paragraph a way is shown to determine a range of optimal capacities of a BESS system in terms of network stability (power generation when needed). Considering the results of the previous analysis, the primal-dual algorithm is chosen to solve this specific problem.

In order to obtain the results, a specific case has been taken. The related information can be seen in the table 2.

Table 2. Wind Farm Data.

Windfarm	Chantada	Number of Wind Turbines	30
Company	GALICIA VENTO, S.L.	Unit Power (kW)	1,670
Town	Chantada and Rodeiro	Wind Turbine Manufacturer	Alstom-Ecotecnia
Province	Lugo and Pontevedra	Model	ECO74
Installed Power (MW)	50	Technology	DFIG

The data of the wind turbines can be found in table 3.

Table 3. Wind Turbine Data.

Rated power	Cut-in wind speed	Rated wind speed
1,670 kW	3.0 m/s	13.0 m/s
Cut-out wind speed	Survival wind speed	Hub Height
25.0 m/s	59.5 m/s	80m

The BESS system used in this paper will be the NAS battery system. The information of this system is provided in table 4.

Table 4. NAS Battery System Data.

Rated Output	2,000 kW
Rated Input	2,000 kW
Rated Capacity	12 MWh

Grouping different modules, the desired power and capacity can be reached. The power (MW) has been used as an index of the capacity of the system, as it is easier to identify and operate within the calculations.

To carry on the research, a representative week per season has been considered and different values of capacities have been evaluated in order to analyze their behavior.

Ten different experiments per week have been done and for the following different power values: 4, 6, 10, 14, 16, 20, 30, 50 MW

This results in a total of 320 experiments. After all, these average values of the indexes have been obtained.

With the aim of rating the good behavior of the batteries the relationship in (31) has been defined:

$$\text{ratio} = \frac{c^t(Pb+Pw)}{c^tPw} \qquad (31)$$

This simply shows the relation between the cases with and without batteries. So the higher it is, the better the behavior of the power plant.

Vector c will be composed with the different weight parameters obtained from demand data. The representative weeks that have been taken can be seen below:

- Spring - 13/04/2015

- Summer - 3/08/2015
- Autumn - 12/10/2015
- Winter - 19/01/2015

The information needed to model the wind resource was taken from the Wind Atlas of the IDAE. Table 5 shows the Weibull parameters used.

Table 5. Seasonal Weibull Parameters at 80 m.

	Spring	Summer	Autumn	Winter
Weibull C (m/s)	10.18	8.73	11.03	12.27
Weibull k ( )	2.36	2.37	2.40	2.42

All these input data gave the results provided in table 6.

Table 6. Average ratios.

4 MW	6 MW	10 MW	14 MW
1.030	1.042	1.065	1.105
16 MW	20 MW	30 MW	50 MW
1.104	1.135	1.159	1.242

Here the positive parameter is represented by the ratio, while the negative one is the power, as it is considered as an expense. If the relation between this ratio and the installed power is plotted against the installed power, the graph presented in figure 2 is obtained.

The behavior is very similar to a decreasing exponential curve. So, as installed power continues being added the amount of improvement achieved is lowered.

Finally, an optimal range of power/capacity can be set between 20 and 30 MW, representing the 40 and 60% of the rated output power.

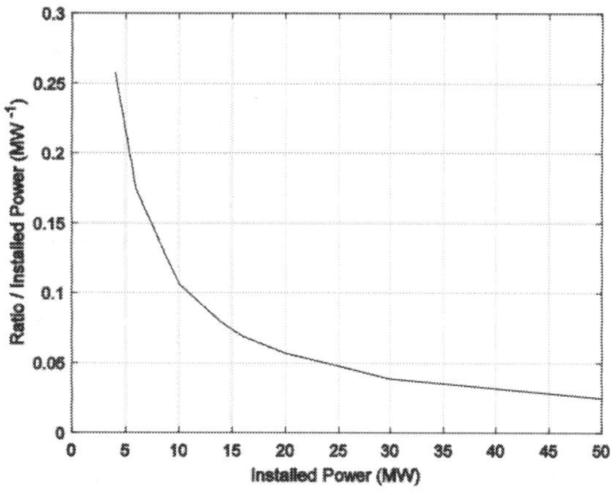

Fig 2. Ratio over installed power vs. installed power.

6 Conclusions

The primal-dual algorithm and the Mehrotra predictor-corrector algorithm have been successfully implemented to optimize the power supplied by a hybrid WF-BESS system. For a period of one day and intervals of 10 min, the primal-dual algorithm performs better in terms of time and, for the same conditions, the predictor-corrector performs better in terms of number of iterations. Therefore, it can be established that the first method is the proper one for the case described here while the second one can be of application for a higher number of intervals.

As a collateral result, using the primal-dual algorithm, it can be concluded that the rated power of a BESS should be around 40-60% of the rated power of the wind farm, because a higher rated power would be underused.

References

1. Protogeropoulos, C., Brinkworth, B.J., Marshall, R.H., Sizing and techno-economical optimization for hybrid solar photovoltaic/wind power systems with battery storage, (1997) International Journal of Energy Research, 21 (6), pp. 465-479.
2. Yao, D.L., Choi, S.S., Tseng, K.J., Lie, T.T., A statistical approach to the design of a dispatchable wind power-battery energy storage system, (2009) IEEE Transactions on Energy Conversion, 24 (4), art. no. 5336289, pp. 916-925.
3. Prasad, A.R., Natarajan, E., Optimization of integrated photovoltaic-wind power generation systems with battery storage (2006) Energy, 31 (12), pp. 1607-1618.
4. Khalid, M., Savkin, A.V., A model predictive control approach to the problem of wind power smoothing with controlled battery storage (2010) Renewable Energy, 35 (7), pp. 1520-1526.
5. Nguyen, M.Y., Nguyen, D.H., Yoon, Y.T., A new battery energy storage charging/discharging scheme for wind power producers in real-time markets (2012) Energies, 5 (12), pp. 5439-5452.
6. Jiang, Q., Wang, H., Two-time-scale coordination control for a battery energy storage system to mitigate wind power fluctuations (2013) IEEE Transactions on Energy Conversion, 28 (1), art. no. 6365250, pp. 52-61.
7. Alimisis, V., Hatziargyriou, N.D., Evaluation of a hybrid power plant comprising used EV-BESS to complement wind power (2013) IEEE Transactions on Sustainable Energy, 4 (2), art. no. 6331027, pp. 286-293.
8. Li, X., Hui, D., Lai, X., Battery energy storage station (BESS)-based smoothing control of photovoltaic (PV) and wind power generation fluctuations (2013) IEEE Transactions on Sustainable Energy, 4 (2), art. no. 6473871, pp. 464-473.
9. Wee, K.W., Choi, S.S., Vilathgamuwa, D.M., Design of a least-cost battery-supercapacitor energy storage system for realizing dispatchable wind power (2013) IEEE Transactions on Sustainable Energy, 4 (3), art. no. 6487428, pp. 786-796.

10. Kou, P., Liang, D., Gao, F., Gao, L., Coordinated Predictive Control of DFIG-Based Wind-Battery Hybrid Systems: Using Non-Gaussian Wind Power Predictive Distributions (2015) IEEE Transactions on Energy Conversion, 30 (2), art. no. 7031952, pp. 681-695.
11. Sun, Y., Zhong, J., Li, Z., Tian, W., Shahidehpour, M., Stochastic Scheduling of Battery-Based Energy Storage Transportation System with the Penetration of Wind Power (2017) IEEE Transactions on Sustainable Energy, 8 (1), art. no. 7501878, pp. 135-144.
12. Zhou, W., Yang, H., Fang, Z., Battery behavior prediction and battery working states analysis of a hybrid solar-wind power generation system (2008) Renewable Energy, 33 (6), pp. 1413-1423.
13. Swierczynski, M., Stroe, D.I., Stan, A.-I., Teodorescu, R., Sauer, D.U., Selection and performance-degradation modeling of limo2/li 4ti5o12 and lifepo4/c battery cells as suitable energy storage systems for grid integration with wind power plants: An example for the primary frequency regulation service (2014) IEEE Transactions on Sustainable Energy, 5 (1), art. no. 6583344, pp. 90-101.
14. Teleke, S., Baran, M.E., Bhattacharya, S., Huang, A.Q., Optimal control of battery energy storage for wind farm dispatching (2010) IEEE Transactions on Energy Conversion, 25 (3), art. no. 5432995, pp. 787-794.
15. Zhang, Y., Dong, Z.Y., Luo, F., Zheng, Y., Meng, K., Wong, K.P., Optimal allocation of battery energy storage systems in distribution networks with high wind power penetration (2016) IET Renewable Power Generation, 10 (8), pp. 1105-1113.
16. Nguyen, C.-L., Lee, H.-H., A Novel Dual-Battery Energy Storage System for Wind Power Applications (2016) IEEE Transactions on Industrial Electronics, 63 (10), art. no. 7473865, pp. 6136-6147.
17. Gamini Jayasinghe, S.D., Mahinda Vilathgamuwa, D., Madawala, U.K., Direct integration of battery energy storage systems in distributed power generation (2011) IEEE Transactions on Energy Conversion, 26 (2), art. no. 5742991, pp. 677-685.
18. Griffin, C., Linear Programming, Penn State Math 484 Lecture Notes.
19. Shu-Cherng Fang, Sarat Puthenpura, Linear Optimization and Extensions: Theory and Algorithms. Prentice-Hall, (1993).
20. Yinyu Ye, Interior Point Algorithms: Theory and Analysis, Wiley-Interscience, (1997).
21. M.J. Atallah et al., Algorithms and Theory of Computation Handbook, CRC Press, (1999).
22. Nicole Lemire, An Introduction to Karmarkar's Method, Windsor Mathematics Report, (1989).
23. Robert M. Freund, Primal-Dual Interior-Point Methods for Linear Programming based on Newton's Method, Massachusetts Institute of Technology, (2004).
24. Sanjay Mehrotra, On Finding a Vertex Solution Using Interior Point Methods, Linear Algebra and its Applications, vol. 152, pp. 233-253, (1991).
25. Hamdy A. Taha, Operations research: An Introduction, Pearson, (1992).
26. Mokhtar S. Bazaara, Linear Programming and Network Flows, Wiley, (2010).

Wind and wave energy resource of Germany reported by ERA-Interim reanalysis data

Eugen Rusu and *Florin* Onea*

Department of Mechanical Engineering, ''Dunarea de Jos'' University of Galati, Domneasca'' Street, 47, Galati 800008, Romania

Abstract. The aim of this work is to identify the most suitable offshore wind farms from Germany that present relevant wave conditions, suitable for the development of a wave energy project. By using the ERA-Interim data (wind and waves) reported for the time interval from 1999 and 2018, was possible to identify the more important areas, by taking also into account the seasonal distributions. Several wave energy converters were considered for assessment, for which a capacity factor located between 2.5% and 14% was reported, better results being accounted by the Seabased system (rated at 15 kW). Finally, we can concluded that the North Sea represent an important area in terms of the marine energy and since at this moment there are operational wave projects, this will represent a suitable area for the development of a mixed wind-wave project.

1 Introduction

On a European energy market, there is a growing interest for the renewable energy solutions, and one of the most promising sector is related to the marine areas. At this moment a significant amount of research and investments are being put into the offshore wind and wave energy, each sector reporting various levels of development [1]. The offshore wind industry is more mature, as in the case of Germany were 25 projects were operational at the end of 2018. A cumulated capacity of 6380 MW is provide by 1305 wind turbines, from which approximately 136 units were connected in 2018, which were mainly related to the projects Borkum Riffgrund II (464.8 MW) and Arkona (384 MW), respectively. From a financial point of view, only in 2018 approximately 0.4 €BN were invested in the offshore wind projects from Germany [2].

In order to accelerate the development of the wave energy sector, one solution will be to use the infrastructure of the existing offshore projects in order to develop mixed wind-wave projects [3–5]. In this way, will be possible to extract two energy source from a single location, being also possible to provide some effective coastal protection in some situations [6]. The North Sea seems to represent an important area for the development of the Wave Energy Converter (WECs), this being the case of the NEMOS system (Germany) [7], Wavestar (Denmark) [8] or Wave Dragon (Denmark) [9]. The SINN Power wave energy converter [10] is another system developed in Germany, being reported two ongoing commercial projects.

At this moment, there are various approaches to evaluate the wind and wave resources from the marine areas, from which we can mention the use of numerical models. This is the case of the marine conditions from the coastal waters of Germany (North and Baltic seas) were throughout some reanalysis databases, different meteorological patterns were identified [11–13].

2 Methods and materials

The target area considered for evaluation is related to the nearshore and offshore waters from the vicinity of Germany. Most of the offshore wind farms operate in the North Sea, which is also defined by more important wave resources than in the case of the Baltic Sea, and therefore for the present work only the projects from this region will be taken into account. A short presentation of the selected project is provided in Table 1.

Table 1. Overview of some German offshore wind projects operating in the North Sea [14].

Site	Project capacity (MW)	Number of turbines
P1 - Alpha Ventus	60	12
P2 - Amrumbank West	302	80
P3 - BARD Offshore 1	400	80
P4 - Butendiek	288	80
P5 - DanTysk	288	80
P6 - Global Tech I	400	80
P7 - Gode Wind 1 & 2	582	97
P8 - Nordsee One	332	54
P9 - Riffgat	108	30
P10 - Sandbank	288	72

* Corresponding author: florin.onea@ugal.ro

The wind and wave parameters considered in this work are: the wind speed reported at 10 m height above sea level (U10); the significant wave height (Hs) and the mean wave period (Te). They are related to the ERA-Interim dataset, which is a project maintained by the European Centre for Medium-range Weather Forecast.

In order to assess the local variations, it was considered for processing two datasets that covers the 20-year time interval (1999-2018), being defined by four data per day (corresponding to 00-06-12-18 UTC) and by a spatial resolution of 0.125o x 0.125o, which is the finest grid available on the ECMWF server (https://www.ecmwf.int/en/forecasts/datasets/reanalysis-datasets/era-interim).

In general for a renewable project, the wind conditions are evaluated at a 80 m height (U80) and therefore the ERA-Interim conditions will be adjusted as follows [15,16]:

$$U80 = U10\frac{\ln(z_{80}) - \ln(z_{10})}{\ln(z_{10}) - \ln(z_0)} \quad (1)$$

where, U80 - wind speed at 80 m, U10 - initial wind speed (at 10 m), z_0 - roughness of the sea surface (0.0002 m), z_{10} and z_{80} - reference heights.

The wave energy flux (Jwave in W/m), associated with to the deep water areas is defined as [17]:

$$J_{wave} = \frac{\rho_{water} g^2}{64\pi} T_e H_s^2 \quad (2)$$

ρ_{water} (kg/m^3) – seawater density, g (m/s) – gravitational acceleration.

Several wave converters will be considered for assessment, their characteristics being indicated in Table 2. The rated powers of the selected WECs cover a larger range of values (from 15 kW to 7000 kW), being possible to estimate in this way the performances of various devices.

Table 2. Technical specifications of the considered WECs [17,18].

WEC (rated capacity)	*Hs* bins (in meters)	*Te* bins (in seconds)
Seabased (15 kW)	1:0.5:7	4:1:16
Pelamis (750 kW)	0.5:0.5:8	5:0.5:13
Wave Dragon (7000 kW)	1:1:7	5:1:17

The expected power generated by a WEC can be estimated as follows [6]:

$$P_E = \sum_{i=1}^{n_T} \sum_{j=1}^{n_H} p_{ij} \cdot P_{ij}, \quad (3)$$

where p_{ij} is related to the energy percentage associated to the bin defined by the line i and column j, where as P_{ij} is the expected electric power output defined in the power matrix of each WEC for the same bin (defined by line i and column j).

Figure 1 presents the power matrix of the Wave Dragon system, from which we can identify the expected power output for different sea states combination.

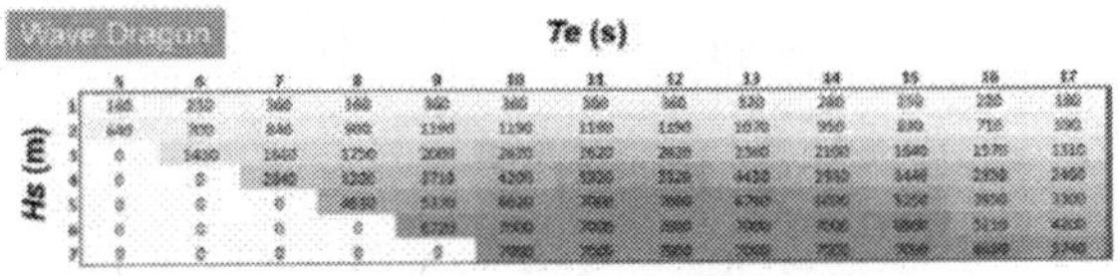

Figure 1. Wave Dragon (7000 kW) power matrix [19]

3 Results

In Figure 2 and Figure 3, are presented the spatial distributions of the wind speed (U80) and wave energy flux, where with yellow was indicated the German land area. From the analysis of the wind conditions, much higher conditions seems to be reported in the North Sea where the wind speed may reach maximum of 11 m/s. Regarding the German areas located in the Baltic Sea, we may expect much lower resources that vary between 7 m/s and 9 m/s, respectively. As for the wave resources, the first thing to notice is that there are reported region with missing data, this aspect being more visible in the case of the Baltic Sea. For the waves associated to the German sector from the North Sea, we may expect an increase of wave energy flux from 4 kW/m to 14 kW/m, as we go from nearshore to offshore.

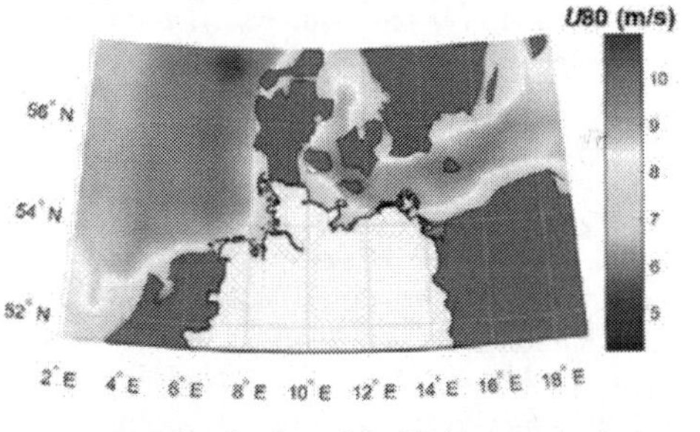

Figure 2. Spatial distribution of the U80 parameter (average values) reported by the ERA-Interim dataset for the 20-year interval (1999-2018).

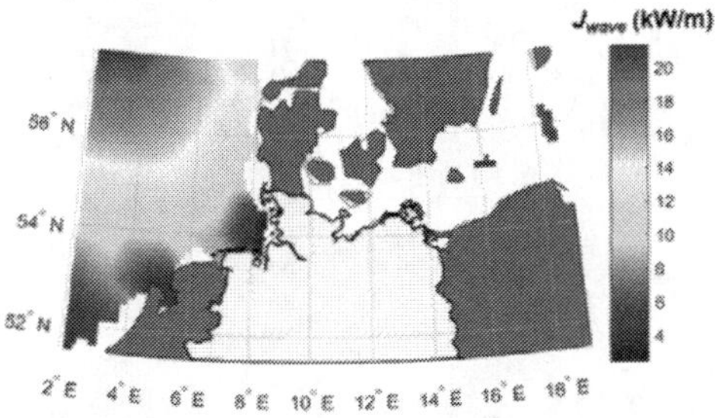

Figure 3. Spatial distribution of the Jwave parameter (average values in kW/m) reported by the ERA-Interim dataset for the 20-year interval (1999-2018).

Figure 4 illustrates the seasonal variations of the wind speed, by taking into account: a) spring (March-April-May); b) summer (June-July-August); c) autumn (September-October-November); d) winter (December-January-February). During spring and summer season, a similar spatial distribution is noticed, with the mention that in the case of the first one the wind speed of 8 m/s-9 m/s can also be found between the longitude of 4° and 6° east. As we go to the winter season, the values start to increase reaching maximum of 13 m/s and minimum of 9 m/s, respectively.

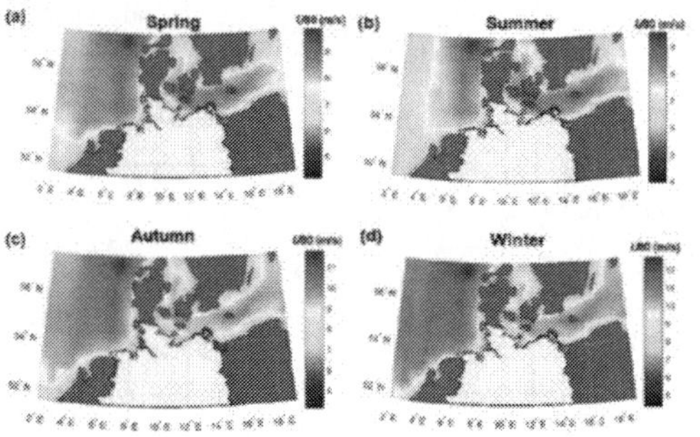

Figure 4. Seasonal distribution of the U80 parameter (average values) reported by the ERA-Interim dataset for the 20-year interval (1999-2018).

In Figure 5 is presented a similar analysis, taking into account this time the wave resources. The percentage of missing data is significantly much lower during the autumn-winter interval, but even so, the accuracy of the ERA-Interim data to predict the local wave conditions can be discussed. For the North Sea area, significantly much higher resources are being reported during winter, being expected in the German sector at an average wave regime of 25 kW/m, compared to only 5 kW/m accounted by the summer season. Regarding the Baltic Sea conditions, it seems that the values located in the range of 5 kW/m and 10 kW/m are representative for the winter time. As for the performances of the WECs, in

Figure 6 presents the power output reported by the systems mentioned in Table 2. At this point, it is important to mention that the project Pelamis is no longer operational, but was taken into account since this was one of the first commercial project. All the selected WECs are designed to operate in offshore areas [20].

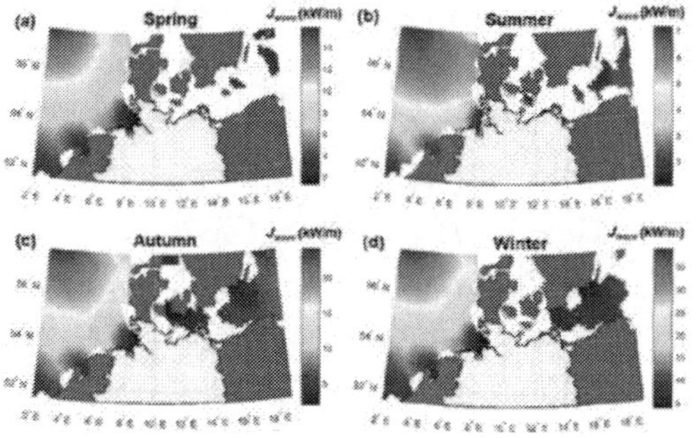

Figure 5. Seasonal distribution of the Jwave parameter (average values in kW/m) reported by the ERA-Interim dataset for the 20-year interval (1999-2018).

From this point of view, the Seabased system report the lowest performances, that are directly associated with the rated power of this device (15 kW). A maximum of 2.08 kW is accounted by the site P10 (project Sandbank), while a minimum of 1.22 kW is reported for P2 (project Amrumbank West). By looking on the values reported by the Pelamis and Wave Dragon, we may notice that some sites stand out in terms of their attractiveness, namely: P1 (Alpha Ventus), P3 (BARD Offshore 1), P5 (DanTysk), P6 (Global Tech I) and P10 (Sandbank).

The capacity factor can be considered as a indicator of efficiency being defined as the ratio between the power produced and the rated power of a particular WEC. These values are presented in Figure 7, were a reverse pattern is noticed compared to the power distribution. A maximum capacity of 14% may be expected from Seabased, that in general reveal an average of 11 % (per site group), compared to 7.8% reported for Pelamis and 5% for Wave Dragon.

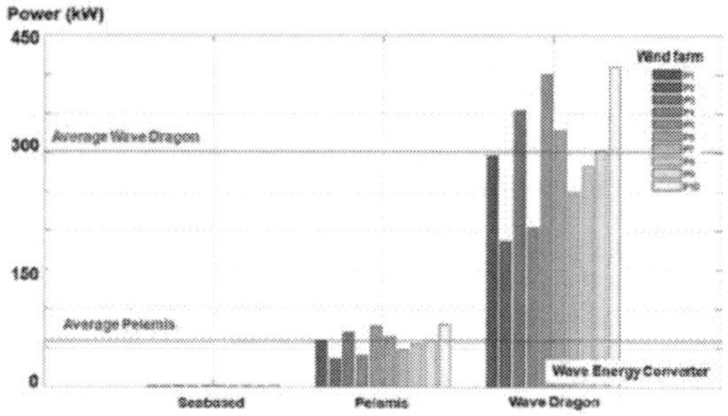

Figure 6. Expected power output (in kW) from the Seabased, Pelamis and Wave Dragon system reported for the locations of the offshore wind farms indicated in Table 1.

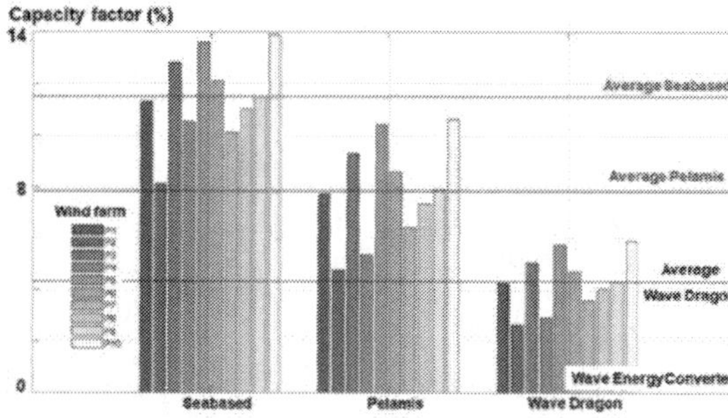

Figure 7. Capacity factor (in %) associated to the Seabased, Pelamis and Wave Dragon systems. Values reported for the locations of the offshore wind farms indicated in Table 1.

4 Conclusions

In the present work a general perspective of the offshore wind and wave resources from the coastal waters of Germany was provided, by taking also into account the performances of some wave energy converters that may operate in the vicinity of some wind projects.

In terms of quality, the wind data reveal better results being not affected by the presence of missing data as in the case of wave, more consistent resources being reported in the North Sea, were an average of 11 m/s

may be noticed (full time distribution). The accuracy of the selected dataset can be discussed, especially in the case of the Baltic Sea, and probably for this type of environment will be more suitable the implementation of a regional wave model. Nevertheless, if we look on the North Sea conditions more promising conditions to implement a mixed wind-wave farm are reported by the sites P10 (Sandbank – 288 MW), P5 (DanTysk – 288 MW) and P3 (BARD Offshore 1 – 400 MW), respectively. Finally, we can concluded that these wind farms can easily support the development of a wave project taken into account their large capacities and also the fact that in the basin of North Sea, there is interest for the development of wave energy sector.

Acknowledgments

This work was carried out in the framework of the research project REMARC (Renewable Energy extraction in MARine environment and its Coastal impact), supported by the Romanian Executive Agency for Higher Education, Research, Development and Innovation Funding – UEFISCDI, grant number PN-III-P4-IDPCE-2016-0017. The ERA-Interim data used in this study have been obtained from the ECMWF data server.

References

[1] C.V.C. Weiss, R. Guanche, B. Ondiviela, O.F. Castellanos, J. Juanes, Marine renewable energy potential: A global perspective for offshore wind and wave exploitation, Energy Conversion and Management. 177 (2018) 43–54. doi:10.1016/j.enconman.2018.09.059.

[2] Offshore wind in Europe – key trends and statistics 2018, WindEurope. (2019). https://windeurope.org/about-wind/statistics/offshore/european-offshore-wind-industry-key-trends-statistics-2018/ (accessed March 8, 2019).

[3] F. Onea, S. Ciortan, E. Rusu, Assessment of the potential for developing combined wind-wave projects in the European nearshore, Energy Environ. 28 (2017) 580–597. doi:10.1177/0958305X17716947.

[4] F. Onea, E. Rusu, Sustainability of the Reanalysis Databases in Predicting the Wind and Wave Power along the European Coasts, Sustainability. 10 (2018) 193. doi:10.3390/su10010193.

[5] E. Rusu, F. Onea, Joint Evaluation of the Wave and Offshore Wind Energy Resources in the Developing Countries, Energies. 10 (2017) 1866. doi:10.3390/en10111866.

[6] F. Onea, E. Rusu, The expected efficiency and coastal impact of a hybrid energy farm operating in the Portuguese nearshore, Energy. 97 (2016) 411–423. doi:10.1016/j.energy.2016.01.002.

[7] Wave Energy, NEMOS. (n.d.). https://www.nemos.org/waveenergy (accessed May 8, 2019).

[8] E.J. Ransley, D.M. Greaves, A. Raby, D. Simmonds, M.M. Jakobsen, M. Kramer, RANS-VOF modelling of the Wavestar point absorber, Renewable Energy. 109 (2017) 49–65. doi:10.1016/j.renene.2017.02.079.

[9] J.P. Kofoed, P. Frigaard, E. Friis-Madsen, H.Chr. Sørensen, Prototype testing of the wave energy converter wave dragon, Renewable Energy. 31 (2006) 181–189. doi:10.1016/j.renene.2005.09.005.

[10] SINN Power | Wave Energy, SINN Power | Wave Energy. (n.d.). https://www.sinnpower.com (accessed May 8, 2019).

[11] T. Lorenz, I. Barstad, A dynamical downscaling of ERA-Interim in the North Sea using WRF with a 3km grid-for wind resource applications, Wind Energy. 19 (2016) 1945–1959. doi:10.1002/we.1961.

[12] N. Groll, R. Weisse, . A multi-decadal wind-wave hindcast for the North Sea 1949-2014: coastDat2, Earth Syst. Sci. Data. 9 (2017) 955–968. doi:10.5194/essd-9-955-2017.

[13] L. Tuomi, A. Sarkanen, Wave forecasts for the Baltic Sea using ECMWF wind fields as forcing data, Ieee, New York, 2008.

[14] 4C Offshore Overview - 4C Offshore, (n.d.). https://www.4coffshore.com/ (accessed March 12, 2019).

[15] A. Raileanu, F. Onea, E. Rusu, Assessment of the Wind Energy Potential in the Coastal Environment of Two Enclosed Seas, Ieee, New York, 2015.

[16] F. Onea, E. Rusu, Efficiency assessments for some state of the art wind turbines in the coastal environments of the Black and the Caspian seas, Energy Explor. Exploit. 34 (2016) 217–234. doi:10.1177/0144598716629872.

[17] E. Rusu, F. Onea, An assessment of the wind and wave power potential in the island environment, Energy. 175 (2019) 830–846. doi:10.1016/j.energy.2019.03.130.

[18] J. Morim, N. Cartwright, M. Hemer, A. Etemad-Shahidi, D. Strauss, Inter- and intra-annual variability of potential power production from wave energy converters, Energy. 169 (2019) 1224–1241. doi:10.1016/j.energy.2018.12.080.

[19] D. Silva, E. Rusu, C. Soares, D. Silva, E. Rusu, C.G. Soares, Evaluation of Various Technologies for Wave Energy Conversion in the Portuguese Nearshore, Energies. 6 (2013) 1344–1364. doi:10.3390/en6031344.

[20] L. Rusu, F. Onea, The performance of some state-of-the-art wave energy converters in locations with the worldwide highest wave power, Renew. Sust. Energ. Rev. 75 (2017) 1348–1362. doi:10.1016/j.rser.2016.11.123.

https://doi.org/10.1051/e3sconf/201912204004

A Strategy for Power Generation Optimization in a Hybrid Wind-BESS Power Plant

Daniel Villanueva[1*], *Andrés* E. Feijóo[1], *and Neeraj* D. Bokde[2]

[1]Electrical Engineering Department, University of Vigo, 36310-Vigo, Spain
[2]National Institute of Technology, Nagpur, India

Abstract. The wind is an uncontrollable primary resource, although its energy can be stored. This fact can be used for the design of strategies for a better management of electric power networks. An option for achieving this goal is to install Battery Energy Storage Systems (BESS) in the wind farms (WF). When dealing with WFs combined with BESSs the most important is to manage the power production in order to meet the requirements of the network or those related with the owner of the plant. Both challenges constitute an optimization problem. This paper proposes an Evolutionary Algorithm (EA) to solve it, where a fitness function must be maximized under the consideration of certain constraints. The fitness function depends on the target of the power production, which may be either to help the network become more stable or to maximize the profit, assessing each scenario and accepting the best one. The constraints of the optimization problem are related to the levels of the BESSs: the maximum power transferred to or from it and the output power of the plant.

1 Introduction

Wind power generation presence in electrical power networks has been growing over the last decades, which has led their control to be more and more critical. Wind speed cannot be directly managed but it can be predicted with high accuracy, for which the efforts should be focused on network management, with the goal of injecting as much electric power from the wind as possible into it. But this action is not enough due to some reasons, such as the variability of the resource and its independence of the daily demand curve. In both cases, an amount of wind power can be wasted and/or used with a low efficiency, i.e., when it is not essential.

If the wind power production can be regulated or adapted to a specific daily demand curve, it can contribute to network stability and the profit can be higher. If it is the case, an interesting alternative of power injection strategy can be established. This consists of storing wind energy when there is a surplus, with the goal of injecting it into the power network when needed, according to the load demand curve. When not needed, it is preserved. This control can be considered as a contribution of the system to the steady-state stability of the network and can improve the profit of the power plant. If such a strategy is possible, then the consideration of WFs as very variable and uncontrollable power plants can be rethought.

Several means are generally used to store energy: different types of BESSs, capacitors, compressed air, liquid air, hydrogen, flywheels, hydroelectricity, superconductors, and others [1]. The intention of this paper is not to assess the most suitable mean to be used in combination with WFs, but several reasons support the use of BESSs:

i) The use of a BESS in combination with a WF arises as a solution applicable to all locations because there is no need of specific conditions in the installation.
ii) The efficiency of the system is very high.
iii) The number and/or size of the BESS can be adapted to the size of the WF, even though the bigger the WF is, the more efficient the installation is.
iv) They are especially valuable for small networks when other types of energy storage fail [2].

When dealing with a system formed by a WF and a BESS, the first issue to be considered is the battery size [3, 4], which is critical. However, it is not dealt with here and the value used in the case study is the existing one in a real power plant. The equations and constraints used in this paper appear in [4-13]. Another issue is the application of the BESS, that can be in an hourly, daily, weekly or monthly term [14], but the objective here is particularized for a daily term. Besides, prediction is not considered here, it is assumed that the wind speed is known with enough anticipation to apply the algorithm. This is realistic due to the daily term being analyzed.

Another issue to be considered is the model used for the BESS. Several models appear in the related literature [9, 15, 16] and they have been simplified here, considering certain efficiency during the charging and discharging processes. Finally, it is also important to decide the optimal allocation of the BESS in the network

* Corresponding author: dvillanueva@uvigo.es

and its configuration [17-19]. In this paper, the BESS is located adjacently to the WF and the chosen configuration is the one called dual inverter [19].

Then, to carry out a strategy based on the adaptation of the wind power produced by a power plant constituted by a WF and a BESS to the power requested along a period, is an optimization problem. The objective should be that the power plant generates more power when it is more valuable. However, the power plant will have some technical constraints that limit its operating capacity, most of them related with the BESS but also with the whole installation, turning the optimization problem into a non-linear one. Here, a specific EA is proposed, adapted to solve this problem. The reasons for using an EA are the constraints and the large amount of variables.

This paper is organized as follows: in section II, the concepts, expressions and constraints considered when dealing with a plant formed by a WF and a BESS are introduced; section III explains the application of the EA to the problem described in this paper; a specific case is shown in section IV and the conclusions are exposed in section V.

2 Power Supplied by a Wind Farm with a BESS

As power plant consisting of a WF with a BESS is analyzed. The equations and constraints that model its behavior are presented in this section.

The power plant output is the sum of the power provided by the WF and by the BESS, as in (1).

$$P_{g_i} = P_{b_i} + P_{w_i} \qquad i = 1, \dots, n_i \qquad (1)$$

where P_{g_i} is the power supplied by the power plant in the interval i. P_{b_i}, when positive, is the power provided by the BESS to the electrical network in the interval i. When it is negative, $|P_{b_i}|$ is the power provided by the WF to the BESS. Finally, P_{w_i} is the power supplied by the WF in the interval i and n_i is the number of intervals.

The charging process of the BESS can be explained by (2). The energy in the BESS in a specific interval can be obtained as the sum of the energy stored in the previous interval and the power supplied to it along the interval, multiplied by the length of the interval and by a coefficient that accounts for the losses due to the efficiency of the process.

$$E_{b_{i+1}} = E_{b_i} - \Delta T \cdot P_{b_i} \cdot \eta_c \qquad i = 1, \dots, n_i \qquad (2)$$

where E_{b_i} is the energy stored in the BESS in the interval i, η_c is the efficiency of the charging process and ΔT is the interval length. Notice that P_{b_i} is negative during the charging process. Therefore, $E_{b_{i+1}} \geq E_{b_i}$ always after applying (2).

If the BESS is discharging, the relationship among the variables is expressed in (3). In this case, the energy in the BESS in a specific interval can be obtained as the subtraction of the power provided by the BESS multiplied by the length of the interval, from the energy in the previous interval. A coefficient due to losses is also applied to the power provided by the BESS due to the efficiency of the process.

$$E_{b_{i+1}} = E_{b_i} - \Delta T \cdot P_{b_i} / \eta_d \qquad i = 1, \dots, n_i \qquad (3)$$

where η_d is the efficiency of the discharging process. Therefore P_{b_i} is the power provided by the BESS to the electrical network and $\Delta T \cdot P_{b_i}/\eta_d$ is the total amount of energy removed from the BESS. In this case, it is always $E_{b_{i+1}} \leq E_{b_i}$ after applying (3).

Besides, there are several constraints that must be taken into account to apply the given equations:

i) Constraints related to the level of the BESS (4): The BESS has a minimum level of charge in order to extend its useful life. It also has a maximum level of charge, i.e., its rated value.

$$E_{b_{min}} \leq E_{b_i} \leq E_{b_{max}} \qquad i = 1, \dots, n_i \qquad (4)$$

where $E_{b_{min}}$ is the minimum level of charge of the BESS and $E_{b_{max}}$ its maximum one.

ii) Constraints related to the charging and discharging processes of the BESS (5): Due to the installation configuration and to the proper BESS, there are limits regarding the maximum power that can be transferred from the WF to the BESS and from it to the electrical network.

$$P_{b_{min}} \leq P_{b_i} \leq P_{b_{max}} \qquad i = 1, \dots, n_i \qquad (5)$$

where $P_{b_{min}} < 0$ is the minimum value of the power provided by the BESS, i.e., the power limit to charge the BESS given in absolute value. $P_{b_{max}}$ is the maximum value of the power provided by the BESS.

iii) Constraints related to the power supplied by the power plant (6): Again, due to the installation configuration, there are limits regarding the power that can be transferred from the power plant to the electrical network, i.e., the power plant cannot consume power and the output power cannot be higher than its rated power. Obviously, in this case, both limits are not critical, but out of that range, the installation has to be specifically designed to work in those situations, and that is not the classic configuration.

$$0 \leq P_{g_i} \leq P_{g_r} \qquad i = 1, \dots, n_i \qquad (6)$$

where P_{g_r} is the rated power of the power plant.

3 Application of the Evolutionary Algorithm

An EA [20] includes processes related to physical or biological evolution such us crossover, mutation,

https://doi.org/10.1051/e3sconf/201912204004

recombination and the selection of the fittest individuals, to obtain the best solution of an optimization problem.

In the case studied in this paper, the application of the EA is very specific, due to the fact that a population of only one individual is needed, where a recombination of its elements is performed on each iteration. Then, the fitness of the new individual is assessed and it is accepted as the only member of the next generation when it is better than the parent's one. If the parent keeps fitting better than the child, the parent is the only member of the next generation. This algorithm does not correspond to a genetic algorithm itself but has some of its features.

More in depth, it can be said that the first individual is the vector formed by the power supplied by the WF in each of the intervals of the period, and, on each iteration, the vector is reorganized in order to obtain a better scenario, where the power supplied by the power plant is more suitable to the needs of the electrical network. In order to reorganize, part of the power provided by the WF during a specific interval is removed and added to another one. Normally, the period considered is one day, and the intervals last for ten minutes.

Therefore, on each iteration, a random value of power supplied by the WF is taken from a random interval and added to another random one. Then, the fitness function of the new vector formed by the P_{g_i} elements is calculated. The fitness function is obtained as in (7).

$$f = \sum_i \left(\omega_i^2 \cdot P_{g_i} \right) \qquad i = 1, \dots, n_i \qquad (7)$$

where f is the fitness function and ω_i^2 is the weight of the interval i. The weight of an interval is related to the importance of supplying power in that specific interval with respect to the rest. The weights can be obtained from the daily demand curve or from the prices of the power along the day, depending on whether the target is the contribution to the stability of the network or to obtain the maximum profit from the power plant, respectively. The reason why a squared value as a weight is included is that the highest values of power production in a specific interval are difficult to reach, so this speeds up the algorithm.

Finally, the fitness function is compared with its parent's one in order to decide which one is the only member of the new generation. The one with highest fitness function is retained and the other, rejected. The process is depicted in the scheme of Fig. 1 where 'ff' stands for fitness function.

In Fig. 1 the intervals last for one hour in order to better appreciate the process. An initial situation (top graph) with a given fitness function (ff) can be seen. Then, a random value of power (15 MW) is suggested to be moved from a random interval (2-3h) to another random interval (15-16h) (middle graph), and the fitness function in this hypothetic situation is obtained (ff'). Finally, two possible situations can be reached (bottom graphs): in case that ff'>ff, the hypothetic situation is accepted (left one) and in other case, discarded (right one).

The process is repeated until either the fitness function does not increase its value in a given number of iterations or a maximum number of iterations is reached. For example, the evolution of the fitness function can be assessed every 100 iterations and the algorithm will have no more than 10 million iterations.

The result of the algorithm is the vector with the new values of power supplied by the power plant and, as a consequence, the exchange of power between the WF and the BESS. In fact, this EA represents the real behavior expected from the power plant itself due to the fact that the BESS is taking energy from the WF along a specific interval, and is giving back that energy along another interval. The algorithm decides, on each iteration, if a concrete scenario is more suitable than the previous one, considering the objective established. This situation, together with the tendency of the EAs to reach the best solution, provides a useful tool to solve the proposed problem.

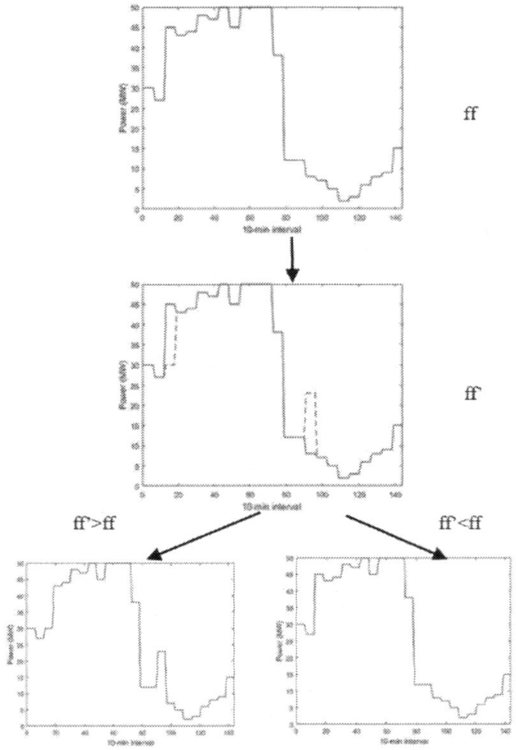

Fig. 1. Reorganization of the power supplied by the Power Plant performed on each iteration of the EA.

Moreover, as the algorithm is representing the desired behavior, the losses must be considered, due to the conversion of the energy from electric to chemical, when charging the BESS, and vice versa for the discharging process. In order to do that, the efficiencies of both processes (η_c, η_d) are taken into account in the algorithm as shown in equations (2) and (3).

Mathematically, the optimization problem is written as follows:

$$\max f = \sum_i (\omega_i^2 \cdot P_{g_i})$$

$$s.t. \begin{cases} E_{b_{min}} \leq E_{b_i} \leq E_{b_{max}} \\ P_{b_{min}} \leq P_{b_i} \leq P_{b_{max}} \\ 0 \leq P_{g_i} \leq P_{g_r} \end{cases}$$

considering the relationships in eqs. (1), (2) and (3).

4 Simulation Results

In order to check the results provided by the algorithm, a case study is proposed in this section. A WF is considered, constituted by $n_{wt} = 25\ WTs$, each one with $P_r = 2000\ kW$, $v_{ci} = 4\ m/s$, $v_r = 15\ m/s$ and $v_{co} = 25\ m/s$. A location is considered, where the wind speed is described by a Weibull distribution of parameters $C = 12\ m/s$ and $k = 2$. In order to simplify the calculations without loss of accuracy, the wind speed value is equal in all WTs.

The wind speed data are simulated using an autoregressive model of order 1, AR(1) [21], as in equation (8).

$$x_i = \varphi \cdot x_{i-1} + \varepsilon_i \qquad i = 1, \dots, n_i \tag{8}$$

where x is a Normal variable, x_i is its value in the interval i, φ is the autoregressive coefficient or lag-one autocorrelation, and ε_i is a white noise signal in the interval i. The autoregressive coefficient used is $\varphi = 0.95$ and the white noise is simulated by using random Normal data with parameters $\mu = 0$ and $\sigma = \sqrt{1 - \varphi^2}$. And converting the Normal data into Weibull data through a change of variables [22] as in (9).

$$u_i = C\left(-log\left(\left(1 - erf\left(\frac{x_i - \mu}{\sqrt{2}\sigma}\right)\right)/2\right)\right)^{1/k} \tag{9}$$

where $log(\,)$ is the natural logarithm and $erf(\,)$ is the error function defined by (10).

$$erf(x) = \frac{2}{\sqrt{\pi}} \int_0^x exp(-t^2)\, dt \tag{10}$$

where $exp(\,)$ is the exponential function.

To obtain the power supplied by each WT the quadratic model for power curves is applied [23], according to (11).

$$P_{w_{ij}} = \begin{cases} 0 & u_i \leq u_{ci} \\ P_r \cdot \frac{u_i^2 - u_{ci}^2}{u_r^2 - u_{ci}^2} & u_{ci} \leq u_i \leq u_r \\ P_r & u_r \leq u_i \leq u_{co} \\ 0 & u_{co} \leq u_i \end{cases} \tag{11}$$

where $P_{w_{ij}}$ is the wind power supplied in the interval i by the WT j, which corresponds to the wind speed u_i. Besides, as the wind speed value is considered equal in all the WTs simultaneously, the power supplied by the WF in the interval i is $P_{w_i} = \sum_j P_{w_{ij}} = n_{wt} \cdot P_{w_{ij}}$.

The parameters of the BESS are: the minimum level, $E_{b_{min}} = 30\ MWh$, the maximum level, $E_{b_{max}} = 150\ MWh$, the charging limit of the BESS, $P_{b_{min}} = -25\ MW$, and its discharging limit, $P_{b_{max}} = 25MW$.

The parameters of the power plant are: the rated power, $P_{g_r} = n_{wt} \cdot P_r = 50\ MW$, the efficiency of the charging process, $\eta_c = 0.90$ and the efficiency of the discharging process, $\eta_d = 0.90$.

In this case, the vector of weights is defined with the objective of contributing to the stability of the network. Then, using a daily demand curve [24], it can be established that the minimum value of demand corresponds to ω_i=0 and the maximum value of demand to ω_i=10. The rest of weights are defined linearly between these two values. Notice the fact that the fitness function includes the squared weights, so the real weights used are those shown in Fig. 2, where the difference between the weights, ω_i^2, and the coefficients, ω_i is depicted. As it has been said, this is done in order to speed up the algorithm: the highest values of power production are delayed to be reached using ω_i instead of ω_i^2.

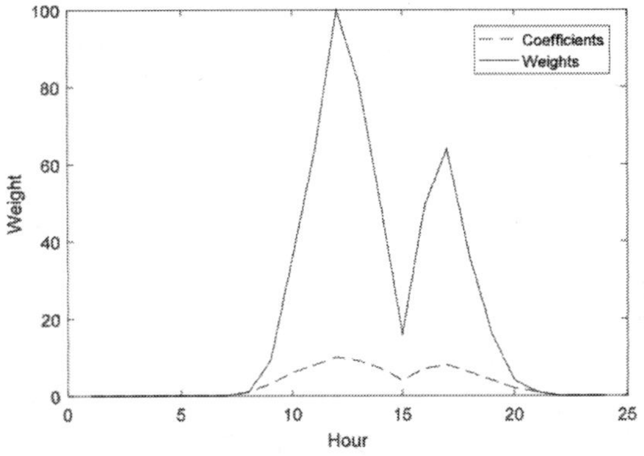

Fig. 2. Coefficients and weights as a function of the hour.

The following additional variables are needed: the total period considered is one day, each interval lasts for ten minutes, the evolution of the fitness function is assessed every 100 iterations, the maximum number of iterations is ten million and the initial state considered for the BESS corresponds to its minimum charge.

Then, once the input data is defined, the algorithm can be applied. In Figs. 3, 4 and 5 there are three applications of the algorithm to different random input data.

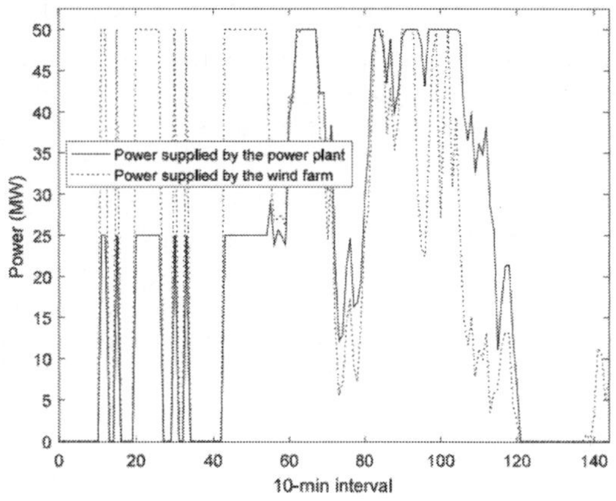

Fig. 3. EA applied to case 1.

In the first case, it can be seen a vector of power supplied by the WF where the rated power is generated during the night (0 to 50, 10-min interval) and also some power is produced during the hours with higher demand (50 to 120 10-min interval). After applying the algorithm, some of the power supplied by the WF during the night is moved to the latter intervals, increasing the production of the power plant in those ones. However, there are some limits to this reassignment: the limit of $25\ MW$ charging the BESS is reached repeatedly during the night, therefore, not all the power supplied by the WF can be used to charge the BESS. Besides, the limit in the power supplied by the power plant ($50\ MW$) is also reached several times during the periods with high demand. Obviously, there are some other cases where there is not a limit reached.

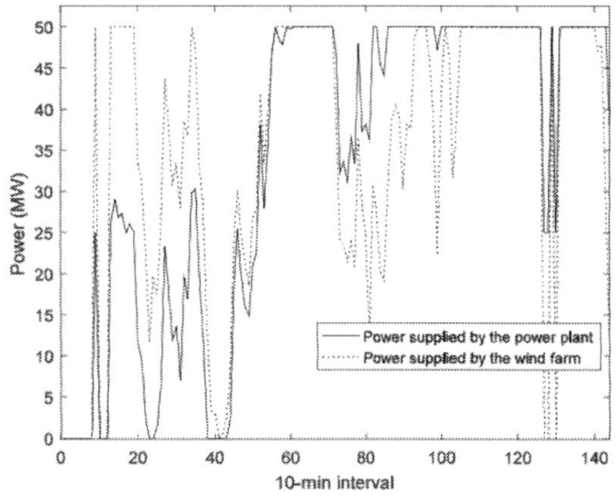

Fig. 4. EA applied to case 2.

In the second case, the scenario is a bit different. There is a large amount of power produced by the WF during the key intervals and a small one generated during the rest. Notice that the total production of energy may be different from one case to another. Here, after applying the algorithm, part of the power produced in the less interesting intervals is moved to the other ones, supplementing them when the rated power of the power plant is not reached. In Fig. 4, it can be seen as some kind of filling of a surface. Although at a first sight it may not be evident, some limits are also reached here, such as the charging limit of the BESS, in the interval 8, discharging, in the interval 100, rated power, in the interval 104 and output power higher than zero, in the interval 0.

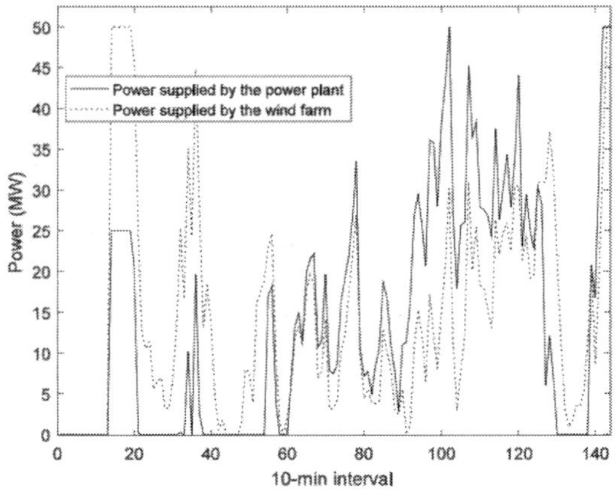

Fig. 5. EA applied to case 3.

In the last case the total energy supplied by the WF is lower than in the second case. In fact, it almost never reaches the rated power. Here, the power produced during the night is used to charge the BESS and it gives it back during the period of high demand. Notice the fact that, in this case, the limits avoid the possibility of no generation during all the night. Besides, the small amount of energy available in the BESS, together with the discharging limits provokes that the rated power is almost never reached. However, the power supplied by the power plant makes tentative attempts to adapt its shape to the daily demand curve.

Moreover, in order to assess the performance of the algorithm, its efficiency is defined by eq. (12).

$$\rho_i = \frac{Energy\ supplied\ by\ the\ power\ plant\ along\ day\ i}{Energy\ supplied\ by\ the\ WF\ along\ day\ i} \quad (12)$$

Efficiencies in the three cases are $\rho_1 = 0.9660$, $\rho_2 = 0.9809$ and $\rho_3 = 0.9377$. Even though the efficiencies when charging and discharging the BESS are lower, not all the power is supplied through the BESS, most of it is provided directly to the electrical network and that is the reason of the higher performance of the power plant. Besides, in Fig. 6, it can be seen the evolution of the fitness function in the three cases.

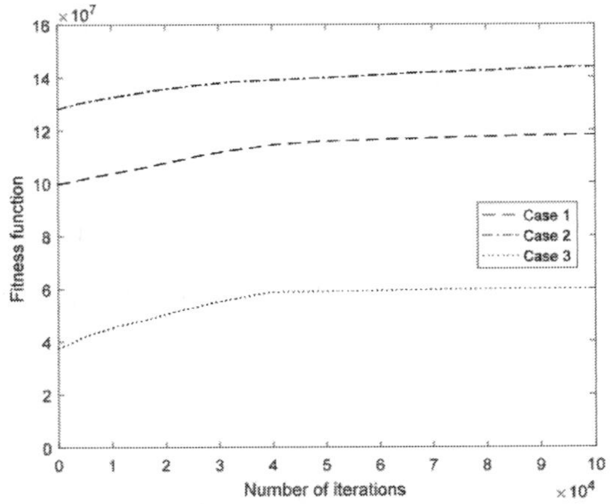

Fig. 6. Evolution of the fitness function in cases 1, 2 and 3 along the algorithm.

In all cases the fitness function behaves in the same manner: first, it increases very fast along the number of iterations 1 to 100 due to evident 'movements of power', then, it enters into a stage of slow increase (from 100 to 100,000 iteration) and, finally, it tends to a steady value (around 100,000 iteration to the end of the algorithm). In Fig. 6 the intermediate stage can be easily seen.

In the three cases, the solution proposed improves the contribution of the power plant to the stability of the network because more power is supplied when the demand is higher. And, obviously, the value of the fitness function is higher than just considering the power provided by the WF.

5 Conclusions

The combined use of WFs and BESS, in order to adjust generation and demand of power, seems to be of much interest, as there is no possibilty of establishing any control over the wind power.

In this paper, an EA is proposed to manage the production of a power plant constituted by a WF equipped with a BESS. The final result is obtained by optimizing a fitness function that prioritizes some periods of generation. The fitness function can be different depending on the main objective, i.e., the stability of the network or the maximum profit. Besides, the following constraints are considered in the algorithm: minimum and maximum levels of energy in the BESS, maximum power transferred when charging or discharging the BESS and minimum and maximum power supplied by the power plant.

If this kind of power plant works in line with the results of the proposed algorithm, the stability of the electrical network will be clearly reinforced and/or the maximum profit will be obtained by the owner. Additionally, in most cases both statements will be achieved simultaneously.

References

1. Banham-Hall, D.D., Taylor, G.A., Smith, C.A., Irving, M.R., Flow BESS for enhancing wind power integration (2012) IEEE Transactions on Power Systems, 27 (3), art. no. 6158621, pp. 1690-1697.
2. Mohod, S.W., Aware, M.V., Micro wind power generator with battery energy storage for critical load, (2012) IEEE Systems Journal, 6 (1), art. no. 6003747, pp. 118-125.
3. Protogeropoulos, C., Brinkworth, B.J., Marshall, R.H., Sizing and techno-economical optimization for hybrid solar photovoltaic/wind power systems with battery storage, (1997) International Journal of Energy Research, 21 (6), pp. 465-479.
4. Yao, D.L., Choi, S.S., Tseng, K.J., Lie, T.T., A statistical approach to the design of a dispatchable wind power-battery energy storage system, (2009) IEEE Transactions on Energy Conversion, 24 (4), art. no. 5336289, pp. 916-925.
5. Prasad, A.R., Natarajan, E., Optimization of integrated photovoltaic-wind power generation systems with battery storage (2006) Energy, 31 (12), pp. 1607-1618.
6. Khalid, M., Savkin, A.V., A model predictive control approach to the problem of wind power smoothing with controlled battery storage (2010) Renewable Energy, 35 (7), pp. 1520-1526.
7. Nguyen, M.Y., Nguyen, D.H., Yoon, Y.T., A new battery energy storage charging/discharging scheme for wind power producers in real-time markets (2012) Energies, 5 (12), pp. 5439-5452.
8. Jiang, Q., Wang, H., Two-time-scale coordination control for a battery energy storage system to mitigate wind power fluctuations (2013) IEEE Transactions on Energy Conversion, 28 (1), art. no. 6365250, pp. 52-61.
9. Alimisis, V., Hatziargyriou, N.D., Evaluation of a hybrid power plant comprising used EV-BESS to complement wind power (2013) IEEE Transactions on Sustainable Energy, 4 (2), art. no. 6331027, pp. 286-293.
10. Li, X., Hui, D., Lai, X., Battery energy storage station (BESS)-based smoothing control of photovoltaic (PV) and wind power generation fluctuations (2013) IEEE Transactions on Sustainable Energy, 4 (2), art. no. 6473871, pp. 464-473.
11. Wee, K.W., Choi, S.S., Vilathgamuwa, D.M., Design of a least-cost battery-supercapacitor energy storage system for realizing dispatchable wind power (2013) IEEE Transactions on Sustainable Energy, 4 (3), art. no. 6487428, pp. 786-796.
12. Kou, P., Liang, D., Gao, F., Gao, L., Coordinated Predictive Control of DFIG-Based Wind-Battery Hybrid Systems: Using Non-Gaussian Wind Power Predictive Distributions (2015) IEEE Transactions on Energy Conversion, 30 (2), art. no. 7031952, pp. 681-695.

13. Sun, Y., Zhong, J., Li, Z., Tian, W., Shahidehpour, M., Stochastic Scheduling of Battery-Based Energy Storage Transportation System with the Penetration of Wind Power (2017) IEEE Transactions on Sustainable Energy, 8 (1), art. no. 7501878, pp. 135-144.
14. Zhou, W., Yang, H., Fang, Z., Battery behavior prediction and battery working states analysis of a hybrid solar-wind power generation system (2008) Renewable Energy, 33 (6), pp. 1413-1423.
15. Swierczynski, M., Stroe, D.I., Stan, A.-I., Teodorescu, R., Sauer, D.U., Selection and performance-degradation modeling of limo2/li 4ti5o12 and lifepo4/c battery cells as suitable energy storage systems for grid integration with wind power plants: An example for the primary frequency regulation service (2014) IEEE Transactions on Sustainable Energy, 5 (1), art. no. 6583344, pp. 90-101.
16. Teleke, S., Baran, M.E., Bhattacharya, S., Huang, A.Q., Optimal control of battery energy storage for wind farm dispatching (2010) IEEE Transactions on Energy Conversion, 25 (3), art. no. 5432995, pp. 787-794.
17. Zhang, Y., Dong, Z.Y., Luo, F., Zheng, Y., Meng, K., Wong, K.P., Optimal allocation of battery energy storage systems in distribution networks with high wind power penetration (2016) IET Renewable Power Generation, 10 (8), pp. 1105-1113.
18. Nguyen, C.-L., Lee, H.-H., A Novel Dual-Battery Energy Storage System for Wind Power Applications (2016) IEEE Transactions on Industrial Electronics, 63 (10), art. no. 7473865, pp. 6136-6147.
19. Gamini Jayasinghe, S.D., Mahinda Vilathgamuwa, D., Madawala, U.K., Direct integration of battery energy storage systems in distributed power generation (2011) IEEE Transactions on Energy Conversion, 26 (2), art. no. 5742991, pp. 677-685.
20. Juste, K.A., An evolutionary programming solution to the unit commitment problem (1999) IEEE Transactions on Power Systems, 14 (4), pp. 1452-1459.
21. P. F. Correia, J. M. Ferreira, and de Jesús, “Simulation of correlated wind speed and power variates in wind parks,”Elect.PowerSyst.Res., vol. 80, no. 5, pp. 592–598, May 2010.
22. Villanueva, D., Feijóo, A., Pazos, J.L. Simulation of correlated wind speed data for economic dispatch evaluation (2012) IEEE Transactions on Sustainable Energy, 3 (1), art. no. 6102285, pp. 142-149.
23. Villanueva, D., Pazos, J.L., Feijóo, A., Probabilistic load flow including wind power generation (2011) IEEE Transactions on Power Systems, 26 (3), art. no. 5682076, pp. 1659-1667.
24. Lin, C.E., Hong, Y.Y., Chuko, C.C., Real-time economic dispatch - A fast base-case approach (1987) Electric Power Systems Research, 12 (3), pp. 167-173.

A holistic approach for the optimization of offshore wind farm layouts considering cable layouts

Ilayda Ulku[1*], *Cigdem* Alabas-Uslu[2]

[1]Istanbul Kultur University, Industrial Engineering Department, Istanbul, Turkey
[2]Marmara University, Industrial Engineering Department, Istanbul, Turkey

Abstract. A wind farm, mainly, is composed of a set of turbines, one or more transmitters and a set of electrical cable connections between turbines and transmitters. Determination of turbine locations within the farm to maximize total power generation is called turbine location (TL) problem. Relative turbine positions affect the amount of overall energy because of wake effects. Determination of cable connections among turbines and transmitters to collect produced energy by turbines at transmitters is called cable layout (CL) problem. While TL problem is directly effective on the total energy production in the farm, CL problem indirectly affects the total energy due to the power losses. In the literature, TL and CL problems are solved sequentially where the layout found by solving of TL is used as an input of CL problem. To minimize wake effects in TL problem, distances between turbine pairs should be increased, however, as the distances are increased the cable cost increases in CL problem. A new mathematical model is developed to deal with simultaneously solving of TL and CL problems. A set of test instances are used to show the performance of the proposed model. The experiments show the practical use of the proposed holistic model.

1 Introduction

Instead of importing energy at high cost, investment cost of renewable energy which consists of research and development, plant installation, technology, materials and manpower can be afforded. Wind energy becomes a key source of renewable energy because of its high potential to provide the required energy capacity.

Selection and determination of the wind farm site, determination of turbine locations (TL) on the wind farm, designing electrical infrastructure and cable layout (CL) are the main design problems in the installation of an offshore wind farm [2, 3, 5, 9, 10, 11]. TL and CL problems, in the literature, are solved in sequentially due to the difficulties in solving both the problems. Wake effects are minimized by increasing the distances between turbine pairs to generate maximum total energy in TL problem. On the other side, as the distances between turbines increase, the total cable cost also increase. Therefore, objectives of TL and CL problems are conflicting and solving of these problems concurrently may results in more cost-effective wind farms. In a wind farm, the region of reduced and changed wind speed behind the wind turbines is called wake region. If a turbine is in the wake region of one or more turbines, energy production of the effected turbine reduces. One of the most popular wake models used in the related literature is developed by Jensen [1] in 1983. This wake model is based on the distance behind the rotor and assumes that the wake linearly expands.

*Corresponding author: i.karabulut@iku.edu.tr

2 Literature Review

TL problem was defined firstly by Mosetti et al. [2] and Jensen's wake decay model was used considering various wind speeds and directions. They developed a genetic algorithm by using a 10x10 grid of possible turbine locations to maximize the ratio of power to installation cost of turbines. Grady et al. [3] obtained better results than Mosetti et al. [2] by using the same example cases and same wake effect. Turner et al. [4] proposed two mathematical models; a quadratic-integer model and a linear-integer programming model for TL problem. They used an interaction matrix prepared in advance to involve wake effects in to the proposed mathematical models. Ulku and Alabas-Uslu [5] developed a new nonlinear mathematical model to solve wind farm layout problem not only for minimization of cost per power under the multiple wake effects but also for maximization of total power production. Their mathematical model has totally unimodularity property which eliminates the binary variables of possible turbine locations. Although their model has nonlinear objective function, the problem becomes to be simpler compared to have both integer variables and nonlinear functions.

CL problem, designing of electrical cable layout between wind turbines and transformer, becomes another major challenge, after designing optimal turbine layouts. Especially for offshore wind farms, 20% of total installation cost is resulted from cable and cable installation cost. Total cable cost is proportional with the length of total cables installed among turbines and transmitter [6-10]. The cable cost can be decreased in overall up to 10% by optimizing the cable layout problem [11]. As the power flow is unsplittable [8], the power (energy) generated by a turbine is transferred to another turbine or transmitter through only one capacitated cable. Therefore, a solution to CL problem must be at least a minimum spanning tree with capacitated cables. Since the minimum spanning tree problem with capacitated arcs is an NP-hard problem, CL problem is also an NP-hard problem [12]. That's why, numerous studies to solve CL problem have used heuristic methods [13-16]. On the other hand, several studies are based on exact optimization models such as mixed integer linear programming (MILP) models [6, 8, 10, 17-20]. Hertz et al. [8] included parallel cables in their study that can be installed in the same dug hole for onshore wind farms. Fischetti and Pisinger [6] focused on the power losses and non-crossing cables of multiple types and presented both exact solutions using a MIP and heuristic solutions using a hybridization of MIP and heuristic rules. As both TL and CL problems are hard to solve optimally, there is no any study that considers turbine location decisions, cable layout decisions and energy flow decisions through the layout simultaneously [21-23]. All of the existing studies in the literature solve these problems consecutively. Pillai [21] proposed a cable layout optimization problem which is a sub-problem of the turbine layout problem. He included cable connections, the placement of the transformer location to show the effect on the annual energy production and the layout cost. Wedzik and Szypowski [22] developed two-step optimization algorithm to solve first cable layout to minimize the total length of the network, then cable crossing is eliminated through the connected turbines where turbine positions are obtained at the earlier stage. Also, Fischetti [23] focused on the optimal design of wind turbine layout and cable layout. The result of optimal turbine positions was used as an input data for solving of cable layout to minimize total cable cost.

In this study, we present a MINLP model to solve TL problem under multiple wake effect condition and CL problem under multiple cable type condition, simultaneously to provide better solutions to these two conflicting problems. Main contributions of this study are two-fold. First of all, offshore wind farms are rather expensive to invest and any improvement in the investment cost will be quite valuable. Since this study aims to develop optimization-based solution approach, there will be cost savings for a planned wind farm comparing to existing methodologies. Last but not least, as our knowledge, there is no any study in the related literature which considers the two main decision problems (TL and CL) synchronically. Therefore, our study will also contribute to the literature from perspective of operational research.

The rest of this paper is organized into the following. The mathematical model approach is introduced in section 2 at first while section 3 specifies computational results of the proposed approach to show performance of the presented model on different problem instances. Finally, section 4 provides the concluding remarks.

3 Proposed Mathematical Programming Framework for the Holistic Approach

The developed model in this study to create a holistic approach for the both TL and CL problems is given through Eq. (1) – Eq. (34). A graph $G = (V, A)$ is used to define the set of turbines under V and a single transmitter in the wind farm while A is the set of possible connections between each pair of nodes in $v \in V$. The wake enlargement, α is calculated in Eq. (1) where the coefficients z symbolizes the hub height and z_0 represents the surface roughness [1].

$$\alpha = \frac{0.5}{ln\frac{z}{z_0}} \tag{1}$$

The wind speed, u_{ij} is determined with Eq. (2) according to Jensen's wake model as u_0 is the ambient wind speed. The velocity deficit, Vd_{ijkl}, in the wind speed due to the turbine located at (k, l) is given in Eq. (3) by using axial induction factor a in Eq. (4) which is dependence of thrust coefficient, C_T and rotor radius r [24]. A_{ijkl} is the overlapping area between the swept area of turbine (i, j) and wake area of turbine (k, l). If turbine (i, j) is completely in the wake of turbine (k, l), then $A_{ijkl} = A_{ij}$ [5].

$$u_{ij} = u_0(1 - Vd_{ijkl}) \tag{2}$$

$$Vd_{ijkl} = \frac{2a}{\left(1 + \frac{\alpha D_{ijkl}}{r}\right)^2} \frac{A_{ijkl}}{A_{ij}} \tag{3}$$

$$a = 0.5\left(1 - \sqrt{1 - C_T}\right) \tag{4}$$

3.1. Multiple Wake Effects

When a turbine is located at position (i, j) and affected by more than one turbine's wake, there occurs a multiple wake effect. The sum of the squares of each single velocity deficits generated by each wake is equal to $Vdef_{ij}$ as given in Eq. (5) [1]. Thus, the residual velocity u_{ij} is calculated in Eq. (6).

$$Vdef_{ij} = \sqrt{\sum_{i=1}^{N} \sum_{j=1}^{N} \sum_{k=1}^{N} \sum_{l=1}^{N} {Vd_{ijkl}}^2} \tag{5}$$

$$u_{ij} = u_0\left[1 - Vdef_{ij}\right] \tag{6}$$

Eq. (7) is used to calculate the power obtained by each individual wind turbine (i, j) where A is cross-sectional area, ρ is air density, and η is turbine efficiency. When velocity deficits are included into the power calculation, Eq. (7) is transformed to Eq. (8).

$$P_{ij} = \eta \frac{1}{2} \rho A\, u_{ij}^3 \tag{7}$$

$$P_{ij} = \eta \frac{1}{2} \rho A \left[u_0 \left[1 - \sqrt{\sum_{i=1}^{N} \sum_{j=1}^{N} \sum_{k=1}^{N} \sum_{l=1}^{N} Vd_{ijkl}^{2}} \right] \right]^3 \quad (8)$$

$$ATC = (TCost)TN \left(\frac{2}{3} + \frac{1}{3} e^{-0.00174(TN)^2} \right) \quad (9)$$

The objective function as defined in Eq. (11) includes financial issues that is considered as cost of investment per power unit. The annual cost of investment includes the annual total cost of cables in Eq. (10) and annual total cost of turbines in Eq. (9) that is widely used in the literature and adapted from Mosetti et al [2]. A grid size of size NxN' is used to represent the wind farm area where each cell (I, j) represents a possible turbine location. An interaction matrix of size $J = (N^2)x\left(N'^2\right)$ is generated a priori by using equations (1-5) where all possible multiple wake effects are included. There is $J - 1$ interaction between each possible turbine location (i, j) and other possible turbine locations. A cable of type $r, r \in R$ has cable capacity Cap_r, and there exist a $Cost_r$ according to the cable capacity between connected turbines at location (i, j) and (k, l). For the given NxN' grid size the location of the transmitter is assumed to be known in advance. Since the generated power by a single turbine can leave the turbine with a single cable, the generated power can be transferred as one power unit from that turbine. In addition, cable connections to the transmitter is restricted with a given maximum cable number C due to the technical specifications of the transmitter. Parameters of the problem which are known in advance are listed below:

Cap_r: Cable capacity of type r in terms of number of turbines

$Cost_r$: Cost of cable of type r per distance unit

C: Maximum number of cables which can be connected to the transmitter

MTN: The number of turbines

tngh : Maximum allowed distance for each possible turbine position

$TCost$: Turbine cost

$x_{SS'}$: Transmitter in location(S, S')

D_{ijkl}: Euclidian distance between turbines in locations (i, j) and (k, l) for $i, j, k, l \in$ I

As the aim of the holistic approach is to find cost-effective wind farm layouts in terms of both turbine locations and cable connections, the following decision variables of the problem as used.

TN: Total number of turbines

$$x_{ij} = \begin{cases} 1, \text{there exists a turbine in location } (i,j) \\ 0, \text{otherwise} \end{cases}$$

$$y^r{}_{ijkl} = \begin{cases} 1, \text{turbines } (i,j) \text{ and } (k,l) \text{ are connected with a cable} \\ 0, \text{otherwise} \end{cases}$$

$flow_{ijkl}$: power flow from turbine (i, j) to turbine (k, l)

In Eq. (12), spanning tree constraint is introduced to obtain the connection between every pair of nodes turbines and transmitter in the wind farm. The incoming flow to the transmitter is represented with Eq. (13) to guarantee the total number of turbines in the wind farm. The number of turbines used in the wind farm must be at least MTN as shown in Eq. (14). It is provided with Eq. (15) that total number of cables exits from a turbine is either 1, if a turbine placed in location (i, j), or zero a turbine is not placed in (i, j). Cable connections are avoided if there is another turbine between two turbines in the same row or the same column of the grid design as provided by Eqs. (16-17). Eqs. (18-19) prevent the presence of overlapping cables in each row and column. The cable connection between a pair of turbines is obtained with the maximum allowed distance for each possible turbine position as shown in Eq. (20). Eqs. (21-22) gives the calculation of outflow from each turbine and similarly inflow to each turbine. Flow conservation is ensured by Eq. (23). The flow amount through a cable is restricted by the capacity of installed cable type as stipulated by Eq. (24). Eqs. (25-26) assure to install at most one cable between a turbine pair. The unidirectionality of the cables is guaranteed with Eqs. (27-28). Neither the cable connections nor flow between (i, j) and (i, j) are available as shown in Eqs. (29-30). Also, there is no outflow from the transmitter and it is stipulated in Eq. [31]. With Eq. (32), maximum number of cables which can be connected to the transmitter is represented where the binary restrictions to the variables are represented in Eqs. (33-34).

$$ACC = \sum_{(i,j),(k,l)\in A} \sum_{rt\in R} cost_r y^t{}_{ijkl} D_{ijkl} \frac{(t)(1+t)^T}{T} \quad (10)$$

$$min = \frac{(ACC + ATC)}{P_{ij}} \quad (11)$$

$$\sum_{(i,j),(k,l)\in A} y^r{}_{ijkl} \geq TN \quad \forall r \in R \quad (12)$$

$$\sum_{k=1}^{N} \sum_{l=1}^{N'} flow_{k,l,S,S'} = TN \quad (13)$$

$$\sum_{(i,j),(k,l)\in A} x_{ij} \geq MTN \quad (14)$$

$$\textstyle\sum_{k=1}^{N} \sum_{l=1}^{N'} \sum_{r=1}^{R} y^r{}_{ijkl} = x_{ij} \quad \forall (i,j) \in V, i \neq S,\ J \neq S' \quad (15)$$

$$y^r{}_{ijil} = 1 - x_{is} \quad \forall (i,j), (i,l) \in V, r \in R, s > j, s < l \quad (16)$$

$$y^r{}_{ijkl} \leq 1 - x_{sj} \quad \forall (i,j),(k,l) \in V, r \in R, s > i, s < k \quad (17)$$

$$\textstyle\sum_{l=1,l\neq j}^{N'} \sum_{r=1}^{R} y^r{}_{ijil} \leq x_{ij} \quad \forall (i,j) \in V \quad (18)$$

$$\textstyle\sum_{k\neq i}^{N'} \sum_{r=1}^{R} y^r{}_{ijkj} \leq x_{ij} \quad \forall (i,j) \in V \quad (19)$$

$$BigM\left(1 - y^{r}_{ijkj}\right) \ge D_{ijkl} - tngh \quad \forall (i,j),(k,l) \in V, r \in R, (i,j) \ne (S,S'), (k,l) \ne (S,S') \tag{20}$$

$$\sum_{k=1}^{N}\sum_{l=1}^{N'} flow_{ijkl} = outflow_{ij} \quad \forall i,j,(i,j) \ne (k,l) \tag{21}$$

$$\sum_{k=1}^{N}\sum_{l=1}^{N'} flow_{klij} = inflow_{ij} \quad \forall (k,l),(i,j) \ne (k,l) \tag{22}$$

$$outflow_{ij} - inflow_{ij} = x_{ij} \quad \forall (i,j),(i,j) \ne (S,S') \tag{23}$$

$$flow_{ijkl} \le \textstyle\sum_{r=1}^{R} Cap_r y^{r}_{ijkl} \quad \forall (i,j),(k,l) \in V \tag{24}$$

$$\textstyle\sum_{r=1}^{R} y^{r}_{ijkl} \le x_{ij} \quad \forall (i,j),(k,l) \in V \tag{25}$$

$$\textstyle\sum_{r=1}^{R} y^{r}_{ijkl} \le x_{ji} \quad \forall (i,j),(k,l) \in V \tag{26}$$

$$y^{r}_{ijkl} + y^{r}_{klij} \le x_{ij} \quad \forall (i,j),(k,l) \in V, r \in R \tag{27}$$

$$y^{r}_{ijkl} + y^{r}_{klij} \le x_{kl} \quad \forall (i,j),(k,l) \in V, r \in R \tag{28}$$

$$y^{r}_{ijij} = 0 \quad \forall (i,j),(k,l) \in V, r \in R \tag{29}$$

$$flow_{ijij} = 0 \quad \forall (i,j) \in V \tag{30}$$

$$flow_{SS'ij} = 0 \quad \forall (i,j) \in V \tag{31}$$

$$\sum_{i=1}^{N}\sum_{j=1}^{N'}\left(\sum_{r=1}^{R} y^{r}_{ijSS'}\right) \le C \tag{32}$$

$$y^{r}_{ijkl} \in \{0,1\} \quad \forall (i,j),(k,l) \in V, r \in R \tag{33}$$

$$x_{ij} \in (0,1) \quad \forall (i,j) \in V \tag{34}$$

4 Computational Results of Proposed Approach

To represent the performance of the proposed nonlinear model a set of test problems are generated for a single wind direction from north to south. Fischetti and Pisinger [6] used real wind farms with different cable costs and cable capacities where the cable specifications used in this study are obtained from their study. The proposed MINLP model is tested for three different real wind farms under one wind direction. The wind regime, one wind direction from north to south with constant wind speed of 12 m/s, is widely used in the wind farm layout literature [2-5, 25]. The cable specifications are given in Table 1. In Table 1, the second column indicates the maximum cable bound to the transmitter. In the third column the number of cable types is shown and it is directly related with the fourth column. Turbine number carried with the cable is represented in fourth column while related cable cost is represented in the last column.

Table 1. Cable Specifications.

Instance #	C	r	Cap_r	$Cost_r$
1	10	3	7	370
			11	393
			13	435
2	4	2	5	407
			10	607
3	4	2	4	382
			9	630

LINGO software package was used as a solver to solve MINLP model. In table 2, performance of the proposed MINLP model is represented in solving of combined TL and CL problems under study.

Table 2. Performance of Holistic Approach.

Instance Number	TN	Power	Annual Cost of Cables	Annual Cost of Turbines	Annual Cost of Investment	Total Cost of Investment /Power
Instance 1	30	15785.29	268959.69	27831128.90	28100088.58	1780.15
Instance 2	30	15785.29	238783.88	27831128.90	28069912.77	1778.23
Instance 3	31	15870.05	432521.90	28484972.89	28917494.79	1822.14

As the cable type is changed the turbine number for the second instance in second column is not changed. The turbine layout as shown in Fig. 1 that is the same for both instance 1 and instance 2. Thus, annual cost of turbines is also the same for both instance 1 and instance 2. As allowed cable types differ, the optimal cable connections also differ as shown in fourth column of Table 2. The last column represents the total cost of investment per generated power. Instance 1 has least cost per power ratio among the instances represented. The cable capacity and cost according to the capacity of the first instance is preferable in comparison of the presented instances. In Fig. 1, the black line represents cable of type 1, blue line for type 2, and red line for type 3. Although instance 1 and 2 has similar layout, the cable layout differs from each other due to the cable types.

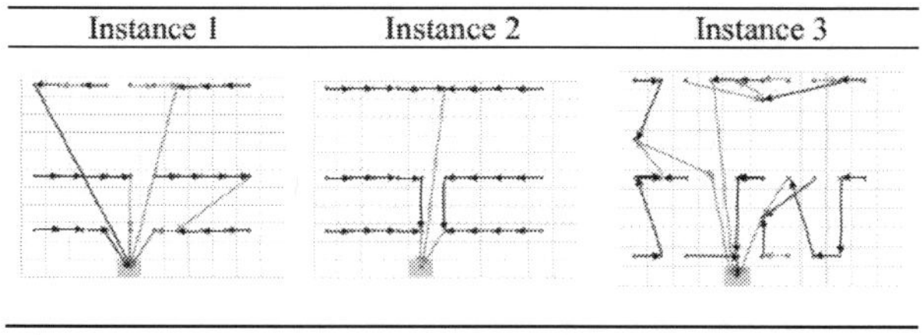

Fig. 1. Cable Connections between Located Wind Turbines.

5 Conclusion

In this study, a holistic approach to solve TL and CL problems is addressed. The combined problem is modelled as a MINLP considering differentiating constraints including economic factors to get close results to the design problems of a real wind farm. Jensen's wake decay model is used for the proposed model to calculate the generated power of each wind turbine. An interaction matrix is used a priori as presented to involve multiple wake effect in the MINLP model. Multiple cable types which differ in capacities and costs are included in the model to provide sending of generated power by the located turbines to a single transmitter assuming that the location of the transmitter is known in advance. Three different case studies were solved and analysed using the MINLP. Although, several constraints to eliminate the crossing cables are imposed in the model to avoid cable damages and costly cable installation, it is observed that the generated results may contain some crossing cables. Therefore, as a future study, we plan to enhance the MINLP to prevent the crossing cables completely.

Acknowledgment

Participation in this conference was supported under grant [ULEP-2019-U2407/52/03] by Istanbul Kultur University.

References

1. Jensen, N.O., *A note on wind generator interaction*. (1983).
2. Mosetti, G., C. Poloni, and B. Daviacco, *Optimization of wind turbine positioning in large windfarms by means of a genetic algorithm*. J Wind Eng Ind Aerod., 1994. **51**: p. 105-116.
3. Grady, S., M. Hussaini, and M.M. Abdullah, *Placement of wind turbines using genetic algorithms*. Renewable energy, 2005. **30**(2): p. 259-270.
4. Turner, S., et al., *A new mathematical programming approach to optimize wind farm layouts*. Renewable Energy, 2014. **63**: p. 674-680.
5. Ulku, I. and C. Alabas-Uslu, *A new mathematical programming approach to wind farm layout problem under multiple wake effects*. Renewable Energy, 2019. **136C** p. 1190-1201.
6. Fischetti, M. and D. Pisinger, *Optimizing wind farm cable routing considering power losses*. European J of Operational Research, 2018. **270**(3): p. 917-930.
7. Qi, W., Y. Liang, and Z.-J.M. Shen, *Joint planning of energy storage and transmission for wind energy generation*. Operations Research, 2015. **63**(6): p. 1280-1293.
8. Hertz, A., et al., *Optimizing the design of a wind farm collection network*. INFOR: Information Systems and Operational Research, 2012. **50**(2): p. 95-104.
9. Fischetti, M. and D. Pisinger, *Optimal wind farm cable routing: Modeling branches and offshore transformer modules*. Networks, 2018. **72**(1): p. 42-59.
10. Cerveira, A., et al., *Optimal cable design of wind farms: The infrastructure and losses cost minimization case*. IEEE Transactions on Power Systems, 2016. **31** (6): p. 4319–4329.
11. Blanco, M.I., *The economics of wind energy*. Renewable and sustainable energy reviews, 2009. **13**(6-7): p. 1372-1382.
12. Jothi, R. and B. Raghavachari, *Approximation Algorithms for the Capacitated Minimum Spanning Tree Problem and Its Variants in Network Design*. ACM Trans. Algorithms, 2005. **1**: p. 265–282.
13. Gonzalez-Longatt, F., et al., *Optimal electric network design for a large offshore wind farm based on a modified genetic algorithm approach*. IEEE Systems Journal 2012: p. 164-172.
14. Pillai, A., et al., *Offshore wind farm electrical cable layout optimization. Engineering Optimization*. 2015. **47**(12): p. 1689-1708.
15. Li, D.D., C. He, and Y. Fu. *Optimization of internal electric connection system of large offshore wind farm with hybrid genetic and immune algorithm. 3rd International Conference on Electric Utility Deregulation and Restructuring and Power Technologies*. 2008. IEEE.
16. Zhao, M., Z. Chen, and F. Blaabjerg, *Optimisation of electrical system for offshore wind farms via genetic algorithm*. IET Renewable Power Generation, 2009. **3** (2): p. 205-216.
17. Bauer, J. and J. Lysgaard, *The offshore wind farm array cable layout problem: a planar open vehicle routing problem*. J of the Operational Research Society, 2015. **66**(3): p. 360-368.
18. Berzan, C., et al., *Algorithms for cable network* design *on large-scale wind farms*. 2011: Tufts University.
19. Lumbreras, S. and A. Ramos, *Optimal design of the electrical layout of an offshore wind farm applying decomposition strategies*. IEEE Trans Power Syst, 2013. **28** (2): p. 1434–1441.
20. Banzo, M. and A. Ramos, *Optimization of AC Electric Power Systems of Offshore Wind Farms*, in *Handbook of Wind Power Systems*. 2013, Springer. p. 747-771.
21. Pillai, A.C., *On the optimization of offshore wind farm layouts*. 2017.
22. Wedzik, A., T. Siewierski, and M. Szypowski, *A new method for simultaneous optimizing of wind farm's network layout and cable cross-sections by MILP optimization*. Applied Energy, 2016. **182**: p. 525-538.
23. Fischetti, M., *Mixed integer programming models and algorithms for wind farm layout*. 2014.
24. Katic, I., J. Højstrup, and N.O. Jensen. *A simple model for cluster efficiency*. in *European wind energy association conference and exhibition*. 1987.
25. Marmidis, G., S. Lazarou, and E. Pyrgioti, *Optimal placement of wind turbines in a wind park using Monte Carlo simulation*. Renewable energy, 2008. **33**(7): p. 1455-1460.

 https://doi.org/10.1051/e3sconf/201912205001

Development of a statistical forecasting model for $PM_{2.5}$ in Macau based on clustering of backward trajectories

Tong Xie[1,*], *Kai Meng* Mok[2], *Ka Veng* Yuen[3], and *Ka In* Hoi[4]

[1]Postgraduate student, Department of Civil and Environmental Engineering, University of Macau, Macau SAR, China
[2]Professor, Department of Civil and Environmental Engineering, University of Macau, Macau SAR, China
[3]Distinguished Professor, Department of Civil and Environmental Engineering, University of Macau, Macau SAR, China
[4]Postdoctoral fellow, Department of Civil and Environmental Engineering, University of Macau, Macau SAR, China

Abstract. A daily $PM_{2.5}$ forecasting model based on multiple linear regression (MLR) and backward trajectory clustering of HYSPLIT was designed for its application to small cities where $PM_{2.5}$ level is easily affected by regional transport. The objective of this study is to investigate the regions that affect the fine particulate concentration of Macau and to develop an effective forecasting system to enhance the capture of $PM_{2.5}$ episodes. By clustering the HYSPLIT 24-hr backward trajectories originated at Macau from 2015 to 2017, five potential transportation paths of $PM_{2.5}$ were found. A cluster based statistical model was developed and trained with air quality and meteorological data of 2015 and 2016. Then, the trained model was evaluated with data of 2017. Comparing to an ordinary model without backward trajectory clustering, the cluster based $PM_{2.5}$ forecasting model yielded similar general forecast performance in 2017. However, the critical success index of the cluster based model was 11% higher than that of the ordinary model. This means the cluster based model has better model performance in $PM_{2.5}$ concentration prediction and it is more important for the health of the public.

1 Introduction

Due to serious adverse effects to respiratory and cardiovascular systems, $PM_{2.5}$ has become a major concern of the public and the Chinese government. In the past decade, China has been experiencing a severe $PM_{2.5}$ pollution due to rapid urban development. Many cities do not comply with the annual $PM_{2.5}$ standard. These cities of nonattainment are mostly distributed in three major city clusters including Beijing – Tianjin – Hebei (BTH), Yangtze River Delta (YRD) and Pearl River Delta (PRD) [1]. Macau a gaming and tourism city located in the southern part of PRD, is having similar situation. In order to protect its citizens, it is important to develop a daily forecasting model of $PM_{2.5}$ for Macau.

Till the third quarter of 2018, this city accommodates a population of 663,400 in an area of 30.8 km^2. Due to the small geographical area, statistical model can be a more attractive alternative to large scale deterministic models when developing operational air quality model [2]. The air quality of Macau is not only governed by the local emissions and the dispersion conditions. Its $PM_{2.5}$ level is also susceptible to the transboundary pollution of neighbouring regions. Therefore, reflecting the regional influence of fine particulates within the forecasting model is necessary. Understanding about the distribution of Macau's upwind cities, where the fine particulates are generated and transported by the atmospheric flow is important.

In view of this, the objective of this study is to develop a cluster based $PM_{2.5}$ statistical forecasting model for performing 1 day ahead forecast of daily averaged $PM_{2.5}$ concentration in Macau. Comparing to deterministic models, the large quantity of historical measured data under a variety of conditions in statistical approaches often have higher accuracy [3]. To achieve this, the historical backward trajectories originated at Macau between 2015 and 2017 are analysed and clustered into several groups by K-means clustering in HYSPLIT. Then, a specific statistical model is developed based on current day local meteorological factors, local time-lagged air pollutant concentrations, hourly $PM_{2.5}$ concentration at mid-night from several upwind cities identified in the classified trajectory clusters. The cluster based $PM_{2.5}$ forecasting model is then trained by using the data of 2015 and 2016, and the dataset of 2017 is used for evaluation of the model. Details of the methodology are described in the following section.

2 Methodology

2.1 Air quality and meteorological data

Local air quality data and meteorological data utilized for model development were provided by the Macau Meteorological and Geophysical Bureau. The air quality

* Corresponding author: mb65481@um.edu.mo

dataset consists of hourly concentrations for $PM_{2.5}$, PM_{10}, SO_2 and NO_2 measured at 5 monitoring stations (http://www.smg.gov.mo/smg/airQuality/c_air_stations.htm) from 2015 to 2017. The meteorological datasets consist of measured data (2015-2016) and forecasted data (2017) at the headquarter of the Macau Meteorological and Geophysical Bureau (22°09′36"N 113°33′54"E). The meteorological dataset contains 5 parameters including the wind speed, wind direction, mean sea level pressure, temperature and relative humidity.

Due to the small area of Macau, its $PM_{2.5}$ level is not only influenced by the local pollution sources. It can be easily affected by the regional transport from its neighbouring upwind cities. Therefore, it is reasonable to adopt the $PM_{2.5}$ concentrations of upwind cities as the model inputs. The data of upwind $PM_{2.5}$ concentrations were obtained from the real-time air quality release platform of the China National Environmental Monitoring Centre (CNEMC). The website reports hourly concentrations of $PM_{2.5}$ measured at 367 cities, with a time lag of 1 hour. However, one difficulty that immediately arises is the choice of the upwind cities, which is addressed in the following subsection.

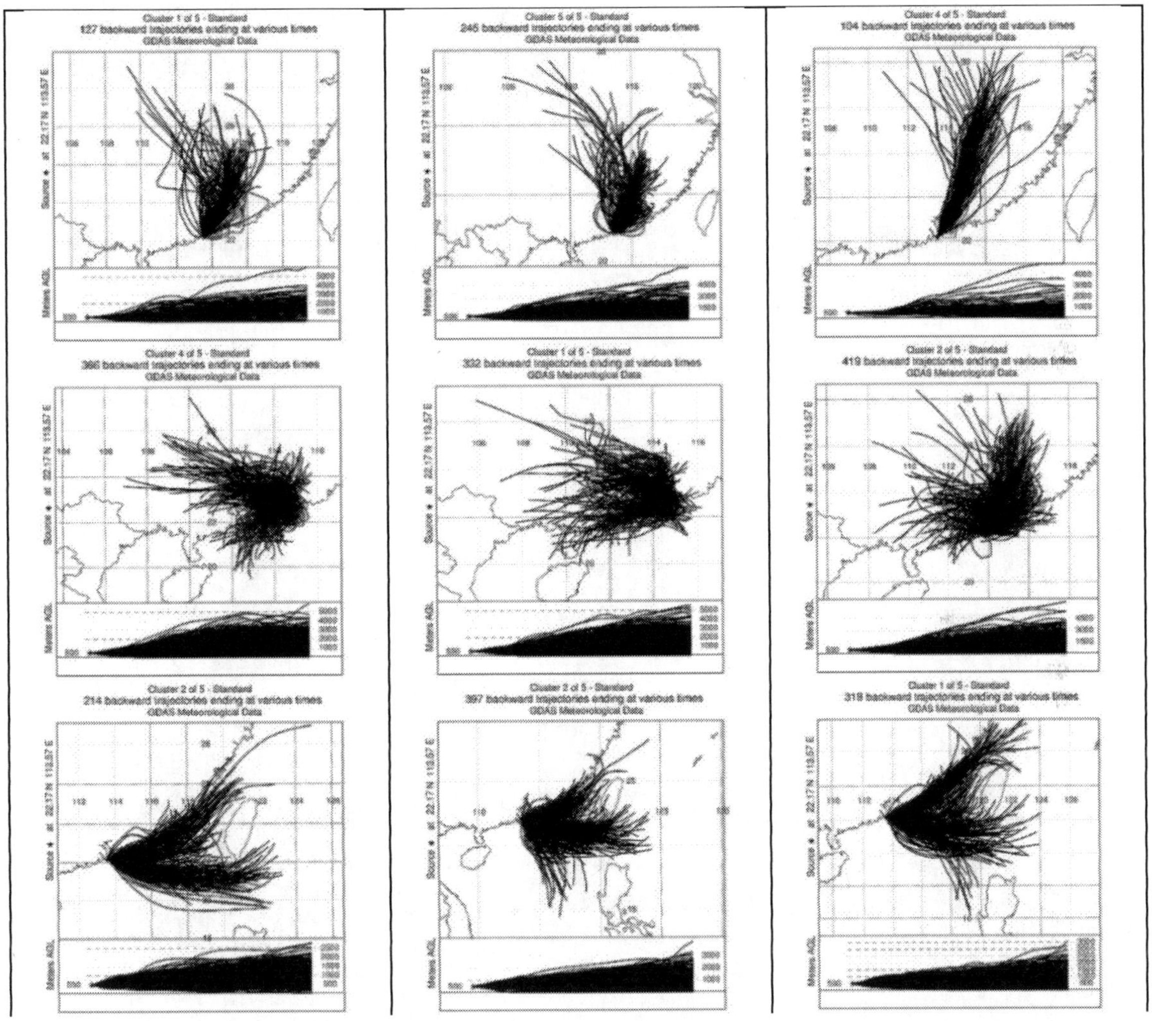

 https://doi.org/10.1051/e3sconf/201912205001

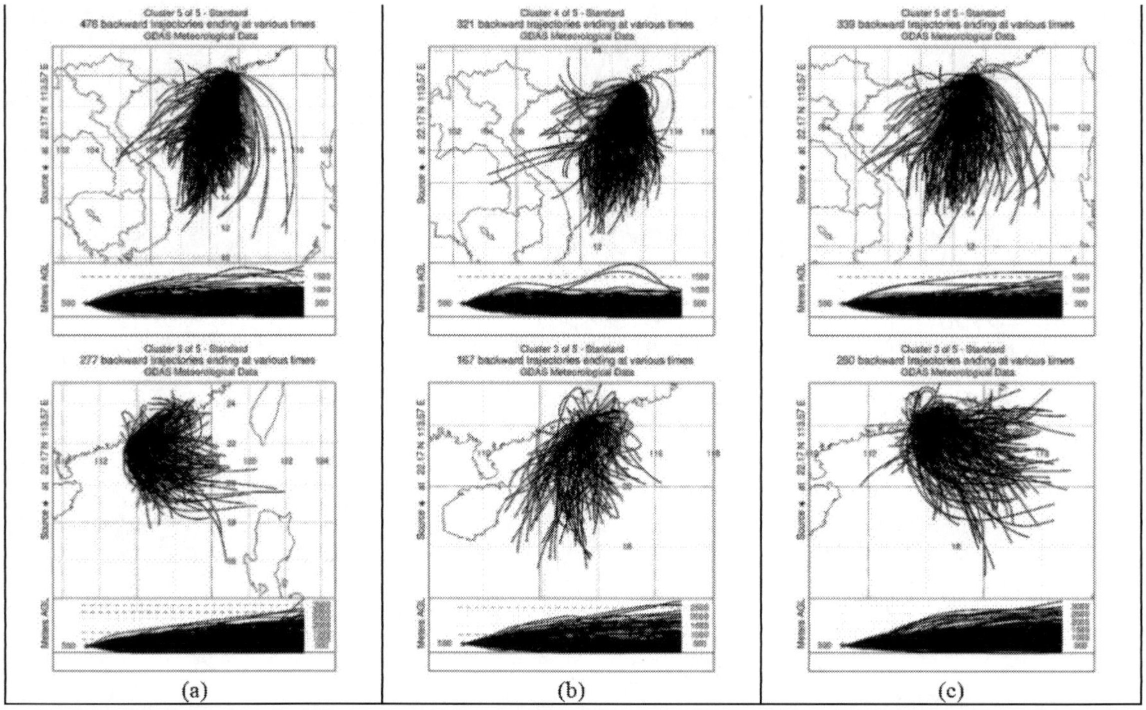

Fig. 1. The HYSPLIT trajectory model computed 24-h backward trajectories originated at 500 meters over Macau at midnight for the period of 2015 (column a), 2016 (column b) and 2017 (column c). (source: ready.arl.noaa.gov/HYSPLIT.php).

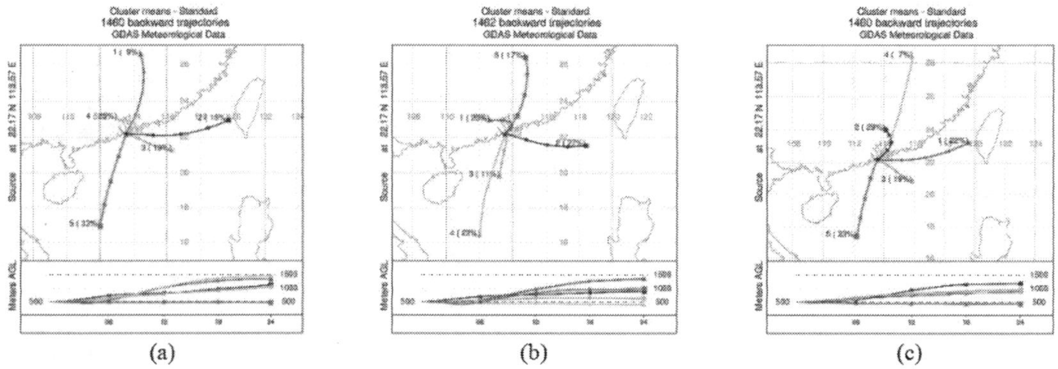

Fig. 2. The mean trajectories of the trajectory clusters computed by the HYSPLIT model for 2015 (a), 2016 (b) and 2017 (c). The squares showed locations of air parcels at 6-h intervals. (source: ready.arl.noaa.gov/HYSPLIT.php).

2.2 Clustering analysis of 24-h backward trajectories

Previous works showed that upwind pollutant concentration of the previous day could influence the performance of air quality forecasting models [4-6], meaning that the choices of upwind cities are important for model development. In this study, the choices for Macau were investigated systematically by using the 24-h backward trajectories with the same starting height of 500 meter for the period between 2015 and 2017. These trajectories were calculated by using the HYSPLIT trajectory model of NOAA. Each trajectory was set to start from Macau at midnight, and the 24-h trajectory end point represented the estimated location of upwind polluted air at midnight on the previous day. All the 24-h backward trajectories for each year between 2015 and 2017 were then classified by K-means into 5 clusters as shown in Fig. 1.

Each column in Fig. 1 shows the clusters of trajectories for a particular year. The general pattern was revealed more clearly by using the mean trajectory of each cluster as in Fig. 2. It shows that the clusters of different years were similar to each other, meaning that this pattern is repeatable over time. The pattern indicates that the regional influence of $PM_{2.5}$ in Macau could be due to (i) the transport of inland fine particulates from the northern part of PRD (north and north-west cluster occurring mostly during winter), (ii) the transport of fine particulates from coastal cities between the Bashi Channel and the Taiwan Strait (northeast cluster

https://doi.org/10.1051/e3sconf/201912205001

occurring mostly during winter and spring), and (iii) the transport of marine aerosols from the South China Sea (southern clusters occurring mostly during summer).

2.3 Cluster based forecasting model of $PM_{2.5}$

The multi-linear regression (MLR) was shown to be an effective tool for air quality forecasting [7]. Here, the MLR was adopted to develop the cluster based forecasting model of $PM_{2.5}$. The general form of the forecasting model is:

$$z_k = \alpha_0 + \boldsymbol{\alpha}_k \boldsymbol{\tau}_{k-1}^T + \varepsilon_k \tag{1}$$

$$\boldsymbol{\alpha}_k = [\alpha_1, \dots, \alpha_m] \tag{2}$$

$$\boldsymbol{\tau}_{k-1} = [x_{1,k-1}, \dots, x_{m,k-1}] \tag{3}$$

where z_k represents the measured $PM_{2.5}$ concentration of the kth day, which is equal to the spatial average of available daily averaged $PM_{2.5}$ concentrations measured at 5 monitoring stations. It is predicted by the linear combination of the input variables with an intercept term while $\boldsymbol{\alpha}_k$ represents the vector of coefficients corresponding to the linear combination. $\boldsymbol{\tau}_{k-1}$ is the vector of input variables available on the (k-1)th day. ε_k is the modelling error associated with the forecasting model. Table 1 shows the compositions of input variables adopted to develop the forecasting model:

Table 1. Compositions of input vector $\boldsymbol{\tau}_{k-1}$ of cluster based forecasting model compared to an ordinary model without upwind city input

<table>
<tr><th rowspan="2"></th><th rowspan="2">$\boldsymbol{\tau}_{k-1}$</th><th colspan="2">Ordinary model</th><th colspan="2">Cluster based model</th></tr>
<tr><th>k</th><th>k-1</th><th>k</th><th>k-1</th></tr>
<tr><td rowspan="10">Local input</td><td>RH [a]</td><td colspan="4" rowspan="5">√</td></tr>
<tr><td>PSEA [a]</td></tr>
<tr><td>TEMP [a]</td></tr>
<tr><td>WSPD [a]</td></tr>
<tr><td>U [a]</td></tr>
<tr><td>[NO_2] [a]</td><td></td><td rowspan="5">√</td><td></td><td rowspan="11">√</td></tr>
<tr><td>[SO_2] [a]</td><td></td><td></td></tr>
<tr><td>[$PM_{2.5}$] [a]</td><td></td><td></td></tr>
<tr><td>[PM_{10}] [a]</td><td></td><td></td></tr>
<tr><td>Macau [b]</td><td></td><td></td></tr>
<tr><td rowspan="6">Upwind city input</td><td>Hongkong [b]</td><td></td><td></td><td></td></tr>
<tr><td>Guangzhou [b]</td><td></td><td></td><td></td></tr>
<tr><td>Shenzhen [b]</td><td></td><td></td><td></td></tr>
<tr><td>Foshan [b]</td><td></td><td></td><td></td></tr>
<tr><td>Dongguan [b]</td><td></td><td></td><td></td></tr>
<tr><td>Xiamen [b]</td><td></td><td></td><td></td></tr>
</table>

[a] Daily averaged data in Macau.
[b] Measured hourly concentration of $PM_{2.5}$ at the last hour of day k-1.
√ means the variable was adopted in the model

In Table 1, the symbols RH, PSEA, TEMP, WSPD and U represent the daily averages of the relative humidity, mean sea level pressure, temperature, wind speed and the north-south component of the wind direction, respectively. These variables were used to reflect the atmospheric stability, the available wind for dilution and the nature (inland or sea) of the trajectory. Besides the meteorological variables, the local daily averaged pollutant concentrations of the (k-1)th day and the hourly $PM_{2.5}$ concentration at 11:00PM of the (k-1)th day in Macau were also incorporated to reflect the initial air quality condition of the kth day. Finally, the hourly $PM_{2.5}$ concentration at 11:00PM of the (k-1)th day for several upwind cities were selected and used to reflect the regional influence of fine particulates.

3 Results and discussion

The model was trained with air-quality and meteorological data of 2015 and 2016. Even though the model involved meteorological inputs of the kth day that were practically not available on the (k-1)th day, the measured meteorological data of the kth day were still used as the model inputs in order to produce better training results. When the trained models were evaluated against the data of 2017, the meteorological forecasts of the kth day were used instead of the measurements. The model accuracy was evaluated according to the measures shown in Eqns. (4) to (9). In these equations, P, O, and $\bar{O}$ represent the prediction, the observation, and the annual average of observed $PM_{2.5}$ concentrations, respectively. For the overall accuracy, the coefficient of determination (R^2) describes how well the model variance explains the observation variance. The normalized mean absolute error (*NMAE*) describes the overall magnitude of the forecast errors relative to the mean of the observations. The index of agreement (*IA*) is a dimensionless indicator between zero and one. $IA = 1$ means perfect prediction, while $IA = 0$ means no agreement at all [4]. This index can detect additive and proportional differences between the observations and the predictions. As for the prediction performance during high $PM_{2.5}$ concentrations, the episode detection rate (*EDR*), represents the probability of successful hits under the observed exceedances ($PM_{2.5} > 35\ \mu g\ m^{-3}$), and the false alarm rate (*FAR*) describes the proportion of false alarms under the forecasted exceedances. In this study, the hit or alarm threshold was $35 \mu g\ m^{-3}$. This value corresponds to the daily limit of the primary $PM_{2.5}$ standard of US-NAAQS, which is equivalent to the grade I $PM_{2.5}$ standard of China (GB3095-2012). The critical success index (*CSI*) is an integrated measure of *EDR* and *FAR* and it can be treated as a discounted *EDR* and the penalty depends on the *FAR* of the model.

$$R^2 = \frac{\sum_{i=1}^{N}(P_i - \bar{O})^2}{\sum_{i=1}^{N}(O_i - \bar{O})^2} \tag{4}$$

$$NMAE = \frac{\sum_{i=1}^{N}|P_i - O_i|}{\sum_{i=1}^{N} O_i} \tag{5}$$

$$IA = 1 - \frac{\sum_{i=1}^{N}(P_i - O_i)^2}{\sum_{i=1}^{N}(|P_i - \bar{O}| + |O_i - \bar{O}|)^2} \quad (6)$$

$$EDR = \frac{no.\,of\ hits}{no.\,of\ observed\ exceedances} \quad (7)$$

$$FAR = \frac{no.\,of\ false\ alarm}{no.\,of\ forecasted\ exceedances} \quad (8)$$

$$CSI = \frac{EDR}{1 + FAR\left(\frac{no.\,of\ forecasted\ exceedances}{no.\,of\ observed\ exceedances}\right)} \quad (9)$$

Table 2 shows the performance of cluster based $PM_{2.5}$ forecasting model and the ordinary $PM_{2.5}$ forecasting model during training period and evaluation period. The ordinary $PM_{2.5}$ forecasting model is a subset of the cluster based model without including the hourly $PM_{2.5}$ concentrations of the upwind cities. During the training phase, it is noted that the cluster based model outperforms the ordinary model due to the extra model complexity to fit the training data. During the evaluation phase, both the ordinary model and the cluster based model achieve similar general performance (R^2, *NMAE*, and *IA*). However, the critical success index (*CSI*) of the cluster based model is significantly higher than that of the ordinary model, meaning that the inclusion of upwind cities is important.

Table 2. Performance of cluster based PM2.5 forecasting model and ordinary model during training period and evaluation period.

Index	Training period (2015-2016)		Evaluation period (2017)	
	Ordinary model	Cluster based model	Ordinary model	Cluster based model
R^2	0.79	0.82	0.75	0.76
NMAE	22.1%	21.5%	25.0%	25.0%
IA	0.94	0.95	0.93	0.94
EDR	78% (173 days)	80% (177 days)	66% (58 days)	85% (75 days)
FAR	19% (41 days)	20% (44 days)	26% (20 days)	34%(39 days)
CSI	65.5%	66.7%	52.4%	63.4%

Fig. 3 shows the measured daily averaged $PM_{2.5}$ concentrations (green line) versus the corresponding predictions (orange line) by the cluster based model and the predictions by the ordinary model (blue line) in 2017. It is noted that predictions by either the cluster based model or the ordinary model are generally in good agreement with the measurements. However, the cluster based model is more capable to capture the peaks. As the most primary objective of air quality forecast model is to correctly predict the episodes, the development of the cluster based model in this study is successful.

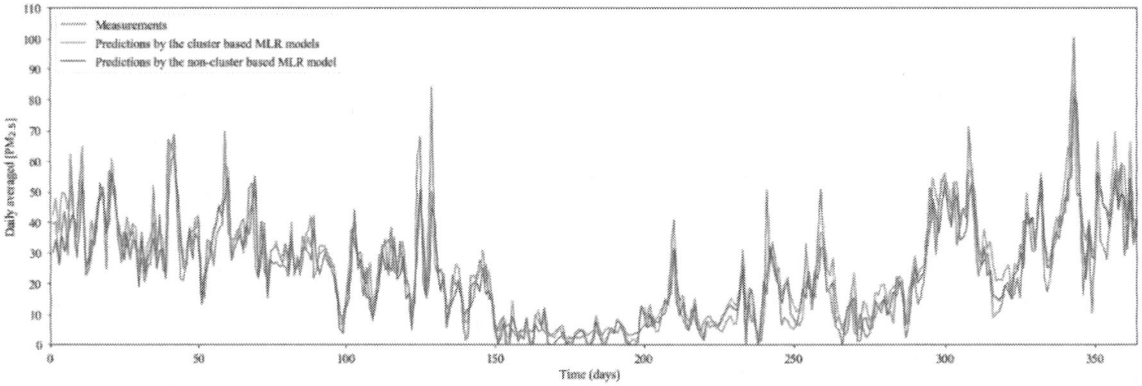

Fig. 3. Plot of measured daily averaged $PM_{2.5}$ concentrations (green line) versus predictions by the cluster based model (orange line) or the predictions by the ordinary model (blue line) in 2017.

4 Conclusions

An empirical $PM_{2.5}$ forecasting model was developed for Macau based on multiple linear regression and backward trajectories clustering of HYSPLIT. The backward trajectories clustering revealed 3 potential regional sources of $PM_{2.5}$ for Macau. A cluster based statistical forecasting model was developed and the regional influence of $PM_{2.5}$ was reflected in the model by using the hourly $PM_{2.5}$ concentrations of the upwind cities before the midnight. Two years of data (2015-2016) were used for model training and one year of data (2017) was used for model evaluation. It was concluded that the cluster based forecasting model overperformed the ordinary $PM_{2.5}$ forecasting model during the episode days, which are of great concern to the general public.

The authors wish to thank the Macau Meteorological and Geophysical Bureau for providing data and Mr. Wenchao Feng for the technical assistance during the process. The authors gratefully acknowledge the NOAA Air Resources Laboratory (ARL) for the provision of the HYSPLIT transport and dispersion model and/or READY website used in this publication (http://www.arl.noaa.gov/ready.php).

References

1. B. Lv, G. Cobourn, Y. Bai, *Atmos. Environ.* **147,** 209-223 (2016)
2. A. Donnelly, B. Misstear, B. Broderick, *Atmos. Environ.* **103,** 53-65 (2015)
3. Y. Zhang, M. Bocquet, V. Mallet, C. Seigneur, A. Baklanov, *Atmos. Environ.* **60,** 632-655 (2012)
4. G.T. Wolff, P.J. Lioy, *J. Air Pollut. Control Assoc.* **28**, 1034-1038 (1978)
5. G. Cobourn, C. Hubbard, *Atmos. Environ.* **33**, 4663-4674 (1999)
6. G. Cobourn, *Atmos. Environ.* **44**, 3015-3023 (2010)
7. A. Vlachogianni, P. Kassomenos, A. Karppinen, S. Karakitsios, J. Kukkonen, *Sci. Total Environ.* **409**, 1559-1571 (2011)

Improving local air quality in cities by reducing nitrogen dioxide pollution from road traffic

Spiru Paraschiv[1,*]

[1]"Dunarea de Jos" University of Galati, Faculty of Engineering, Galati, Romania

Abstract. Trucks and buses play a major role in our lives, transporting goods and thousands of people to cities every day. But these vehicles, although in a much smaller number than the car generates a significant amount of air pollutants. The daily NO_2 concentrations measured by a traffic monitoring station over a period of two years are used to identify the temporal variation of NO_2 pollution as a result of measures to ban the circulation of trucks that do not meet the EURO 6 standard on Stresemannstraße Street in Hamburg. The data shows a decrease in NO_2 concentration due to the measure taken so that in January 2017 the maximum daily NO_2 concentration was 86 μg/m^3 compared to 63 μg/m^3 in 2019. There was also a difference between the daily minimum concentrations during the same period, being approximately 28 μg/m^3 in 2017 and 10 μg/m^3 in 2019. The daily NO_2 observations show a significant decrease in concentration since May 2018 when the non-EURO 6 trucks were banned. The largest decrease in daily concentrations was recorded in March 2019 compared with levels in March 2018, with a lower concentration for 28 days. A different situation was observed in October 2018, when compared to October 2017, showed an increase in concentration for 23 days.

1 Introduction

The urban air in many European cities has become inadequate for breathing due to the high levels of pollution caused mainly by vehicles, especially by diesel cars. Urban citizens are forced to breathe this highly polluted air despite EU legislation that limit ambient air-pollution levels, thus leading to hundreds of thousands of premature deaths every year at European level [1].

Vehicles are the main source of pollution because of their ubiquity and the proximity of gas emitted to people. Some progress has been made to reduce particulate emissions by introducing Euro 6 limits for diesel cars, but NOx remains a key issue, especially from diesel engines. NOx convert in the air to toxic nitrogen dioxide and finally into nitrate aerosol particles and ozone. A stringent problem is the particulate and pollutant emission from older diesel engines and from diesel vehicles with particle filters that have been damaged or removed illegally.

Breathing air with a high concentration of NO_2 it may irritate the airways of the human respiratory system and even for short exposures may worsen respiratory diseases, especially asthma. Longer exposure to high concentrations of NO_2 can contribute to the development of asthma and the most exposed to the effects of this pollutant are children and the elderly.

Local concentrations of NO_2 are largely determined by road traffic emissions from passenger cars, light trucks and heavy duty vehicles [2, 7, 9].

The Euro standards for exhaust emission gases are important tools in Europe to reduce NOx emissions from road traffic in order to reduce NO_2 concentration levels below the limit value and to improve air quality in cities and along highways.

The German government has adopted an ambitious air quality program aimed to ensure that the air quality limit is accomplished throughout Germany over time.

Low emission zones can be an effective way to reduce NOx emissions in cities. An increasing number of large cities implement low emission zones, which ban circulation either for all vehicles or just for heavy goods vehicles only. Low emission zones must be large in size, meet severe emission standards (Euro 6 or better) and be strictly enforced [3, 4].

The ban on circulating certain types of more polluting vehicles in low emission zones could lead to the relocation of pollution sources instead of reducing them, so they do not necessarily lead to a good result.

Low emission zones are areas - usually in cities with different restrictions on the operation of more polluting vehicles, usually older (vehicles with higher emissions cannot enter into the area or have to pay a higher tax if they enter into the area with low emissions) [5, 6, 8].

Cities and governments have adopted low emission zones as a measure to reduce air pollution to comply with EU air quality standards. Low emission zones are often considered the most effective measure that cities can take to improve air quality and the number of these zones has increased steadily and they have been implemented in many EU countries [10, 11].

* Corresponding author: sparaschiv@ugal.ro

Low emission zones requirements are not harmonized in the EU, they are rather governed by different regulations at local and / or national level.

Typical requirements of low emission zones includes: prohibition of higher emission vehicles and of vehicles below a certain EURO emission standard. Some low emission zones have different standards for petrol and diesel cars to take into account different levels of NO_2 emissions.

On 27 February 2018, a decision of the highest court of the federal administration in Germany confirmed that the German cities could introduce restrictions for diesel cars with immediate effect and clarifies that the right of citizens to breathe clean air takes precedence over the rights of private car owners to drive polluting vehicles. As more cities introduce restrictions for polluting diesel cars, the air pollution problem could become more severe for Central and Eastern European countries due to second-hand cars imported from Western Europe.

2 Data

Starting from 31st May 2018, the circulation of diesel lorries that did not comply with the Euro 6 standard on Stresemannstrasse street in Hamburg was banned (Figure 1).

For our study, we obtained data on NO_2 concentrations collected at the traffic monitoring station located on Stresemannstrasse street in Hamburg (the sampling height of the monitoring station: 4 m above ground). The monitoring station is operated by the Hamburg Monitoring System for Air Quality, Germany. Our objective, of this study was to evaluate the air quality improvement by applying the measure to ban the circulation of diesel trucks [12, 13].

In our analysis, we compared the NO_2 concentrations measured prior to the implementation of the diesel car ban, with the NO_2 levels measured after the measures came into force. For the analysis, the measurements in the period prior to the application of the prohibition measure from 1 January 2017 to 31 May 2018 were used. The data from the second analysis period from 1 June 2018 to 30 March 2019 covered the period when the trucks transit ban was effective.

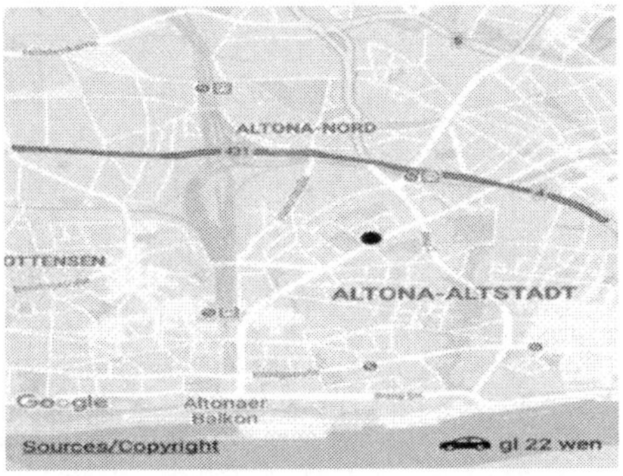

Fig. 1. The environmental zone on Stresemannstrasse street in Hamburg

3 Results and discussions

As shown in Figure 2, during the study period in the evolution of NO_2 concentrations in Hamburg, the average NO_2 concentrations have exceeded the limit value of the EU air quality standard of 40 µg / m³. In January, February and March, it is noted that most exceedances of the limit value are recorded in 2017 and 2018 when no traffic restriction measures were applied.

Following of road traffic restrictions for diesel cars (starting on 31 May 2018), in 2019 there is a downward trend in NO_2 concentrations below the limit value, the highest decrease has been registered in March, when the number of the exceeding days was only 10, compared to 28 days recorded in 2017 and 25 days in 2018. The number of days in January in which exceedances of NO_2 concentrations were recorded were 25 days in 2017, 17 days in 2018 and 15 days in 2019. In February, the number of days with exceedances of the limit value it was 22 days in 2017, 23 days in 2018 and 16 days in 2019.

Figure 2 shows the trend and the value of the increase or decrease of the NO_2 concentration in the case of the restrictive measures application (2019) and in the base case without measures (2018), the red bars representing the absolute value of the increase in concentration in the days when it exceeds occurred in 2019 compared to 2018, and with green bars the value with which the concentration was reduced in the days when the decreases occurred.

The negative numbers represent the absolute value of the decrease in NO_2 concentration and the positive numbers the absolute value of the increase of the concentration compared to the previous year.

Figure 2 shows that the NO_2 concentration has a decreasing trend due to the measures applied, the most obvious decrease being registered in March when the concentrations measured in 2019 exceeded the values from 2018 for only two days. Daily average NO_2 concentrations over the three months and for the three years ranged from 10 µg / m³ to 98 µg / m³. The maximum NO_2 concentration was observed on March 27, 2017, and the lowest NO_2 concentration was detected on January 2, 2019.

Figure 3 shows the daily average NO_2 concentration variation in the pre-restrictive period of 2017 and 2018.

The data monitored in April and May 2017 and 2018 show a general trend of exceeding the limit value of NO_2 concentrations, with exceeding being recorded for 20 days in April in 2017 and 24 days in 2018 respectively.

In May, the days in which the limit value of NO_2 concentrations was exceeded was 28 in 2017 and 22 in 2018 respectively. In April, there was an increase in NO_2 concentrations in 2018 compared to 2017 for 16 days, while in May the exceedance was registered for 13 days.

https://doi.org/10.1051/e3sconf/201912205002

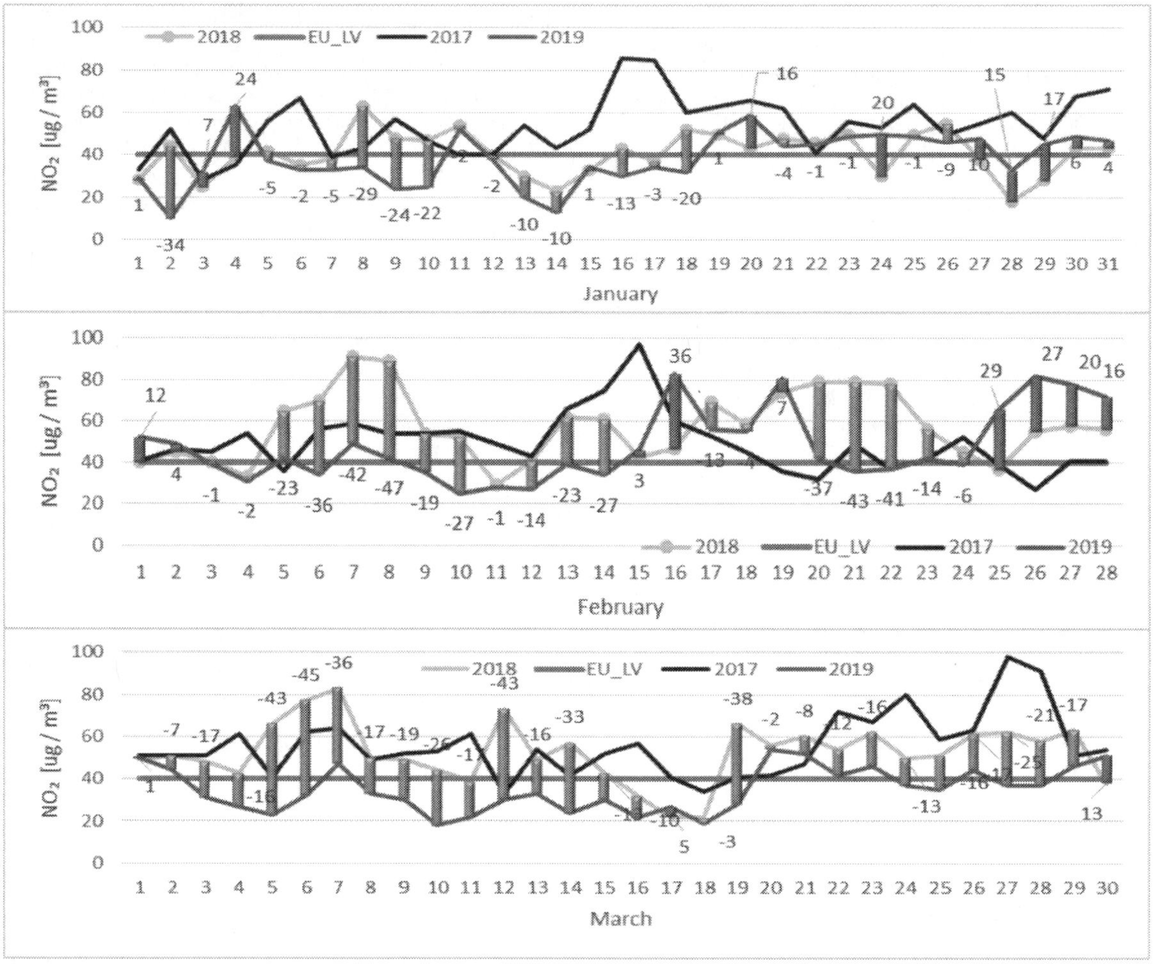

Fig. 2. Daily average of NO_2 concentrations during the pre-restriction and restriction period in the years 2017-2019

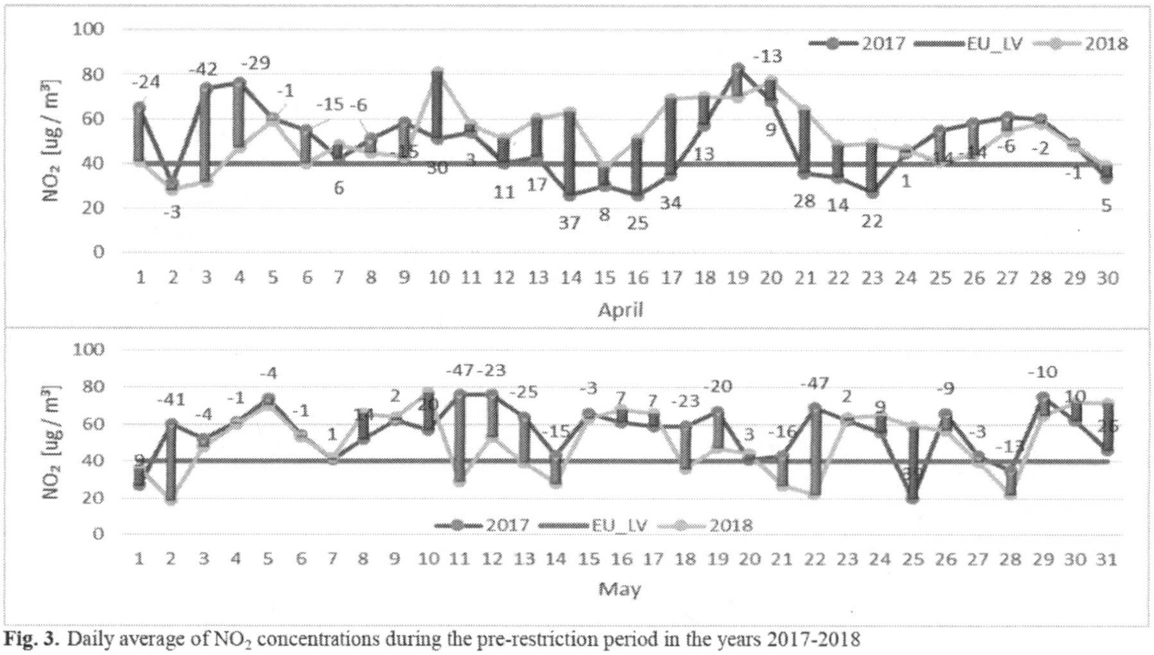

Fig. 3. Daily average of NO_2 concentrations during the pre-restriction period in the years 2017-2018

https://doi.org/10.1051/e3sconf/201912205002

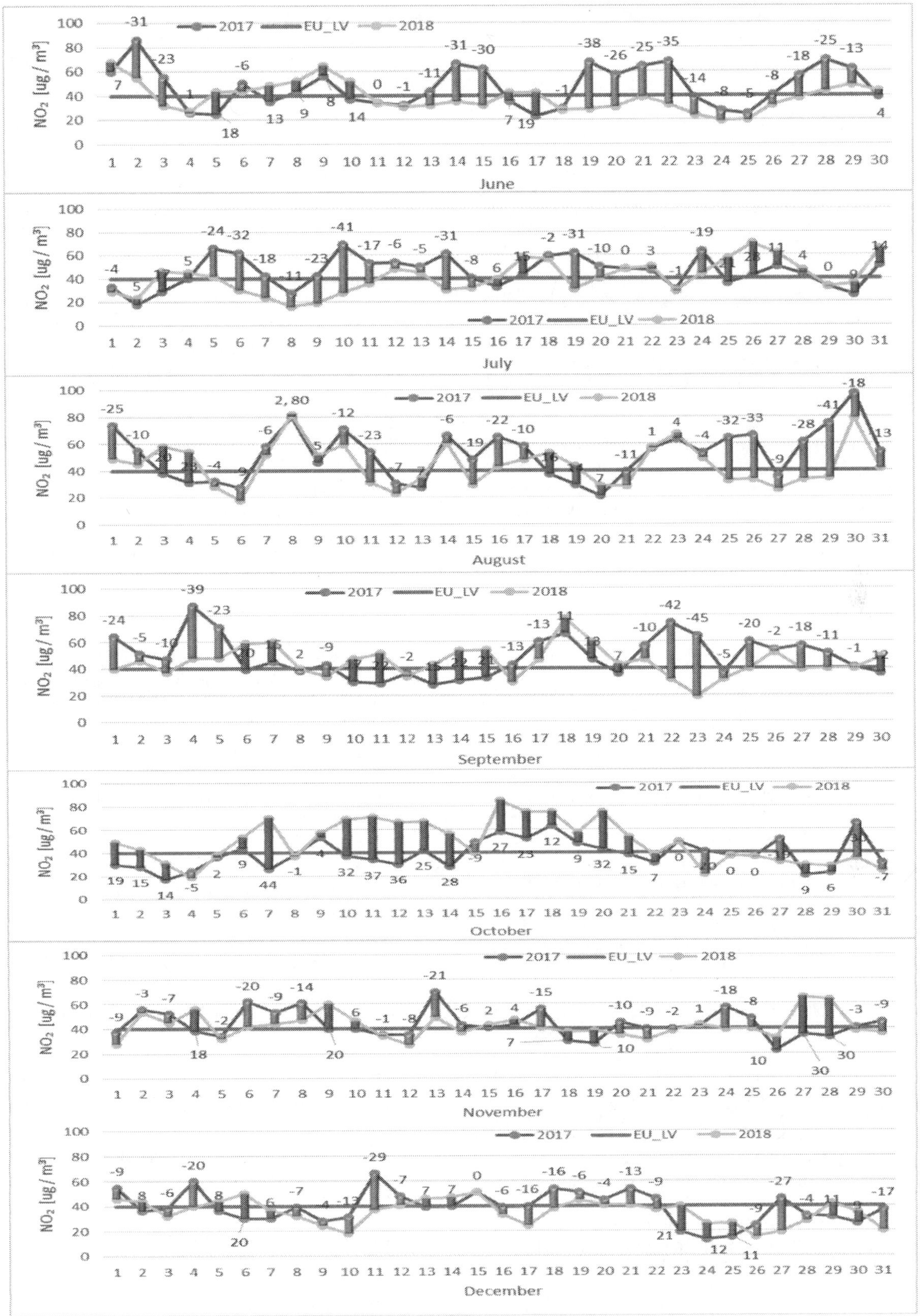

Fig. 4. Daily average of NO_2 concentrations during the pre-restriction and restriction period in the years 2017-2018

The results of the analysis performed for different time periods (June to December) are shown in Figure 4. The average daily variation of the NO_2 concentration shows a decreasing trend during the period of restriction application compared to the time before restriction introduction. The greatest value of the NO_2 concentration was 97 µg / m3 recorded on August 30, 2017.

From Figure 4 it can be seen that in the months before the restriction, the days in which the limit value were exceeded are higher, except October when the trend was reversed and the exceedances were higher during the period of restriction application (2018) and have recorded higher concentrations than in the same period of 2017.

The results show that the daily average concentration of NO_2 has a downward trend after application of traffic restriction on diesel cars, registering higher concentrations than the limit value for 10 days in January 2019 to 18 days in September 2018.

In the period before restriction introduction, the limit value of NO_2 concentrations was exceeded for 11 days in December 2017 and for 28 days in March and May 2017.

Lower values of NO_2 concentrations in December may be due to lower traffic, especially during the holidays (the Christmas) when it is observed a greater reduction in the NO_2 concentrations than the usual values.

4 Conclusion

NO_2 concentrations fluctuate visibly day by day, with daily average concentrations ranging from 10 to 98 µg / m3.

During the nine months before restrictions apply to diesel lorries (June 2017 - March 2018), 66% of days exceeded the NO2 concentration limit. After the restriction application has been a decrease of 12% of days in which there were breaches of the limit on concentrations of NO_2. Due to the measures implemented, the daily average concentration of NO_2 decreased, the strongest effect was in March when except two days all the daily average values were lower than in the pre-ban period and at the same time there were exceedances of the limit value of the NO_2 concentration for only 10 days as shown in Figure 2.

The results presented in this paper clearly confirm the beneficial effect of the measure taken on Stresemannstrasse Street in Hamburg. However, the limit value for NO_2 concentration is still exceeded. In order to comply with the minimum level, important measures need to be taken further and interpretation of past air quality trends can generate relevant results for planning beneficial actions.

Since the ban applies only for lorries and not to all diesel vehicles, in a future analysis, we must study the influence of the background concentrations due to ban on all diesel cars that do not comply with Euro 6 standard applied on nearby Max-Brauer-Allee and the possible relocation of these cars on the Stresemannstrasse street. When the ban does not apply to a large area, the expected effects could be minimal due to influence of the background concentrations that could increase due to increasing traffic on adjacent streets.

The analysis showed that in some days, even after the introduction of the traffic restrictions, there were exceedances of concentration compared with the previous period this may have been due to the influence of atmospheric factors (solar radiation, relative humidity, wind speed) on the lifetime of NO_2.

References

1. K. Gładyszewska-Fiedoruk, E3S Web of Conferences **101**, 03002 (2019)
2. P. Spiru, P.L. Simona, Energy Procedia **157**, 1553-1560 (2017)
3. C. Johansson, B. Lövenheim, P. Schantz, L. Wahlgren, P. Almström, A. Markstedt, M. Strömgren, B. Forsberg, J.N. Sommar, Science of the Total Environment **584–585**, 55–63 (2017)
4. V. Fensterer, H. Küchenhoff, V. Maier, H.E. Wichmann, S. Breitner, A. Peters, J. Gu, J. Cyrys, Int. J. Environ. Res. Public Health, **11**, 5094-5112 (2014)
5. I.S. Mudway, I. Dundas, H.E. Wood, N. Marlin, J.B. Jamaludin, S.A. Bremner, L. Cross, A. Grieve, A. Nanzer, B.M. Barratt, S. Beevers, D. Dajnak, G.W. Fuller, A. Font, G. Colligan, A. Sheikh, R. Walton, J. Grigg, F.J. Kelly, T.H. Lee, C.J. Griffiths, Public-health, **4**, e28-e40 (2019)
6. S. Alahmadi, K. Al-Ahmadi, M. Almeshari, Science of the Total Environment, **676**, 131-143 (2019)
7. S. Paraschiv, D.E. Constantin, S.L. Paraschiv, M. Voiculescu, Int. J. Environ. Res. Public Health **14** (11), 1415 (2017)
8. T.V. Russkova, P.N. Zenkova, Modern problems of remote sensing of the earth from space, **15** 7 221–235 (2018)
9. P Spiru, PL Simona, Energy Procedia **128**, 179-186 (2017)
10. D.M. Agudelo–Castaneda, E.C. Teixeira, F.N. Pereira, Atmospheric Pollution Research, **5**, 411–420 (2014)
11. S. Alahmadi, K. Al-Ahmadi, M. Almeshari, Science of the Total Environment, **676**, 131-143 (2019)
12. The Hamburg Climate Action Plan https://luft.hamburg.de/
13. The ambient air pollution network of Hamburg https://luft.hamburg.de/

https://doi.org/10.1051/e3sconf/201912205003

Quantifying Total Environmental Impact of the Power Sector Using Input-Output Life Cycle Assessment: a case study for Thailand

Isara Muangthai[1]*and *Sue* J. Lin[2]

[1]Warash School of Maritime Science and Engineering, Solent University, East Park Terrace, Southampton, United Kingdom
[2] Department of Environmental Engineering, National Cheng Kung University, Tainan 701, Taiwan

Abstract. The electricity generation is vital to industries and economic development in Thailand. In this study, the input-output life cycle assessment (IO-LCA) is applied to estimate the direct and indirect impacts from the power generation sector for the years 2005 and 2010. Based on the input-output analysis, more than 90% of the total environmental impact of Thailand's power sector involves ten relevant sectors. Results reveal that the most significant environmental damage was on natural resources followed by human health, climate change, and ecosystem quality. The most dominant environmental impacts were non-renewable energy, global warming and respiratory inorganic effects. Furthermore, the power sector, which accounts for 80% and 61% of total each impact in 2010 respectively, had a large direct impact on climate change and human health. On the contrary, the coal and lignite, and metal ore sectors contributed significantly to indirect impacts on ecosystem quality and resources. Regarding the results, some additional suggestions can be made to improve current policies in Thailand, including the implementation of green manufacturing in the iron and steel production, and installing control devices in all power plant units. Consequently, IO-LCA can be applied to other industries for assessing their total environmental impacts, and planning CO_2 mitigation strategies.

1 Introduction

Electricity is a crucial factor in daily life, transportation, households, and all industrial sectors of a country. The electricity generation in Thailand significantly rose from 95,977 Gwh in 2000 to 184,350 Gwh in 2015. Of this total, thermal power contributed 73% of the total national grid generation in 2015, hydro power provided 7%, and other alternative energy (solar, wind, biogas) supplied 20% [1]. This reveals that Thailand's power generation still heavily depends on fossil fuel. In addition, 42% of total CO_2 emissions from the energy consumption sector emitted from the power generation sector in 2015, while transportation, industry, agriculture and mining, and residential and commercial sectors contributed 30%, 19%, 6% and 3%, respectively to total CO_2 emissions [2]. The electricity generation sector plays an essential role in the country's economy, and environmental emissions; hence, it is important assess the potential environmental impacts caused by the electricity generation sector.

Life cycle assessment (LCA) is a tool to quantify the environmental impacts and resource utilization of a product's life cycle, process, or from cradle-to-grave [3]. There are two main approaches for conducting LCA: Process-based (P-LCA) and Input-Output life cycle assessment (IO-LCA). P-LCA is a comprehensive method that provides detailed results. However, P-LCA is a time consuming process, and it can have the problems of defining the system boundary and the circularity effect [4]. Conversely, IO-LCA is an integral method of input-output analysis (IOA) and LCA. IOA is an economic discipline that focus the inter-relationship between industries and households through producing and consuming commodities [5]. It has been used to extend the boundary of the product or process system in LCA; thereby, the economic-wide of embodied (direct and indirect) environmental burdens of a product or process can be traced. Moreover, the IO-LCA approach overcomes the disadvantage of P-LCA because the boundary includes transactions and emissions of all industry sectors. The result of IO-LCA can potentially be used as indicators in comparing products or sectors for strategic policy decision.

This study aims to evaluate the environmental impacts of Thailand's electricity for the years 2005 and 2010 by the IO-LCA. IO-LCA can be used to calculate the potential of environmental impacts from the electricity generation sector and other relevant sectors throughout the supply-chain of Thai's economy. This study focus not only on the direct environment impacts from the power generation process itself, but also the indirect impacts contributed from other related industries in the economy.

*Corresponding author: isara.muangthai@solent.ac.uk

https://doi.org/10.1051/e3sconf/201912205003

2 Methodology

2.1 Input-Output Analysis (IOA)

The IO model presents the financial flows of goods and services within an economy at a specific time [6]. A set of linear equation serves as the interdependency across a different sector of the economy. The standard representation of the IO model defines as follows [7] :

$$\sum_{j=1}^{n} x_{ij} + F_i = X_i \quad (1)$$

$$\sum_{j=1}^{n} a_{ij} X_j + F_i = X_i \quad (2)$$

$$a_{ij} = x_{ij}/X_j , \quad A = [a_{ij}] \quad (3)$$

Where X_i = the total output of every sectors;
x_{ij} = the intermediate demand that represent amount of input from sector i used by per unit of output sector j.
F_i = the exogenous change in final demand for the output.
A = matrix of technical coefficient, calculated by dividing the industry-by-industry direct requirements of sectorial inputs by the sectorial production,
a_{ij} = the dollar value of input required from sector i to produce one dollar worth output of sector j

Eq. (2) can be rewritten as;

$$X = AX + F$$

$$X = (I - A)^{-1}F \quad (4)$$

In this study, the domestic IO table was used to calculate the Leontief inverse matrix; therefore, A table is replaced by D table. Eq.(4) can be rewritten as

$$X = (I - D)^{-1}F \quad (5)$$

Where X = the (n x 1) vector of total economic outputs of the sectors,
I = the identity matrix (n x n),
D = the (n x n) inter-sectoral direct requirements (technical coefficients) matrix
$(I - D)^{-1}$ = the domestic Leontief inverse matrix, representing the total amount of goods or service i directly and indirectly needed to deliver a unit of final demand of good or service j.
F = the (n x 1) vector of exogenous change in final demand

2.2 Input-Output Life Cycle Assessment (IO-LCA)

Suppose that R is a $k \times n$ matrix of environmental burden coefficients, where R_{kj} is environmental burden k (e.g. CO_2 emissions) per dollar output of sector j. The economy-wide of total environmental burden associated with an exogenous demand can be calculated as follows, [8].

$$Q = RX$$

$$Q = R(I - D)^{-1}F \quad (6)$$

Where Q = the $(k \times n)$ vector of total (direct and indirect) environmental burden per monetary unit of output of each sector, total emission million US^{-1}

R = the $(k \times n)$ vector of direct environmental burden or pollutant emissions per monetary unit of output in each sector, direct emission million US^{-1}
F = the (n x 1) vector of exogenous change in final demand

2.3 Life Cycle Impact Assessment

The scope of IO-LCA in this paper is an economic-oriented life cycle assessment of Thailand's electricity generation that focuses on one industry to another within the economy. The functional unit in this study was million US dollars of total output of each economic sector for baseyear 2010. This study uses SimaPro 7.3.3 software along with IMPACT 2002+ version 2.10. Input-Output USA database 2002 was chosen as environmental database inventory in the analysis. The Missing Inventory Estimation Tool (MIET) 3.0 version was employed to estimate the environmental emissions of US database through SimaPro 7.3.3.

The IMPACT 2002+ is a damage oriented impact assessment model, which combines the benefits from both mid-point and end-point methods. Asides from to the aggregation of midpoint/damage structure, the compara- tive assessment of human toxicity and ecotoxicity are estimated in IMPACT 2002+. The assessment of this model involves characterization, damage assessment, normalization, weighting and single score. The method- logy incorporates fourteen midpoint impact categories: human toxicity, respiratory effects, ionizing radiation, ozone layer depletion, photochemical oxidation, aquatic ecotoxicity, terrestrial ecotoxicity, aquatic acidification aquatic eutrophication, terrestrial acidification/nitrifi- cation, land occupation, global warming, non-renewable energy and mineral extraction [9]. All fourteen midpoints categories were linked to four damage categories which is human health, eco-system quality, climate change, and resources. The respective share of each impact to overall damage of the considered category are analyzed through normalization. The details information of IMPACT 2002+ method could be viewed in [9].

2.4 Data consolidation

National IO tables was compiled and published by The National Economic and Social Development Board (NESDB) of Thailand every five years. IO table in 2010 that launched in 2015 by NESDB is the latest IO table. This study employed two economic IO tables with a 180 sectors in years 2005, and 2010 [10,11]. The domestic IO table (D-table); excluding the import values, was adopted for the analysis to obtain more precise results of environmental impacts caused by industrial sectors in Thailand. Besides, the economic activity output were adjusted to the Gross Domestic Product (GDP) base year of 2010 in U.S. dollar prices to avoid fluctuating currency values. GDP data was obtained from World Bank Group [12], and the exchange rate of the Thai currency to the US dollar ($) was adopted from The Bank of Thailand [13].

3 Result and discussion

The results herein shows ten most important sectors which totally accounts for more than 90% of the environmental impacts of the electricity generation sector in Thailand.
These sectors are given in Table 1

Table 1. Definition of ten important sectors

Sector	Definition
Coal and lignite	Mining of coal and lignite
Chemical fertilizer	Extraction or mineral mining
Electricity generation	Electricity generation, transmission and distribution of electric for sale to household, industrial, commercial and public users
Insulated wire and cable	The production of insulated wire and cable
Metal ore	Extraction of metal ore such as tin , iron ore
Pipeline and gas distribution	Gas distribution such as liquid petroleum gasoline
Petroleum and natural gas extraction	The exploration activities for crude petroleum and natural gas, the drilling and operation of wells
Petroleum refineries	Oil-processing refineries.
Railways	Transportation of both passengers and cargo
Sanitary and similar services	Operation of drainage systems and the purification of water for consumption.

3.1 Environmental Damage assessment

The environmental damages from ten sectors relevant to Thailand's electricity generation sector in 2000, 2005, and 2010 are given in Table 2. Results show that electricity generation sector was the main contributor to human health damages. The damage increased from 1.21E+08 in 2005 to 1.58E+08 DALY in 2010 which accounts for 53.3% and 65% of each year's total human health damage, respectively. Moreover, the metal ore sector was the major contributor to ecotoxicity. The total damage slightly decreased from 8.31E+12 PDF*m^2*yr in 2005 to 3.08E+12 PDF*m^2*yr in 2010 which accounts for 62.5% and 54.6% to total ecotoxicity value in 2005 and 2010, respectively (Table 2). The metal ore sector is the primary source of iron and steel. According to the IO tables for 2005 and 2010, total demand of the iron and steel sector was 3,619,685 million USD, and 1,996,407 million USD, respectively. This reveals that total demand for iron and steel declined 1.8 times from 2005 to 2010. For this reason, high demand for iron and steel in 2005 directly affected the metal ore sector. Results also indicate that the electricity generation sector was the largest CO_2 emission source contributing to climate change. The share of each year's total CO_2 mission increased from 78.5% in 2005 to 80.3% in 2010. For the resources, coal and lignite were major factors affecting resources and their share amounted to 70% and 15% of total resources in 2005 and 2010, respectively (Table 2). Coal and lignite increased theirs shared in power generation from 17% in 2005 to 21% in 2010 of the total fuel mix [1]. Results from the IO table reveal that the coal and lignite sector furnished a huge support of 310,300 million USD to the electricity generation sector in 2005, whereas the support from these coal and lignite sector dramatically decreased to 30,309 million USD in 2010. That is, the electricity generation sector reduced the use of domestic coal and lignite approximately 30% from 2005 to 2010. The electricity generation and petroleum and natural gas extraction sectors were two important contributors to the resources category.

Table 2. Environmental damages from ten sectors contributing to the Thai's electricity generation sector in 2005 and 2010

Sector	Year	Human Health (DALY)	Ecotoxicity (PDF*m^2*yr)	Climate change (kgCO_2 eq)	Resource (MJ primary)
Electricity generation	2005	1.21E+08	1.33E+12	1.14E+14	2.20E+15
	2010	1.64E+08	1.67E+12	1.44E+14	2.76E+15
Coal and lignite	2005	2.18E+07	2.93E+12	3.63E+13	1.17E+16
	2010	2.03E+06	2.73E+11	3.38E+12	1.09E+15
Chemical fertilizer	2005	6.09E+06	6.10E+10	3.59E+12	1.45E+14
	2010	4.11E+06	4.12E+10	2.42E+12	9.77E+13
Petroleum and natural gas extraction	2005	2.00E+07	4.84E+10	1.60E+12	1.24E+15
	2010	2.26E+07	7.08E+10	34E+12	1.81E+15
Petroleum refineries	2005	3.01E+06	2.04E+10	1.99E+12	2.87E+14
	2010	3.24E+06	2.19E+10	2.15E+12	3.09E+14
Pipeline and gas distribution	2005	3.54E+07	1.46E+11	1.44E+13	6.39E+14
	2010	4.02E+07	1.93E+11	1.91E+13	8.46E+14
Metal ore	2005	9.46E+06	8.31E+12	6.79E+12	1.64E+14
	2010	3.51E+06	3.08E+12	2.52E+12	6.10E+13
Insulated wire and cable	2005	8.82E+05	2.44E+11	6.96E+11	1.65E+13
	2010	3.08E+05	8.49E+10	2.43E+11	5.77E+12
Railways	2005	6.32E+06	1.13E+10	7.80E+11	1.59E+13
	2010	7.94E+06	1.42E+10	9.80E+11	2.00E+13
Sanitary and similar services	2005	3.78E+05	8.63E+09	3.29E+11	7.15E+12
	2010	5.64E+05	1.29E+10	4.91E+11	1.07E+13
Total impact	2005	2.28E+08	1.33E+13	1.82E+14	1.64E+16
	2010	2.52E+08	5.64E+12	1.79E+14	7.05E+15

3.2 Normalization of total environmental impacts

Normalization of total environmental impacts from Thailand's electricity generation and other related sectors in 2005 and 2010 show in Figure 1. The most important environmental damage from normalization step was resources, followed by human health, climate change, and ecosystem quality. Resources impact dramatically decreased from1.08E+11 in 2005 to 4.64E+10 in 2010. he reduction of domestic coal and lignite consumption in the power sector was a reason for the decrease of resources from 2005 to 2010. In addition, the impact value of resources decline in 2010 because the domestic natural gas shared in grid generation rose 5% from 2005 to 2010 [1]. An increasing share of natural gas in the Thai's power generation in 2010 contributed a less effect to resources when compare with coal and lignite because natural gas is a cleaner fossil fuel,

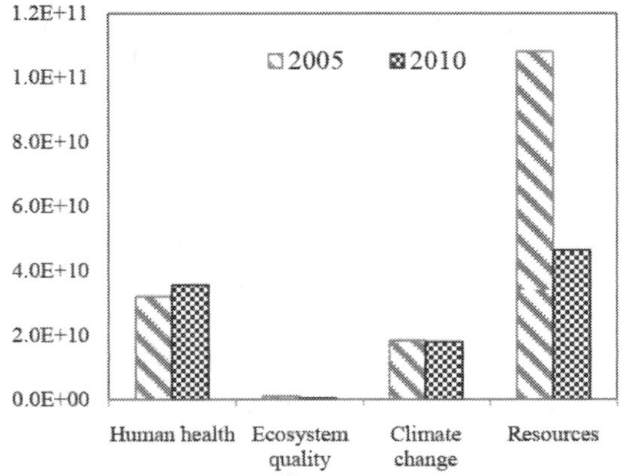

Fig. 1. Normalized environmental damage for 2005 and 2010

The damage value of human health increased from 3.21E+10 DALY in 2005 to 3.56E+10 DALY in 2010, while the damage value of climate change slightly decreased from 1.84E+10 in 2005 to 1.81E+10 in 2010 (Fig 1). For the climate change normalized score, it is the same category as the midpoint category global warming. Results reveal that the electricity generation was the key contributor to global warming in two years. However, the ecosystem quality had the less effect to environmental damage because normalization factor for ecosystem quality is determined with some modification [9].

In order to understand the details of the impact categories, the normalized environmental impact per midpoint level categories for years 2005 and 2010 from 10 sectors are illustrated in Figure 2. Results show that non-renewable energy, global warming and respiratory inorganics effects were three dominant environmental impacts. The normalized value of global warming and respiratory inorganics also increased from 2005 to 2010. On the other hand, the normalized value of non-renewable energy decreased from 2005 to 2010. This result accordance with previous result of normalization of total environmental impact, that is decreasing of non-renewable energy impact from 2005 to 2010.

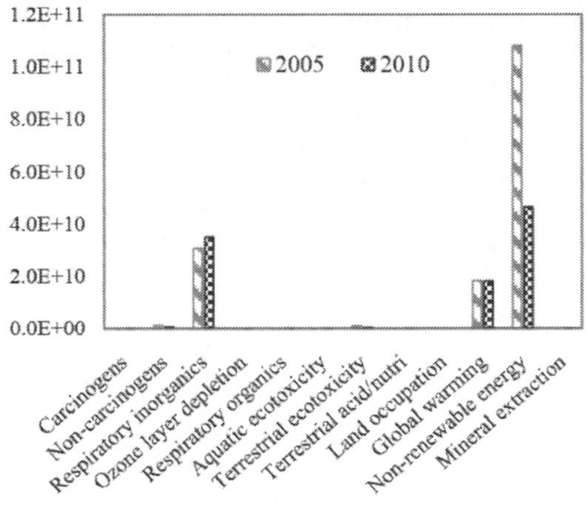

Fig. 2. Normalization of total environmental impacts for 2005 and 2010

3.3 Direct and indirect environmental impacts

Table 3. The percentage of direct and indirect environmental impact of Thailand's electricity generation sector

Year	Nature of impact	Human health	Ecosystem quality	Climate change	Resources
2005	Direct	53%	10%	63%	13%
	Indirect	47%	90%	37%	87%
2010	Direct	61%	30%	80%	39%
	Indirect	39%	70%	20%	61%

The proportion of direct and indirect impacts for each damage category are given in Table 3. Results indicate that the climate change category had the largest direct effects over other impacts, and its proportion accounts for 63%, and 80% in 2005, and 2010, respectively. In addition, the estimated carbon dioxide (CO_2) from Thai's power sector gradually increased every year, especially in 2010 (Fig. 3) [1,2].

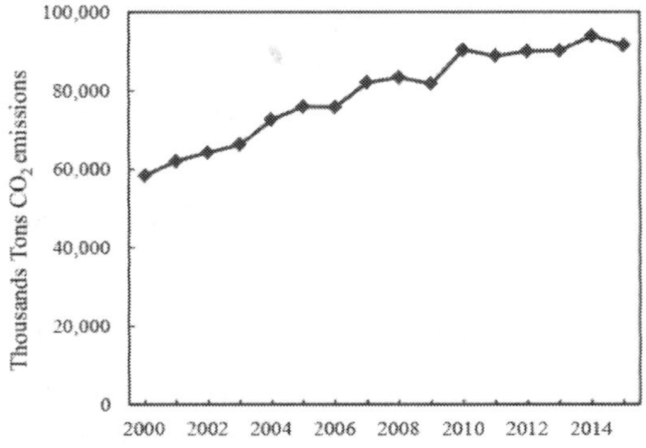

Fig. 3. CO_2 emissions from the Thai's power sector 2000-2015

CO_2 emission from the power sector increased along 2000 to 2015 with average growth 3.3% per year due to the increasing of electricity consumption and population. However, the direct impact on human health from the electricity generation sector itself increased from 53% in 2005 to 61% in 2010. On the other hand, ecosystem quality and resources mainly came from indirect effects. The percentage of ecosystem quality from an indirect effect was 90% in 2005, and 70% in 2010 (Table 3). The metal ore sector is considered as one of the indirect sources that contributes to ecosystem quality. Moreover, the indirect proportion of resources were higher than the direct proportion because the power sector needs to reply on fossil fuel from upstream sectors like the coal and lignite and petroleum and natural gas extraction sectors. Therefore, these sectors emerge as significant sources affecting resources.

4 Conclusion

This study utilized IO-LCA to explore the direct and indirect environmental impacts of Thailand's electricity generation sector. Regarding the damage assessment results, the coal and lignite sector caused major damages in 2005. The normalization results shows that the most crucial environmental damages were resources, followed by human health, climate change, and ecosystem quality. The reduction of domestic coal and lignite in the fuel mix of grid generation in 2010 resulted in a dramatic decreased of resources from 2005 to 2010. In addition, non-renewable energy, global warming, and respiratory inorganic effects were the most dominant normalized environmental impacts from the electricity generation and its related sectors. Additionally, results of the direct and indirect effects reveal that the electricity generation sector had the highest direct impact on climate change because of it was the major contributor to CO_2 emissions. In contrast, other relevant sectors played significant roles regarding ecosystem quality and resources. Since the electricity generation sector links with various industrial sectors, there could be an underestimation of indirect impacts if the other related sector are excluded from the calculation.

In summary, IO-LCA is a useful method that estimate the direct and indirect environmental impacts from the power sector which includes the indirect impacts from the associated sectors in demand-supply system. The main function of Sima Pro software offer the calculation of environmental impacts. The methodology used herein can be applied to other important industries to estimate the embodied environmental impacts, to plan mitigation strategies to lowering CO_2 emissions and other environmental emissions.

5 Recommendation

The Thai government needs to pay attention to the emissions from both direct source from electricity generation and indirect sources such as coal and lignite, petroleum and natural gas extraction and metal ore sectors. Regarding the ecosystem quality, the metal ore sector was the main contributor to ecotoxicity. The iron and steel sector is a major sector that consumes metal ores. The government should adopt a more sustainable development policy for the iron and steel industry. Green manufacturing should focus on investigating energy saving techniques and reducing CO_2 emission from producing steel. Recycling of steel scrap would help to reduce the need for extracting the metal ore. Regarding the result, the electricity generation sector had a tremendous affected global warming, non-renewable energy and respiratory inorganics. A control device such as flue-gas desulfurization and fabrics filter collectors should be installed in all power plant units, and there should regular monitoring the efficiency of the devices. Furthermore, the government of Thailand and other relevant agencies should focus not only the energy-related CO_2 emission from the power sector, but also the concern of human health from its pollution.

Reference

1. Department of Alternative Energy Development and Efficiency, Thailand Alternative Energy 2015. Ministry of Energy, Bangkok, Thailand. pp. 28-29. (2015a).
2. Department of Alternative Energy Development and Efficiency. Energy Balance of Thailand 2015. Ministry of Energy, Bangkok, Thailand. pp. 22-23,48, 60. (2015b).
3. ISO 14040 International Standard, Environmental Management-Life Cycle Assessment-Principles and Framework. International Organization for Standardization, Geneva, Switzerland. (2006)
4. CT. Hendrickson, LB. Lave, HS. Matthews. Environ- mental Life Cycle Assessment of Goods and Services: *An Input-Output Approach*. Resources for the Future, Washington, DC. (2006)
5. G. Finnveden, M. Z. Hauschild, T. Ekvall, J., Guinée, R. Heijungs, S. Hellweg, A. Koehler, D. Pennington, and S. Suh. Recent Developments in Life Cycle Assessment.J Environ Manag. **91**, 1-21 (2009).
6. RE.Miller, PD. Blair. *Input-Output Analysis:* Foundations *and Extensions*. Cambridge University Press, New York. (2009).
7. SJ. Lin, YF. Chang. Linkage Effects and Environmental Impacts from Oil Consumption Industries in Taiwan. J Environ Manag. **49**, 393-411 (1997).
8. CH. Liu, SJ. Lin, C. Lewis. Environmental Impacts of Electricity Sector in Taiwan by Using Input-Output Life Cycle Assessment: The Role of Carbon Dioxide Emissions. Aerosol Air Qual Res. **12**, 733-744. (2012)
9. Jolliet, O., Margni, M., Charles, R., Humbert, S., Payet, J., Rebitzer, G. and Rosenbaum, R. Impact 2002+: A New Life Cycle Impact Assessment Methodology. The International Journal of Life Cycle Assessment, **8**, 324-330. (2003).
10. National Economic and Social Development Board. Input-Output Table of Thailand 2005. Retrieved 20

https://doi.org/10.1051/e3sconf/201912205003

July 2014,from Office of the National Economic and Social Development Board (2005).

11. National Economic and Social Development Board. Input-Output Table of Thailand 2010. Retrieved 24 October 2015, from Office of the National Economic and Social Development Board (2010).
12. World Bank Group.World Development Indicators. Last access 5 December, 2015 (2015).
13. Bank of Thailand Rates of Exchange of Commercial Bankshttps://www.bot.or.th/English/Statistics/ FinancialMarkets/Pages/StatExchangeRate.aspx. Accessed 12 June 2016 (2016).

https://doi.org/10.1051/e3sconf/201912205004

Assessment of Natural Radioactivity Level and Annual Effective Dose of Amber Rice Samples Cultivated in the South of Iraq

Abdulridha S. Younis[1,*] and *Nada F.* Tawfiq[2]

[1]Al-Kitab University, Kirkuk, Iraq
[2]Al-Nahrain University, Baghdad, Iraq

Abstract. In this work, the activity concentrations of naturally occurring radioactive material including ^{238}U, ^{232}Th, and ^{40}K were measured of some amber rice samples cultivated in southern Baghdad and Al Najaf governorate and used in Baghdad governorate, Iraq in 2018, using gamma-ray spectrometer with NaI (Tl) detector. The results show that the activity concentrations for ^{238}U ranged from (2.68 to 10.81) Bq/kg with average 5.94 Bq/kg, ^{232}Th ranged from (B.D.L to 3.37) Bq/kg with average 2.65 Bq/kg, and for ^{40}K ranged from (4.48 to 35.7) Bq/kg with average 16.84 Bq/kg. The annual effective dose from rice consumption by adults for ^{238}U, ^{232}Th, and ^{40}K ranged from (0.41 to 1.6) x10^{-5} Sv/y with average 0.9 x10^{-5} Sv/y, (B.D.L to 0.42) x10^{-5} Sv/y with average 0.2 x10^{-5} Sv/y, and (0.17 to 1.2) x10^{-5} Sv/y with average 0.45 x10^{-5} Sv/y respectively. All values of the average specific activities are less than the global average values of ICRP, and the annual effective dose from rice consumption by adults was lower than the permissible limit of (1 m Sv/y) recommended by the International Commission on Radiological Protection.

1 Introduction

Radionuclides are found throughout nature and exist in the soil, water and food. These radionuclides have half-lives that are approximately the earth's age or older (i.e. about 4 to 5 billion years) [1]. Natural radioactive decay series such as ^{238}U and ^{232}Th as well as singly occurring radionuclides such as ^{40}K exist in the earth and atmosphere in varied levels. The radioactivity present in air or in the agricultural land and in soil may transfer to the crops grown on it. It happens, however, that an amount of some radioactive elements find their way into human bodies [2]. When plants are grown in the contaminated soil, the radioactivity is transferred from the soil to the roots. In the end, the radioactivity is shifted to the human diet [3]. These radionuclides will get transferred into plants together with the nutrients throughout mineral uptake and accumulation in varied components and even reach edible portions [4].
It is well known that traces of radionuclides are found in air, soil, water and human bodies, we inhale and ingest radionuclides every day of our lives and radioactive materials have been ubiquitous on earth since its creation [1,5].

Natural Uranium is considered as one of the most important primordial radionuclides in the earth's crust. It is found in varying soil but small amounts, for example, rocks, soils, water, air, plants, animals and in all human beings. On average, about (90 µg) of uranium exist in the human body which comes from the normal intakes of food, water, and air. Approximately (66%) was found in the skeleton, (16%) in liver, (8%) in kidneys and (10%) in other tissues of the body [6, 7].

The physiological behaviour of uranium compounds depends mainly on their solubility, when ingested, uranium might be concentrated within the bones; it increases the probability of bone cancer, or in the red bone marrow (leukemia).[8,9]

The kidney is considered the target for uranium chemical toxicity, which causes irreversible damage to the kidney leading to growth of tumours. A large quantity of uranium in the kidney can impair renal function. [10, 11, 12,13,14].

The Aim of the study is to determine the specific activities of natural 0radionuclides (^{238}U, ^{232}Th, and ^{40}K) in amber Iraqi rice samples cultivated in the middle and the south of the country using Gamma spectroscopy technique with NaI(Tl) scintillation detector.
And to evaluate the annual effective dose from the rice consumption by individual consumers due to intake of the concerned the above radionuclides.

2 Materials and Methods

Three rice samples (1 kg weight of each sample) were collected from fives farming fields (two from south Baghdad and three samples from Al-Najaf governorate farming fields area are shown in table (1), these samples are kept in Marinally beakers and stored for 30 days before gamma ray analysis is done to allow ^{226}Ra and its short-lived progenies to reach secular equilibrium between uranium and thorium with respective progenies.

* Corresponding author: ars.younis@gmail.com

Table 1. Rice Samples collected locations

Sample	Location
S1	southern Baghdad
S2	southern Baghdad
S3	AL Najaf governorate
S4	AL Najaf governorate
S5	AL Najaf governorate

Measuring the Specific activity of natural Radionuclides is done by using NaI(Tl) scintillation detector (The sodium iodide NaI(Tl) (CANBRA-USA) to PC-MCA (4096 channel) operating voltage (750V), (3×3inch), and 60% efficiency and the delectability of the energy detector values (1332)MeV. The detector is surrounded by a lead shield to reduce the background radiation.

The energy calibration of gamma spectrometer were carried out using standard sources in one litter marinelli beaker mixed radio nuclides with energies (59.53, 88.34, 661.7, and 1178) keV for ^{214}Am, ^{109}Cd , ^{137}Cs , and ^{60}Co respectively (figure :1)

The software (GIENE 2000) program is used to quantify the measurements and to analyze the results.

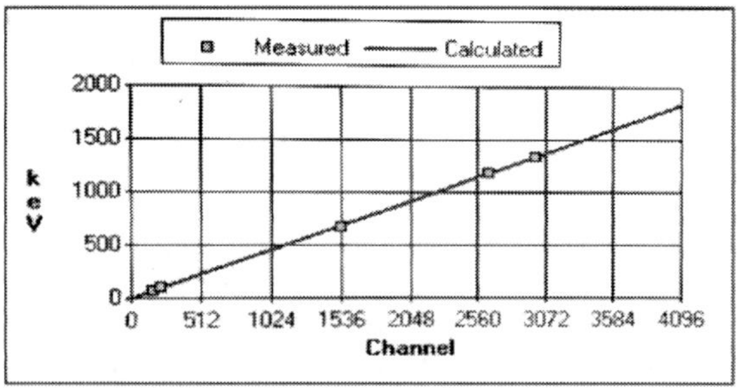

Fig. 1. Energy Calibration Curve.

The efficiency e_f at energy (Eγ) was calculated according to equation:1 below.

$$e_f = \frac{N}{A * I\gamma * T} * 100\% \quad (1)$$

Where:

e_f: The efficiency at energy (Eγ).

N: the net area under the peak at specific peak at energy (Eγ).

A: is the activity of the standard source in (Bq).

Iγ: is the abundance at energy (Eγ).

T: measurement time which equal to (1800s).

To find the efficiencies of other energies of any radioactive isotopes by the efficiency calibration curve as shown in Fig. 2.

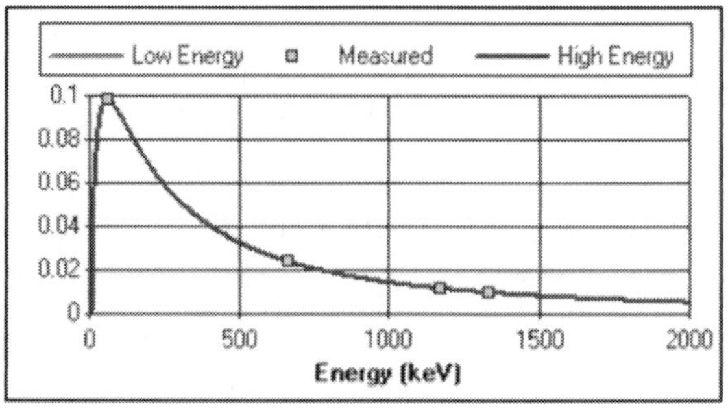

Fig 2. Efficiency Calibration curve

The specific activity of the radionuclides, in (Bq/kg) units, in rice samples were obtained by using equation: 2 [15,16]:

$$A = \frac{N - B.G}{T \times I\gamma \times \mathcal{E}\text{ff} \times M} \quad (2)$$

Where:

A: is the specific activity concentration of radionuclide in (Bq/kg), N: is the area under the peak .

B.G: is the peak of background with shielding.

M: is the mass of the sample (kg).

Iγ : is the intensity at energy (Eγ).

Ɛff: is the efficiency at energy (Eγ).

T: is the time of measurement which is equal to (1800s)

The Annual Effective Dose

The annual effective dose was calculated by the following formula [17]:

$$E = C * R * F_C \quad (3)$$

Where: E is the annual effective dose (Sv/y),

C = Concentration (Bq/kg),

R = Consumption rates of food (kg/year) which is equal to 55kg/y.

Fc =Dose conversion factor for ingestion is equal to ($4.5x10^{-5}$ mSv/Bq) for uranium, IAEA 1996 [18, 19].

3 Results and Discussions

The specific activity of natural radionuclides' ^{238}U, ^{232}Th and ^{40}k in rice samples were given in Table 2.

The range of specific activity concentration for ^{238}U varied from (2.68 to 10.81) Bq/kg with average 5.94 Bq/kg, the highest specific activity levels was found in sample S1 (from southern Baghdad), which can be referred to as the cumulative of this radionuclide in the rice and the minimum activity level found in sample S3 from (Al-Najaf governorate) as shown in table 2 and Fig. 3. The range of specific activity for ^{232}Th varied from (B.D.L to 3.37) Bq/kg with average 2.65 Bq/kg, the highest specific activity levels was found in sample S1 (southern Baghdad), and the minimum activity level found in sample S2 and S4 (southern Baghdad and Al-Najaf governorate), as shown in Fig. 4.

The range of specific activity for ^{40}K varied from (4.88 to 35.7) Bq/kg with average 16.84 Bq/kg, the highest specific activity levels were found in sample S5 (Al-Najaf governorate) and the minimum activity level found in sample S3 (Al-Najaf governorate) as shown in Fig. 5.

The average specific activities of ^{238}U, ^{232}Th and ^{40}K were shown in Fig 6, all values of the average specific activities are less than the global average values [8,22].

As seen in Table 3, the annual effective dose from rice consumption by adults indicates that the annual effective dose in all samples were lower than the maximum permissible limit of (1 mSv/y) recommended by the International Commission on Radiological Protection (ICRP). [17, 18].

Table 2. The average Specific activities of natural radionuclides in rice samples

Sample	Specific Activity Concentration (Bq/kg)		
	^{238}U	^{232}Th	^{40}K
S1	10.81	3.37	12.2
S2	4.42	B.D.L	13.9
S3	2.68	3.18	4.88
S4	6.23	B.D.L	17.5
S5	5.55	1.41	35.7
Max.	10.81	3.37	35.7
Min.	2.68	B.D.L	4.48
Average	5.94	2.65	16.84

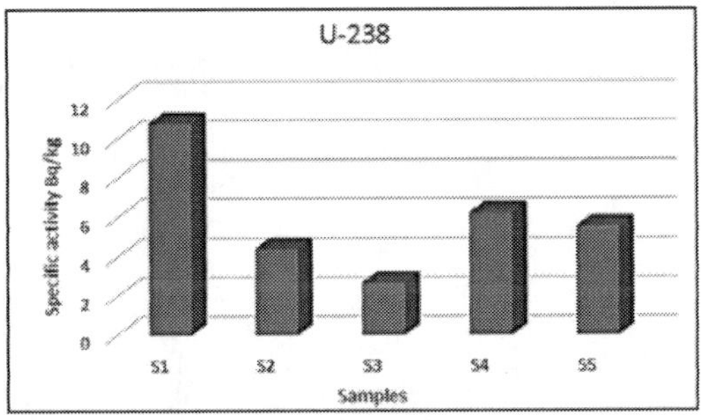

Fig. 3. The specific activity of 238U in rice samples

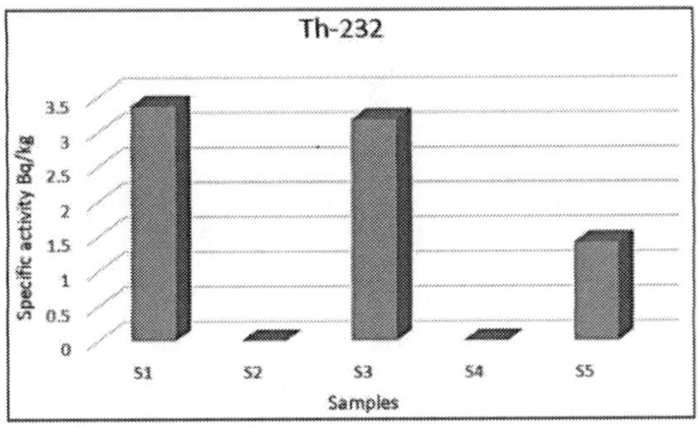

Fig. 4. The specific activity of ^{232}Th in rice samples

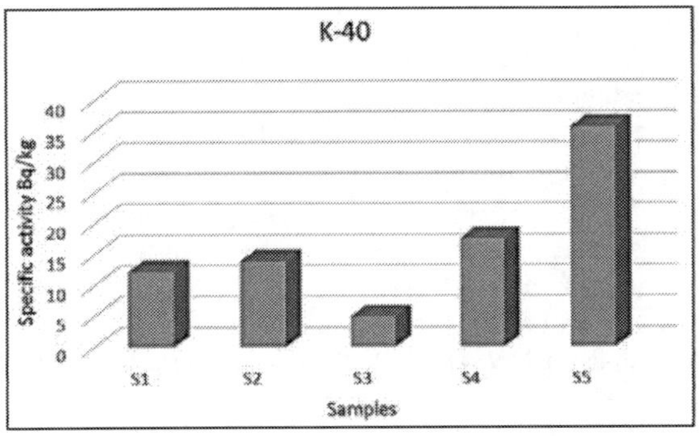

Fig. 5. The specific activity of ^{40}K in rice sample

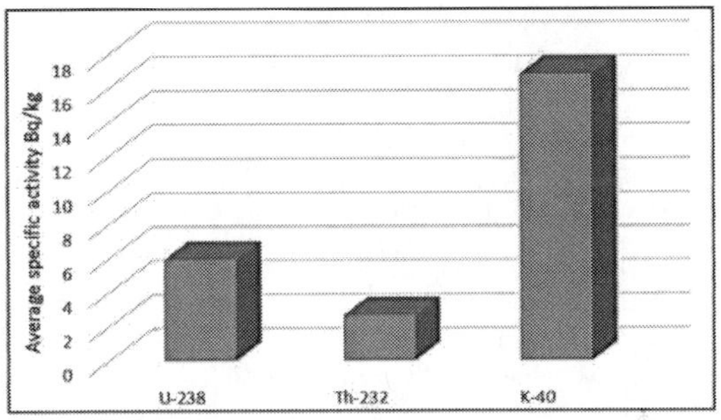

Fig. 6. The average specific activities of ^{238}U, ^{232}Th and ^{40}K in rice samples

Table 3. The annual Effective Dose in rice samples

Sample	Annual Effective Dose $\times 10^{-5}$ (Sv/y)		
	^{238}U	^{232}Th	^{40}K
S1	1.6	0.42	0.42
S2	0.68	B.D.L	0.47
S3	0.41	0.40	0.17
S4	0.96	B.D.L	0.60
S5	0.85	0.18	1.2
Max.	1.6	0.42	1.2
Min.	0.41	B.D.L	0.17
Average	0.9	0.2	0.45

4 Conclusions

The Specific activities of Natural radionuclides (^{238}U, ^{232}Th and ^{40}K) in all samples are lower than the worldwide median values reported by UNSCEAR.[8]

Also it was found that the annual effective doses due to the ingestion of all three natural radionuclides by adults are below the recommended limit by the International Commission on Radiological Protection for radiological safety ICRP.[11, 12,13,14]

References

1. Raymond Murray L. Understanding Radioactive Waste. Battelle Press. Columbus. Ohio: Fourth Edition 1994.
2. U.S. Environmental Protection Agency Radiation Protection Program-Uranium, 2015.
3. The Governorate of the Hong Kong Special Administrative. 'Radiation and food Safety 2013.
4. Laith A. Najam, Nada F. Tawfiq, Fouzey H. Kitah, (2015), Estimation of Natural Radioactivity of Some Medicinal or Herbal Plants Used in Iraq, Detection, 3, 1-7.
5. Cember H., Thomas E. Johnson, Introduction to Health Physics, Fifth Edition, McGraw Hill Education, USA, 2017.
6. Peter van Calsteren, Louise Thomas, Uranium-series dating applications in natural environmental science, Earth-Science Reviews, Vol. 75, Issues 1–4, 155-175, 2006.
7. Harigel G.: "Depleted Uranium Weapons – A threat to Human Health?" 2002.
8. UNSCEAR, United Nations Scientific Committee on the Effects of Atomic Radiation, Report to the General Assembly: Sources and Effects of Ionizing Radiation, New York, Vol.1, 2000.
9. Asaf, D., "Medical Consequences of Contamination with Depleted
10. Uranium", In "Metal of Dishonor", International Action Center, New York, 1999.
11. Deitzy, "Uranium Battle Fields", Progressive Alliance for Community, New Mexico, 1993.
12. ICRP, Recommendations of the International Commission on Radiological Protection. ICRP Publication 60. Ann. ICRP 21 (1–3), 1991.
13. Linnea E. Wahl, Environmental Radiation, Health Physics Society Specialists in Radiation Safety January, 2010.
14. ICRP, "Radiation Protection", Report of Committee IV on Evaluation of Radiation Dose to Body Tissues from Internal Contamination due to occupational Exposure, Pergamon Press, Oxford, 1968.
15. Ridha, A. A., "Determination of radionuclides' concentrations in construction materials used in Iraq " Ph.D. Thesis, College of Science, Al-Mostansrya University, 2013.
16. Hussain R.O. and Abbas E.K. " Measurement of Natural Occurring Radio nuclides (NORMs) in Soil using Gamma- ray Spectrometry " Journal of AL-Kufa ,2, No.2,15-22, 2010.
17. Karim M.S, Chiad,S. S., Habubi, N.F, Mansour,H.L., and Mishjil, K.A." Study of specific activities of some biological samples for selected Iraqi governorates" Fondazione Giorgio Ronchi, ANNO, LXVIII, No.1, 2013.
18. ICRP, 2009. Application of the Commission's recommendations for the protection of people in emergency exposure situations. ICRP Publication 109. Ann. ICRP 39 (1).
19. IAEA, 1996, International basic safety standards for protection against ionizing radiation and safety of radiation sources, Safety Series 115, Vienna.

E3S Web of Conferences **122**, 05005 (2019) https://doi.org/10.1051/e3sconf/201912205005
REEE 2019

Bioelimination of sulfur from high-sulfur coal by selected strains of microorganisms

Raushan Samet[1,2,*], *Azhar* Zhubanova[1,2], *Nuraly* Akimbekov[1,2], *Xiaohui* Qiao[1,2], and *Anel* Kadyrzhanova[2]

[1]Al-Farabi Kazakh National University, Faculty of biology and biotechnology, Biotechnology Department, 71 al-Farabi Ave., Almaty, Republic of Kazakhstan
[2]Scientific Research Institute of Biology and Biotechnology Problems, 71 al-Farabi Ave., Almaty, Republic of Kazakhstan

Abstract. In this study, low-rank lignite coal sample collected from Lenger coal deposit (Turkestan province) in Kazakhstan was subjected to desulfurization by using three bacterial strains isolated from soil with silt and coal itself. The molecular identification of the 16S rRNA gene revealed that the isolated bacteria were *Atlantibacter* sp., *Pseudomonas* sp., *Bacillus* sp. denoted as S1, S2, and T1, respectively. *Pseudomonas* sp. showed the best result in removing organic sulfur (93%) and total sulfur (52%), while *Bacillus* sp. was effective in removing pyritic sulfur (19%) compared to other strains. However, *Atlantibacter* sp. had no significant influence on sulfur content after treatment, thereby reducing its chances to be used in decreasing sulfur content in lignite in future investigations. Additionally, this research would be valuable to develop an innovative biotechnological method for producing an environmentally friendly briquetted smokeless fuel from lignite.

1 Introduction

The coal industry is one of the main sectors of the economy in many countries of the world. Thus, coal is used as a fuel, for the manufacture of building materials, in medicine and the chemical industry. However, its extraction, processing and use lead to environmental pollution.

Therefore, the development of high-tech coal business and the implementation of new growth points, such as coal enrichment, coal gasification, coal hydrogenation, methane extraction from coal beds, catalytic processing of methane, etc. are becoming important for the future of the coal industry.

For example, the enrichment of coal is always aimed at not only removing ash but also reducing sulfur in coal, since sulfur is considered a harmful component of coal for many reasons.

With an increase in sulfur content, the heat of combustion of coal decreases, the amount of sulfur dioxide in flue products released into the surrounding atmosphere increases. Emissions of sulfur compounds cause morbidity and mortality in humans; the death of coniferous trees; acid rain; decrease in crop yields in agriculture, etc.

In order to improve the marketability of the fine coal produced during mining, it is desirable to remove as much sulfur and ash as possible [1].

Sulfur in coal is present in both inorganic and organic forms. The inorganic sulfur in coal consists predominantly of sulfides and sulfates. Sulfide minerals include pyrite, sphalerite, galena, arsenopyrite and others. The sulfate minerals include barite, gypsum, anhydrite, and a number of iron sulfates and others. The organic sulfur in coal is covalently bound into its large complex structure and is difficult to remove physically or chemically, in contrast to pyritic or inorganic sulfur [2].

In this regard, research is being conducted on new physical methods for cleaning coal, such as multistage flotation, electrostatic separation, and oil agglomeration. According to preliminary data, these methods can remove up to 90% of all pyritic sulfur and up to 65% of the total amount of sulfur from coal. However, complete desulfurization is possible as a result of the removal of bound organic sulfur. Therefore, at present, methods of chemical and microbiological purification of coal from organic sulfur are being developed.

The method of chemical cleaning involves the treatment of coal with special chemicals or solvents under pressure. It is possible to remove up to 95% of all sulfur. However, the method is much more expensive compared to the physical, as a result, the cost of coal increases.

Methods of microbiological cleaning, green-based approaches are based on the fact that certain bacteria and fungi absorb sulfur, while coal liquefaction is also possible. Microbial coal desulfurization before combustion has low capital and operating costs and is more energy efficient compared with high-temperature chemical processes [3]. There are diverse representative microbial groups responsible for sulfur oxidation and reduction in different ecosystems [4].

Kazakhstan is among the top ten countries with the largest coal deposits in the world. Brown coal which has

* Corresponding author: mraushans@gmail.com

low calorific value and high ash content is piled up as rubbish causing a serious waste of resources as well as environmental pollution. Kazakhstan needs to acquire new technologies for coal processing, especially green-based approaches [5].

Relatively little studies have been conducted to evaluate properties of Kazakhstan brown coal, specifically reduction of sulfur content of coal by the bacterial means. The present work is intended to show comparative analysis before and after desulfurization of lignite by microorganisms.

2 Materials and methods

2.1 Coal samples

Brown coals of the Lenger coal deposit (South Kazakhstan region) (42°10'51.7"N 69°52'58.8"E) (Fig.1.), which are characterized by medium ash content and significant sulfur content, were used. Coal sampling was carried out in accordance with ISO 18283: 2006 "Hard coal and coke - Manual sampling and ISO 13909–4: 2016 Preview Hard coal and coke - Mechanical sampling - Part 4: Coal - Preparation of test samples". The top layer of 1,5-2,0 cm coal removed with a sterile knife and 500-600 grams of lignite samples gathered with a sterile spatula to the depths of 30 cm. The samples were placed in a sterile container and transported to the laboratory. Each sample was labeled indicating the date and sample number. During transportation and storage of coal samples the rules have been followed in order to prevent the possibility of secondary pollution [5].

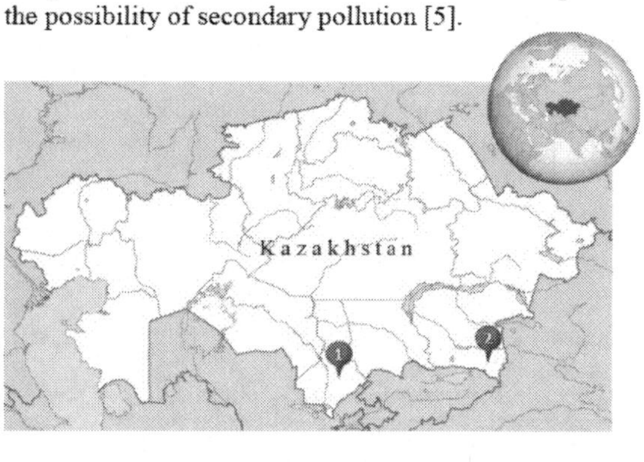

Fig.1. The coal (1 – "Lenger" coal deposit) and microbial (2 – "Oikaragai"coal deposit) sampling points

2.2 Isolation and identification of native microorganisms

Strains were isolated from different ecological niches like local rivers, flowing water, stagnant soil with silt, rusty pipe, soil near the road, sewer cover and coal itself (Oikaragai coal deposit, Kazakhstan) (42°10'51.7"N 69°52'58.8"E) (Fig.1.). Identification of the isolated bacteria was carried out by determining the direct nucleotide sequence of the 16S rRNA gene fragment, followed by determining the nucleotide identity with the sequences deposited in the international GeneBank database, and also constructing phylogenetic trees.

2.3 Nutrient media composition

The following nutrient media for the growth of microorganisms were used:

1) The SOB medium of the following composition: bacto-peptone - 10 g, K_2HPO_4 - 1.5 g, iron-ammonium citrate - 0.75 g, $Na_2S_2O_3{*}5H_20$ - 1 g per 1000 ml of distilled water. Agar with a concentration of 15 g/l was added to the SOB agar medium. The inoculums were incubated for 24 hours at a temperature of 30°C.

2) Thiobacillus Broth of the following composition: $(NH_4)_2SO_4$ - 0.400 g/l, KH_2PO_4 - 4.000 g/l, $CaCl_2$ - 0.250 g/l, $FeSO_4$ - 0.010 g/l, $MgSO_4$ - 0.500 g/l, $Na_2S_2O_3$ - 5.000 g/l. Sterilize by autoclaving at 15 lbs pressure (121°C) for 15 minutes. Samples are inoculated into Thiobacillus Broth. After incubation at 25-30°C for about 7 days or more, turbidity or sulfur precipitation on the surface of the liquid or against the walls of the flasks indicates growth of bacteria. Isolation is subsequently done on Thiobacillus Agar (Agar - 12.500 g/l).

2.4 Treatment conditions

The ability of pure cultures of bacteria to grow on brown coal was studied. Conditions for treatment: pH – 3; Incubation time – 12 days; Pulp density – 5%; Particle size – 2 mm; Temperature – 30°C. Reactions were conducted in a shaking incubator with a rotational speed of 180 rpm. After optimizing the pH, incubation time, pulp density, particle size, and incubation temperature for the microorganisms, the desulphurization of the lignite coal sample was further analyzed. At the end of the incubation time, the coal samples were separated by centrifugation and then washed with distilled water. The coal samples were dried at 45 °C overnight for further analysis [6].

2.5 Determination of sulfur forms

Methods for determining the forms of sulfur are based on the different solubilities of sulfur compounds in hydrochloric and nitric acid solutions according to ISO 157-75 Solid mineral fuel. Determination of sulfur forms (Fig.2.).

The essence of the method: pyrites do not dissolve in hydrochloric acid, but dissolve in nitric acid; sulfates are dissolved in hydrochloric and nitric acids, and organic sulfur compounds do not dissolve in hydrochloric or nitric acids.

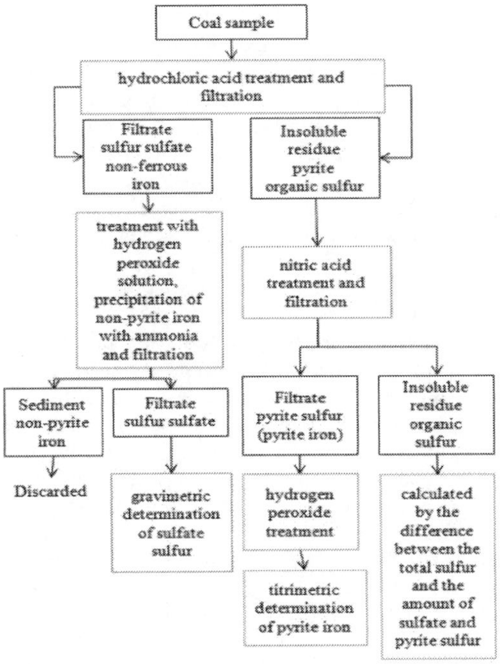

Fig. 2. The general scheme for determining the forms of sulfur in coal.

3 Results and discussion

As a result of the research, 3 cultures of microorganisms were isolated (S1, S2, T1) for further identification.

The isolates were genotyped by the method of determining the direct nucleotide sequence of the fragment of the 16S rRNA gene and the phylogenetic trees were constructed with the nucleotide sequences of the 16S rRNA gene of the reference strains of these species (Fig.3-5).

S1 is identified as *Atlantibacter* sp. and this sample was isolated from soil with silt.

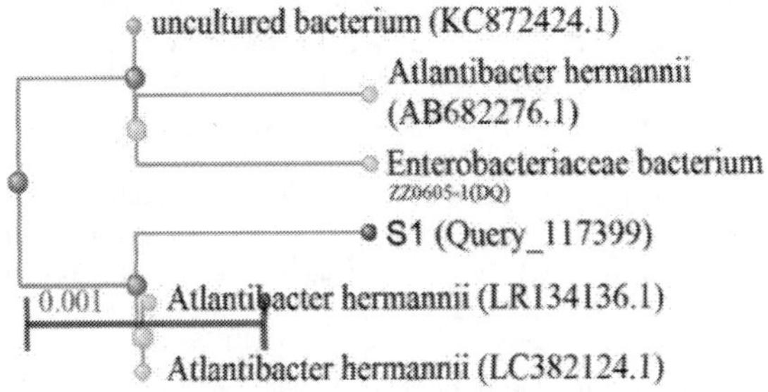

Fig.3. Phylogenetic tree, based on the analysis of a fragment of the 16SrRNA gene of the group *Atlantibacter* sp.

S2 is identified as *Pseudomonas* sp. and this sample was taken from soil with silt.

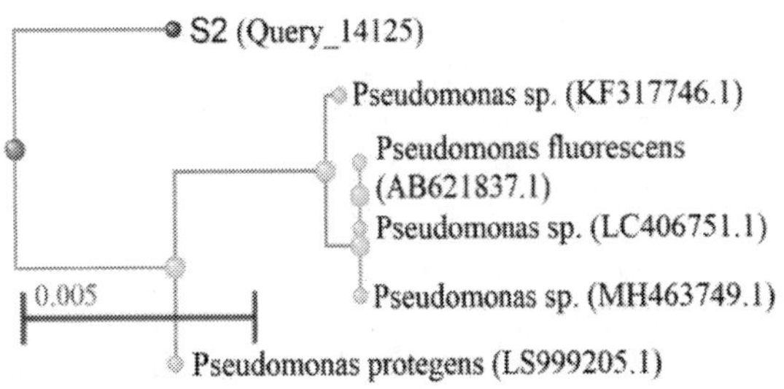

Fig.4. Phylogenetic tree, based on the analysis of a fragment of the 16SrRNA gene of the group *Pseudomonas* sp.

T1 is identified as *Bacillus* sp. and this sample was isolated from coal itself (Oikaragai coal deposit).

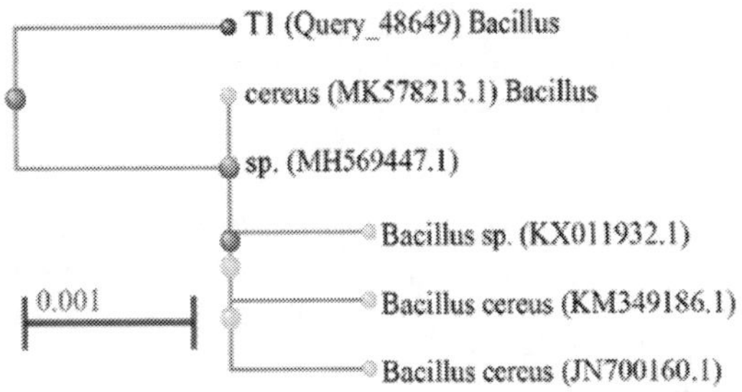

Fig.5. Phylogenetic tree, based on the analysis of a fragment of the 16SrRNA gene of the group *Bacillus* sp.

The results show that the strain S1 is located on the same clade with the nucleotide sequences from the reference strain *Atlantibacter hermannii*, while S2 is located on the same clade with the nucleotide sequences from the reference strain *Pseudomonas* sp. As shown in Fig.5, the T1 strain is located on the same clade with *Bacillus cereus*.

The data obtained as a result of a comparison of genetic characteristics made it possible to determine the species of the strains S1, S2, and T1 in the genera *Atlantibacter, Pseudomonas and Bacillus*, respectively.

According to the total sulfur content, the coal is divided into low-sulfur (up to 1.5% sulfur), medium-sulfur (1.5-2.5%), sulfur (2.5-4%) and high-sulfur (more than 4%). Pyrite, organic, sulfate and elemental sulfur are distinguished in coal [7].

Sulfur compounds in coal are present mainly as organic sulfur, pyritic sulfur, and sulfate sulfur. Pyritic sulfur occurs in coal as mineral matter whereas the organic sulfur is present as an integral part of the coal matrix covalently bounded to its complex structure [6].

One of the indicators characterizing the technical nature of coal - total sulfur (S,%) indicates the total content of this element in all compounds, converted conditionally to elemental sulfur (%) in relation to the analyzed sample.

According to the results obtained (Table 1), sulfur content in the coal of the Lenger coal deposit is about 3.14%.

Table 1. Sum of various forms of sulfur in a sample of brown coal from the Lenger coal deposit

Sulfur forms	Content, %
Total sulfur $-S_{total}$	3,14
Sulphate sulfur – S_S	<0,01
Pyritic sulfur – S_P	1,61
Organic sulfur – S_O	1,53

The lignite coal was desulfurized using the newly isolated bacteria *Atlantibacter* sp., *Pseudomonas* sp., *Bacillus* sp.

Conditions for treatment:

- pH – 3;
- Incubation time – 12 days;
- Pulp density – 5%;
- Particle size – 2 mm;
- Temperature – 30°C.

Our team conducted an experiment to study the process of desulfurization, and the results of the analysis of the effects of microorganisms on total sulfur and sulfur forms of brown coal were analyzed and demonstrated on Table 2.

Table 2. Sulfur concentration before and after microbial treatment of coal sample from Lenger coal deposit.

Sulfur, in %	Initial coal	Biodesulfurized coal with		
		S1	S2	T1
S_{Total}	3,14	2,65	1,52	1,94
$S_{Sulphate}$	<0,01	<0,01	<0,01	<0,01
S_{Pyrite}	1,61	1,56	1,41	1,3
$S_{Organic}$	1,53	1,09	0,11	0,64

As can be seen from Table 2, the total sulfur content of the initial coal sample was 3,14%, where 1,53% corresponded to the organic sulfur and less than 1,62% to the inorganic sulfur (1,61% was pyritic sulfur and less than 0,01% was sulfate sulfur).

It was found that 16% of the total sulfur was removed from the lignite coal after *Atlantibacter* sp. (S1) treatment under optimal conditions, the total sulfur content was reduced to 2,65%, organic sulfur was reduced to 1,09% and inorganic sulfur to less than 1.57% (where 1.56% was pyritic sulfur and less than 0.01% was sulfate sulfur), which was a reduction of 29% in organic sulfur and 3% in pyritic sulfur, while sulfate sulfur had no significant changes.

After the treatment with *Bacillus* sp. (T1), 38% of the total sulfur was removed from the lignite coal, the total sulfur content was reduced to 1,94%. More than half of the organic sulfur (58%) was extracted, reaching 0,64%. Inorganic sulfur in the form of pyrite decreased till 1,3%, which is the 19% from initial coal, however, sulfate sulfur remained stable (<0,01%).

For *Pseudomonas* sp. (S2), the increase of the percentage of the removed total sulfur showed the best results, thus 52% sulfur removal was achieved. The total sulfur content was reduced to 1,52%. Organic sulfur was almost fully removed, reducing to 0,11%, and inorganic sulfur to 1,42% (where 1,41% was pyritic sulfur and less than 0,01% was sulfate sulfur). The reduction was of 93% for organic sulfur and 12% for pyritic sulfur.

Based on these studies and results, a literature review was conducted on these strains. Thus, there are relatively few studies on *Pseudomonas* sp. and *Bacillus* sp. used in desulfurization. For example, Charanjit Rai and Jon P. Reyniers stated that the results from the preliminary microbial desulfurization of Illinois and Texas lignite by the organisms of the genus *Pseudomonas* show that *P. putida* was much more effective than *P. aeruginosa*. The *P. putida* reduced the pyritic sulfur content by 76% in 5 to 7 days. Whereas, *P.aeruginosa* was hardly effective (26 to 32.5%) in reducing the pyritic sulfur content [8].

The other study demonstrated the mutant strain *Pseudomonas stutzeri* LH42 (Mutant ZW-15) which was used for the biodesulfurization of coal. After 15 days' processing, it degraded 93.25% of organic sulfur and 41% of total sulfur. Thus, we can conclude that Mutant ZW-15 can be used as an efficient strain in the coal biodesulfurization [9].

A.A. El-Midany, M.A. Abdel-Khalek stated that *B. subtilis* can remove more than 70% of sulfur and ash content in the feed. The final concentrate, using *B. subtilis*, contains 0.9% sulfur and 1.95% ash with a recovery of 90–92% [10].

4 Conclusion

Recently, a number of entrepreneurs have become interested in the problem of reducing the sulfur content in coal. This is due to the sale of coal for export to other countries, where there are rather strict requirements for the sulfur content of concentrates [11].

Also, the current environmental situation in urbanized regions tightens the requirements for the quality of municipal solid fuel that is burned, along with that, the municipal fuel that is burned must have a low sulfur content, minimal opacity and the required particle size distribution.

To date, the Association of Mining and Metallurgical Enterprises is the largest industry association in Kazakhstan, which is composed of more than 100 companies of ferrous and non-ferrous metals, uranium and coal mining. Experts of this Association predict that coal will remain the main most reliable strategic type of fuel, ensuring the development of electric power. Therefore Kazakhstan needs to acquire new technologies for coal processing, especially green-based approaches.

As a result of the studies, three cultures of microorganisms were identified and comparisons of genetic characteristics made it possible to determine the species of these strains in the genera *Atlantibacter, Pseudomonas,* and *Bacillus*, respectively. The experiment to study the desulfurization process was conducted and the results of the effect of the microorganism treatment on total sulfur and sulfur forms of lignite coal were analyzed

Thus, *Pseudomonas* sp. showed the best result in removing organic sulfur (93%) and total sulfur (52%), while *Bacillus* sp. was effective in removing pyritic sulfur (19%) compared to other strains. However, *Atlantibacter* sp. had no significant influence on sulfur content after treatment, thereby reducing its chances to be used in decreasing sulfur content in lignite in future investigations. Additionally, this would be valuable information to consider while conducting the development of a biotechnological method of producing an environmentally friendly briquetted smokeless fuel from brown coal.

We would like to express our gratitude to the laboratory of Molecular Ecology in Inner Mongolia University, Huhhot, China and Xiaohui Qiao, PhD student of al-Farabi Kazakh National University for assistance with particular techniques, and thus helped us to get results of better quality.

References

1. F. Kargi, J. Robinson, *Applied and environmental microbiology*, **44**, 878 (1982)
2. P. Prayuenyong, Songklanakarin J. Sci. Technol., **24**, 493-507 (2002)
3. P.R. Dugan, W.A. Apel, *Metallurgical applications of bacterial leaching and related microbiological phenomena*, 223 (1978)
4. U. Kumar et al. *Advances in Soil Microbiology: Recent Trends and Future Prospects*, 65 (2018)
5. N.Sh. Akimbekov, X. Qiao, K.T. Tastambek,, *Experimental Biology*, **76**, 28 (2018)
6. T. Liu, J. Hou, Y. Peng, International Journal of Mineral Processing, **162**, 6 (2017)
7. V.B. Kuskov, Y.V. Kuskova, N.V. Nikolaeva, Mining information and analytical bulletin, 180 (2013)
8. C. Rai and J.P. Reyniers, Biotechnology Progress, **4**, 225 (1988)
9. T. Hu, Y. Yang, M. Zhang, Q. Cheng, Y. Gao, *3rd* International *Conference on Electromechanical Control Technology and Transportation*, 41 (2018)
10. A.A. El-Midany, M.A. Abdel-Khalek, Fuel, **115**, 589 (2014)
11. E.I. Nazimko, *Proceedings of the Donetsk mining institute*, **34**, 60 (2014)

 https://doi.org/10.1051/e3sconf/201912205006

A study on heavy metals contamination of surfacial materials by environmental magnetism and chemical analysis in Antaibao Open Pit Coal Mine, Shanxi Province, China

Manrong Chen[1,*], *Shangyi* Wang [2], *Binbin* Yu[1] , *Weiguo* Zhang [3], and *Lizhong* Yu [3]

[1] Environmental Science and Engineering College, Yangzhou University, Jiangsu province, Yangzhou, 225127,China.
[2] Research Center for Scientific Development in Fenhe River Valley, Taiyuan Normal University,Taiyuan，Shanxi, 030012, China.
[3] State Key Lab. of Estuarine and Coastal Research, East China Normal University (ECNU), Shanghai, 200062,China.

Abstract: Antaibao Open Pit Coal Mine(AOPCM)'s mining activities have caused heavy metals contamination of surface, so it was urgent task to find a suitable method and survey full and prompt and long-term monitor on the heavy metals pollution. We chose methods of environmental magnetism and chemical analysis to analyze the surface materials on heavy metals contamination, by correlation of magnetic parameters and contents of chemical analysis, we can greatly reduce workload of chemical analysis and then achieve greater, faster, better, non-destructive, less chemical pollution and more economical results. The magnetic parameters and its groups act as proxy for the contents from chemical analysis. Three different sample areas sediments: loess or loess-like sediments, sedimentary rocks and alluvium, samples have different values of magnetic parameters. The values of magnetic susceptibility in this study were lower than others sediments and soils, secondly, the base rock and sedimentary rocks from the FD were of lower values of magnetic susceptibility than others of AOPCM. Contamination of Pb and Cd are more serious than other heavy metals. we can infer the contaminated level of anthropogenic heavy metals, DS > OD > FD. The mechanism may be mainly by the origin of the magnetic fraction in the anthropogenic particulate pollution and connected with the high-temperature technological processes during production and/or processing materials which have significant Fe content. Another point, when we want to study mechanism of magnetic properties used as proxy of heavy metals, the weak magnetic samples were measured and analyzed, we should be carefully.

1 Introduction

Heavy metals pollution in open pit coal mine attracted the attention of academic researchers for long periods around the world.

The Antaibao Open Pit Coal Mine (AOPCM) (see Fig. 1)cooperated with Western Petroleum Company, USA, till 1991, is one of the largest open pit coal mines in the world, total area about 376 km^2, constructed in 1985, and coal production in September 1987, raw coal production 15.33 Mt/a (Hao zhi, 1999). This mine service period is 1985-2077, the stripping-to-ore-ratio was about 5.5 m^3/t (Hong Yu, 1995), roughly total about 2,107,875,000 m^3 overburden materials were produced to September 2012 from then, parts of tremendous overburden materials piled on the surface with elements from deep underground, included emissions of heavy metals from fossil fuel combustion, and atmospheric particulates fallout, and life wastes etc., heavy metals concentration of the surface materials may be changed in the mining areas. Scientists from Beijing Normal University completed research on heavy metals pollution of AOPCM (Wang S. et al., 1987). For purpose of cleaning coal, some authors studied the elements (heavy metals, major elements and rare earth elements) in coal of AOPCM, but all-round monitoring and investigation reports on the heavy metals contamination of surface materials have not been read by authors.

Environmental magnetism (R. Thompson & F. Oldfield, 1986) was applied to study heavy metals contamination in soils, dusts, peats and sediments from 1980's. The magnetic parameters were used as proxies to monitoring the concentration of heavy metals, by correlation between magnetic parameters, and the concentrations of Cu, Pb, Zn, Cd, Cr, Hg, etc., which be analyzed by ICP-AES (atomic emission spectroscopy with inductively coupled plasma), and AAS (Atomic Absorption Spectrophotometer) and samples pretreatment by HNO_3-HF-$HClO_4$ decomposition method. Which opened up the possibility that magnetic measurement might have a role to play in pollution monitoring.

2 Material and methods

* Corresponding author: mrchen@yzu.edu.cn

2.1 Sampling strategy

To understand the state of heavy metals contamination in the AOPCM, sampling strategy not only focus on mine area, but also the downstream of the river system. (DS , OD and FD,3 sampling areas)

2.2 Chemical analysis on Sediments and soils

The samples were dried in oven at 40℃, for metal analysis, and be dissolved by concentrated HCl-HF-$HClO_4$ acid attack. The total contents of Cd, Cu, Pb, Fe, As, Cr, Hg were analyzed in 44 samples.

2.3 Environmental magnetic measurement

All the magnetic measurements were carried out in the State Key Laboratory of Estuarine and Coastal Research at East China Normal University. The instruments include Bartington MS2 magnetic susceptibility meter, Molspin demagnetizer, pulse magnetizer and Minispin magnetometer.

3 Results and discuss

3.1 The characteristics of magnetic properties and heavy metal concentration of sediments in AOPCM

3.1.1 The characteristics of magnetic properties of surfacial sediments in AOPCM

3.1.1.1 The general magnetic characteristics of surfaciel materials in AOPCM

The magnetic properties of surface materials in AOPCM showed in the Table 1 to Table 3 .we can roughly understand the magnetic characteristics of the surface materials.

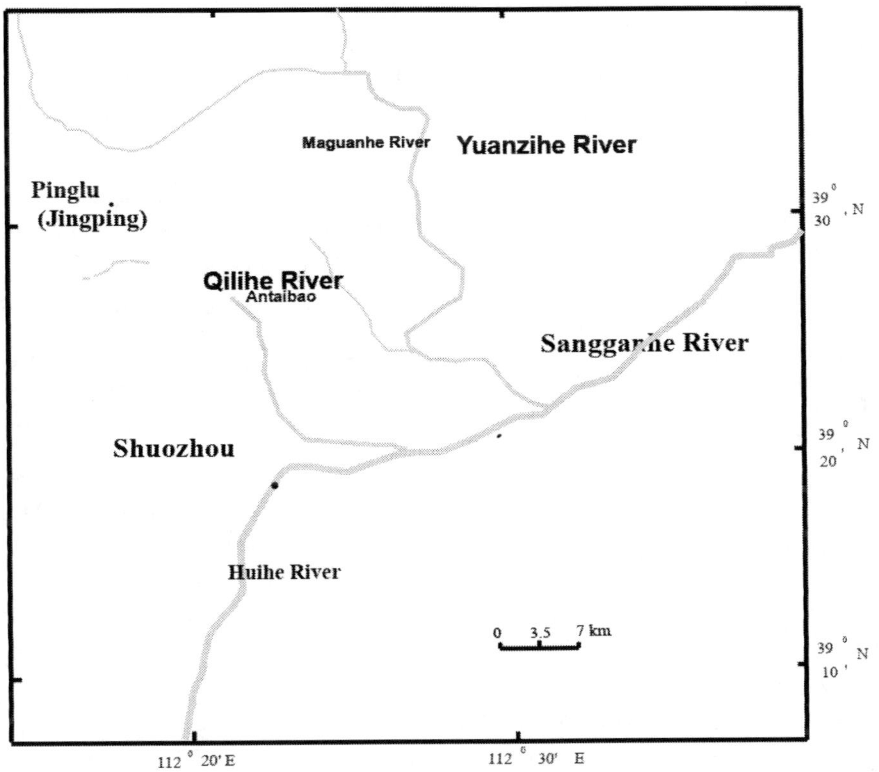

Fig. 1 The location of Antaibao Open Pit Coal Mine (AOPCM)

Table 1 Statistics of magnetic parameters of sediments in AOPCM (n=40)

parameters	Mean	Range	Standard Dev.	parameters	Mean	Range	Standard Dev.
χ ($10^{-8}m^3/kg$)	41.28	0.66-118.37	24.65	χ_{fd} ($10^{-8}m^3/kg$)	3.43	-11.29-33.43	7.07
SIRM ($10^{-6}Am^2/kg$)	6555.36	55.66-50266.03	8000.73	ARM ($10^{-6}Am^2/kg$)	109.85	1.95-290.61	68.43

χ_{arm} ($10^{-8}m^3/kg$)	150.16	6.13-512.3	102.91	HIRM ($10^{-6}Am^2/kg$)	435.62	-732.19-2900.45	707.52
IRM_{-100} ($10^{-6}Am^2/kg$)	-2436.46	-8611.50-2639.21	2157.49	IRM_{-300} ($10^{-6}Am^2/kg$)	-5734.11	-51390.08--25.50	8575.00
S_{-100}	70.82	21.63-87.99	13.56	S_{-300}	89.68	52.24-104.54	13.21
χ_{arm}/χ	6.10	0.88-99.12	15.19	χ_{arm}/SIRM(10^{-2}m/A)	58.48	3.17-1079.30	167.12
L-ratio	0.38	-0.20-2.91	0.54	SIRM/χ （kA/m）	16.18	1.27-100.89	16.62
SIRM/ARM	65.78	2.91-396.64	68.72				

Comparing the lower arithmetic mean value in this study, so samples may be dominated by magnetic mineral haematite, and low χ in this study may be affected by much calcite in sample. see Table 2.

The χ vs. SIRM showed that weak positive correlation and lower values, the concentration of ferrimgnetic mineral may be lower, and the χ_{ARM} vs. χ for magnetic granulometry, samples with higher values of χ_{ARM} have SD and PSD grain of magnetite, from J. King et al. (1982), a new method for identifying relative grain size variations in magnetite involves the parameter anhysteretic (ARM) or anhysteretic susceptibility (χ_{ARM}), which is particularly sensitive to the single domain (SD) and small pseudo-single domain (SPD) grains of the finer magnetite fraction, similarly the χ is sensitive to larger PSD and smaller multidomain (MD) .

3.1.1.2 The different of magnetic characteristics in three sampling areas

As the three different areas sediments: loess or loess-like sediments, sedimentary rocks and alluvium, samples have different values of magnetic parameters. (see Table 2)

Table 2 Statistics of main magnetic parameters in different sampling area in AOPCM

Position	χ_{lf} ($10^{-8}m^3kg^{-1}$)		SIRM ($10^{-6}Am^2kg^{-1}$)		ARM ($10^{-6}Am^2kg^{-1}$)		S_{-300mT} (%)	
	Range	Mean	Range	Mean	Range	Mean	Range	Mean
OD	1.36-67.31	41.08	338.37-12287.79	5940.634	8.53-290.61	120.99	83.38-103.07	76.53
FD	0.66-77.33	23.02	55.66-7317.6	2169.585	1.95-163.12	45.53	56.19-98.03	87.35
DS	22.56-118.37	35.50	3808.27-50266.03	1488.21	58.55-197.15	112.27	93.85-104.54	97.50

OD n=27, FD n=6, DS n=7.

3.1.1.3 Low values of magnetic susceptibility of sedimentary rock in AOPCM

We compared sedimentary rocks samples in this study with relative researches, which showed that conclusions: the firstly, the values of magnetic susceptibility in this study were lower than others sediments and soils, secondly, the base rock and sedimentary rocks from the FD were of lower values of magnetic susceptibility than others of AOPCM.

According to the researches by many authors, sedimentary rocks are of low magnetic susceptibility, because the magnetic minerals were dissolution.

3.1.2 The heavy metal concentration of surfacial sediments in AOPCM

3.1.2.1 The general characteristics of heavy metals concentration in sediments in AOPCM

According to the Table 3 , the concentrations of Cu, Pb, Cr, Hg, Cd, As, Fe are different greatly in 44 samples, the range of values of Cu, Pb, Cr, Hg, Cd, As, Fe and Al are showed in statistic, the ratios (maximum/minimum) are 7.65 (Cd) > 4.07 (Hg) > 3 (Fe) > Cr (2.54)> Cu (2.38)> 2.07 (Pb) > 1.88 (As) respectively. The order of standard dev. are Hg > Cd > Fe > As >Cr > Pb > Cu > Al. Antaibao open pit coal mine make geomorphological changes, and similarly to the heavy metals concentration of surface sediments. (see Table 3 and Table 4).

Table 3 The concentration (mg/kg, but % for Fe and Al) and EF of heavy metals of sediments in AOPCM

Sample	Cu	Pb	Cr	Hg	Cd	As	Fe	Al
PS01	20.24(1.03)	31.22 （1.34）	80.2(1.33)	0.039(0.99)	0.073(1.00)	9.57(0.99)	1.92(0.80)	6.53
PS02	15.33(0.91)	34.55 （1.75）	45.1(0.88)	0.028(0.84)	0.092(1.48)	8.34(1.02)	1.17(0.57)	5.55
PS03	13.99(0.88)	30.97 （1.65）	50.3(1.04)	0.018(0.57)	0.023(0.39)	8.16(1.05)	1.09(0.56)	5.25
PS04	29.68(1.63)	30.89 （1.44）	81.9(1.48)	0.039(1.07)	0.085(1.26)	10.05(1.13)	2.48(1.12)	6.02
PS05	28.55(1.35)	30.68 （1.23）	70.2(1.09)	0.038(0.90)	0.098(1.25)	8.86(0.86)	2.28(0.88)	6.99
PS06	25.48(1.28)	38.99 （1.66）	59.7(0.98)	0.018(0.45)	0.089(1.21)	9.97(1.02)	2.19(0.90)	6.59
PS07	26.26(1.45)	37.69 （1.76）	68.9(1.24)	0.028(0.77)	0.067(1.00)	8.34(0.94)	2.26(1.02)	6.01
PS08	20.36(1.06)	30.62 （1.35）	87.6(1.50)	0.017(0.44)	0.058(0.82)	8.12(0.86)	2.64(1.13)	6.35
PS09	24.99(1.41)	35.68 （1.71）	80.3(1.48)	0.034(0.96)	0.088(1.34)	8.66(1.00)	2.34(1.08)	5.87
PS10	28.25(1.34)	37.83 （1.52）	90.1(1.40)	0.028(0.67)	0.176(2.26)	9.38(0.91)	2.49(0.97)	6.96
PS11	26.12(1.31)	35.35 （1.50）	75.8(1.25)	0.041(1.03)	0.167(2.27)	8.22(0.84)	2.64(1.09)	6.59
PS12	24.02(1.25)	37.69 （1.66）	64.2(1.10)	0.038(0.99)	0.077(1.08)	8.85(0.94)	2.12(0.91)	6.35
PS13	22.69(1.21)	30.11 （1.36）	90.1(1.57)	0.038(1.02)	0.077(1.11)	9.72(1.06)	2.57(1.13)	6.19

PS14	29.02(1.63)	31.97 (1.52)	80.2(1.47)	0.028(0.78)	0.099(1.50)	8.84(1.01)	2.07(0.95)	5.91
PS15	20.69(1.11)	22.34 (1.01)	74.1(1.30)	0.025(0.67)	0.062(0.90)	8.68(0.95)	1.56(0.68)	6.18
PS16	15.68(0.83)	45.65 (2.05)	90.2(1.57)	0.036(0.95)	0.058(0.83)	8.67(0.94)	2.57(1.12)	6.25
PS17	12.69(0.69)	22.58 (1.03)	60.4(1.07)	0.029(0.78)	0.057(0.83)	8.43(0.93)	1.37(0.61)	6.12
PS18	24.65(1.08)	23.68 (0.88)	68.4(0.98)	0.038(0.83)	0.088(1.04)	9.12(0.81)	3.11(1.11)	7.58
PS19	20.68(0.99)	23.69 (0.96)	56.2(0.88)	0.037(0.88)	0.058(0.75)	7.33(0.71)	2.06(0.81)	6.94
PS20	16.39(0.86)	46.23 (2.05)	66.2(1.13)	0.037(0.97)	0.054(0.76)	8.76(0.94)	2.35(1.01)	6.33
PS21	28.36(1.53)	34.36 (1.57)	57.4(1.02)	0.038(1.03)	0.084(1.23)	8.57(0.95)	2.38(1.06)	6.12
HH22	25.25(1.35)	30.02 (1.36)	50.3(0.88)	0.025(0.67)	0.067(0.97)	6.65(0.73)	2.19(0.96)	6.18
HH23	28.62(1.26)	30.41 (1.13)	56.9(0.82)	0.024(0.53)	0.147(1.75)	6.38(0.57)	3.01(1.09)	7.52
HH24	26.45(1.42)	30.36(1.38)	75.6(1.33)	0.031(0.83)	0.169(2.45)	9.08(0.99)	2.05(0.90)	6.17
YZ25	24.05(1.25)	36.01(1.59)	82.3(1.41)	0.037(0.96)	0.064(0.90)	10.05(1.07)	2.03(0.87)	6.35
YZ26	24.35(1.21)	38.32(1.61)	114.6(1.86)	0.061(1.51)	0.088(1.18)	12.01(1.22)	3.27(1.33)	6.68
YM27	20.15(1.24)	30.36(1.58)	86.3(1.74)	0.015(0.46)	0.078(1.29)	8.76(1.10)	1.98(1.00)	5.39
YM28	25.36(1.41)	29.66(1.39)	74.2(1.35)	0.024(0.67)	0.083(1.24)	10.34(1.17)	2.56(1.16)	5.97
PS29	30.24(1.53)	36.66(1.57)	68.5(1.14)	0.027(0.68)	0.081(1.11)	9.16(0.95)	2.33(0.97)	6.54
PS30	25.36(1.20)	34.33(1.38)	56.9(0.88)	0.031(0.74)	0.088(1.13)	7.81(0.76)	2.08(0.81)	6.98
PS31	28.69 (1.56)	35.37(1.63)	64.2(1.15)	0.028(0.76)	0.064(0.94)	8.64(0.96)	2.45(1.10)	6.07
PS32	22.63 (1.18)	31.02(1.37)	51.2(0.88)	0.037(0.96)	0.042(0.59)	8.28(0.88)	2.19(0.94)	6.35
PS33	26.69 (1.37)	28.47(1.23)	67.6(1.13)	0.024(0.61)	0.068(0.94)	8.68(0.91)	1.35(0.57)	6.47
PS34	14.32 (0.80)	30.54(1.44)	47.6(0.87)	0.027(0.75)	0.102(1.54)	8.48(0.97)	1.46(0.67)	5.93
PS35	26.58 (1.26)	30.58(1.23)	84.2(1.31)	0.038(0.90)	0.083(1.07)	10.26(1.00)	2.25(0.88)	6.97
PS36	26.22 (1.39)	24.36(1.09)	49.5(0.86)	0.041(1.09)	0.167(2.39)	9.43(1.02)	2.14(0.93)	6.25
PS37	25.87 (1.42)	27.02(1.26)	58.2(1.05)	0.017(0.47)	0.076(1.13)	8.71(0.98)	2.07(0.93)	6.02
PS38	28.85 (1.69)	35.65(1.77)	82.5(1.59)	0.043(1.26)	0.068(1.08)	8.75(1.05)	2.08(1.00)	5.64
PS39	29.56 (1.58)	34.67(1.57)	98.6(1.73)	0.025(0.67)	0.079(1.14)	9.18(1.00)	2.41(1.06)	6.18
PS40	29.69 (1.49)	31.08(1.32)	80.2(1.32)	0.047(1.18)	0.087(1.18)	10.09(1.04)	2.56(1.06)	6.58
PS41	25.56 (1.38)	29.37(1.34)	64.1(1.14)	0.035(0.95)	0.078(1.14)	8.64(0.95)	2.09(0.93)	6.12
PS42	12.89 (0.69)	25.65(1.17)	70.2(1.23)	0.027(0.72)	0.025(0.36)	7.95(0.87)	2.17(0.95)	6.17
PS43	22.97 (1.22)	28.67(1.30)	87.5(1.53)	0.026(0.69)	0.078(1.12)	8.64(0.94)	2.34(1.02)	6.21
PS44	16.39 (0.77)	30.34(1.20)	61.2(0.94)	0.044(1.03)	0.029(0.37)	8.69(0.83)	1.98(0.76)	7.08
Mean EFs	1.24	1.43	1.23	0.83	1.17	0.95	0.94	

Note: Data in the parentheses are enrichment factor (EF).

Table 4 Statistics of the heavy metals and EFs in the AOPCM (n=44) (mg/kg, but % for Fe and Al)

Sample	Cu	Pb	Cr	Hg	Cd	As	Fe	Al
Arithmetic Mean	23.66 (1.23)	32.08 (1.00)	71.1(1.22)	0.03(0.78)	0.08(1.13)	8.85 (1.01)	2.20(0.89)	6.33
Minimum	12.69 (0.80)	22.34 (1.14)	45.1(0.93)	0.015(0.47)	0.023(0.39)	6.38 (0.87)	1.09(0.53)	5.25
Maximum	30.24(1.33)	46.23(1.69)	114.6(1.64)	0.061(1.33)	0.176(2.08)	12.01 (1.14)	3.27(1.10)	7.58
Median	25.31(1.37)	30.93(1.38)	70.2(1.22)	0.031(0.82)	0.078(1.12)	8.70 (1.00)	2.19(0.90)	6.23
Standard Dev.	5.06	6.66	22.9	27.27	26.99	25.82	26.88	0.49
B.V. of soil, China	20	23.6	61.0	0.040	0.074	9.2	2.44	6.62

Note: B.V. of soil, China is background value of soil in China; the data from NEMCC (1990). Data in the parentheses are enrichment factor (EF).

3.1.2.2 Enrichment Factors (EFs) of heavy metals in AOPCM

In order to identify the anthropogenic source of metallic elements, enrichment factor (EF) is widely employed. The formula was:

$$EF=(Me/Al)_{sample}/(Me/Al)_{soil} \qquad [3\text{-}1]$$

Where $(Me/Al)_{sample}$ is the metal to Al ratio in the samples; $(Me/Al)_{soil}$ is the average ratio in soils in China (NEMCC, 1990). Based on Wang et al., (2012), EF values between 0.5-1.5 (i.e. $0.5 \leqslant EF \geqslant 1.5$) suggest that trace metals may be entirely from natural weathering processes and greater than 1.5 (i.e. $EF > 1.5$) suggest that a significant portion of trace metal is delivered from non-natural weathering processes. The mean values of EFs in this study are Pb > Cu > Cr > Cd > As > Fe > Hg. The enrichment factors of Pb, Cu, Cr, and Cd greater than 1.5(see Table3, Table4), suggesting obvious anthropogenic inputs of these elements. Contamination of Pb and Cd are more serious than other heavy metals.

In 3 Sampling areas (FD, DS and OD), as formula [3-1] and arithmetic mean of EFs of Cu, Pb, Cr, Hg, Cd, Al and Fe, then calculated mean EI of the samples from different area, the formula EI:

$$EI_{FD}=\Sigma EFs_{FD}/\text{number of samples in the FD} \qquad [3\text{-}2]$$

$$EI_{DS}=\Sigma EFs_{DS}/\text{number of samples in the DS} \qquad [3\text{-}3]$$

$$EI_{OD}=\Sigma EFs_{DS}/\text{number of samples in OD} \qquad [3\text{-}4]$$

The results of $EI_{FD}=1.038$, $EI_{DS}=1.187$, and $EI_{OD}=1.116$, so we can infer the contaminated level of anthropogenic heavy metals, DS > OD > FD.

3.2 Relationship between magnetic properties of materials and heavy metals contamination

The correlation coefficient of concentrations of heavy metals and magnetic parameters were analyzed (see Table 5), in this study.

Table 5 Correlation coefficient of concentrations of heavy metals and magnetic parameters

	Cu	Pb	Cr	Hg	Cd	As	Fe	Al		Cu	Pb	Cr	Hg	Cd	As	Fe	Al
χ_{lf}	0.64	0.13	-0.02	0.05	0.62	-0.18	0.46	0.38	χ_{fd}	0.26	0.15	-0.15	0.02	0.22	0.03	0.02	0.14
ARM	0.54	0.08	-0.08	-0.06	0.49	-0.14	0.38	0.24	χ_{ARM}	0.52	0.05	-0.07	0.01	0.41	-0.10	0.32	0.25
SIRM	0.25	-0.10	-0.19	-0.11	0.22	-0.37	0.11	0.08 H	IRM	0.06	-0.32	0.06	0.12	0.10	0.16	-0.01	0.05
IRM-100	-0.54	0.01	-0.05	-0.09	-0.55	0.29	-0.42	-0.26	IRM-300	-0.23	0.04	0.02	0.12	-0.18	0.38	-0.11	-0.07
SIRM/χ	-0.01	-0.43	-0.07	-0.19	-0.12	-0.22	-0.18	-0.07	S-100	0.25	0.14	0.18	0.24	0.12	-0.06	0.26	0.05
S-300	0.23	0.33	0.02	-0.12	0.08	-0.04	0.11	-0.03	L- ratio	-0.03	-0.25	0.00	0.16	0.02	0.04	0.14	0.18
SIRM/ARM	-0.10	-0.39	-0.02	-0.09	-0.06	-0.17	-0.18	-0.03									

Notice: n=40, $r_{0.10}$=0.2573, $r_{0.02}$=0.3578

A lot of studies on the magnetic parameters were used as proxy concentration of heavy metals, in China urban, Nanjing , Lanzhou , Shanghai , Beijing , etc. Earlier study on the 1977, Beckwith et al.(1986) analyses of heavy metals showed positive correlations (99%) between χ_{lf} and Zn, Pb, and Cu. (Urban soils 0-30 cm).

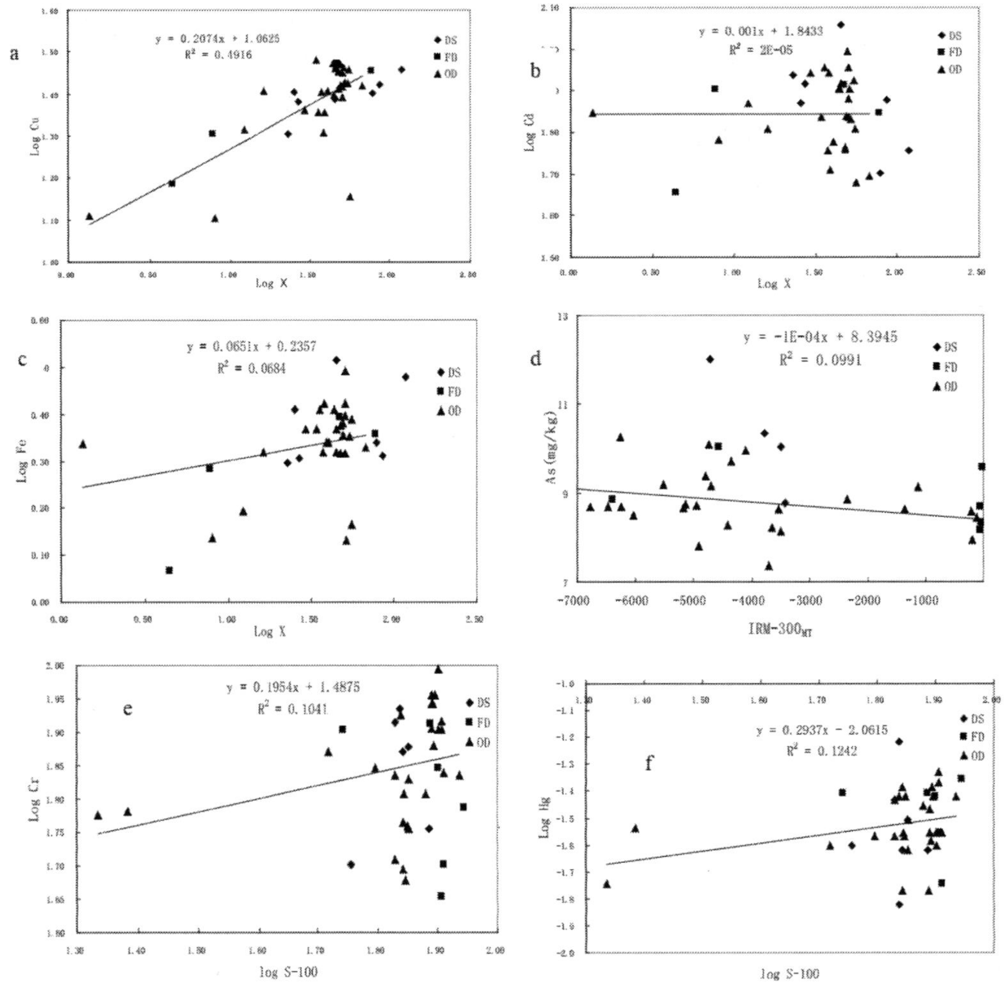

https://doi.org/10.1051/e3sconf/201912205006

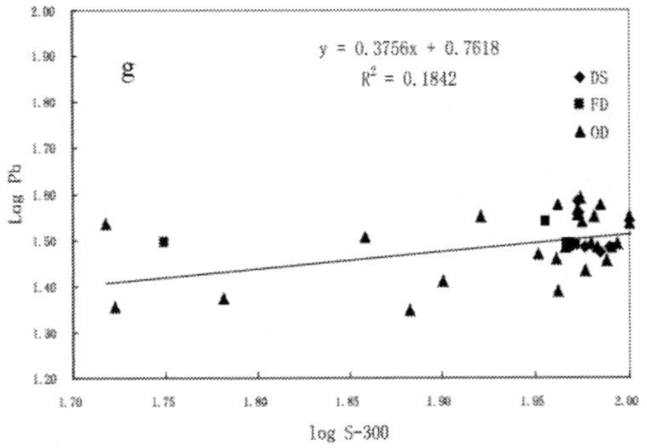

Fig.2 Relations Log-Log for all samples (n=40) between Cu (a), Cd (b), Fe (c) &χ, Cr (e), Hg (f) & S-100, Pb (g) & S-300, but As (d) & IRM-300mT, and the correlation coefficients (R2) are calculated only for samples with OD (n=27).

The linear relation (Fig. 2) between Log Cu (a), Cd (b), Fe (c) & Log χ, Log Cr (e), Hg (f) & Log S-100$_{mT}$, Log Pb (g) & Log S-300$_{mT}$, and As (d) & IRM-300$_{mT}$, suggest that these magnetic parameters can reflect the concentration variations of heavy metals.

3.3 Mechanism of magnetic properties be used as proxy of heavy metals

Great effort has been made in understanding the relationship between magnetic properties (particularly magnetic susceptibility) and the content of heavy metals, magnetite or hematite for the purposes of interpreting magnetic anomalies and rock magnetism study, and investigation of pollution.

The measured susceptibility of weakly ferromagnetic samples, in which water, carbon, calcium carbonate or silica are abundant, will be reduced by diamagnetism. According the book "Pinglu Soil", the concentration of calcite in soil profile in Antaibao is 11.3-12.5% from 0-150cm.

Magnetite and hematite concentration are more relatively in oxidation than in reduction condition in sediments, because the reduction and dissolution of magnetite and hematite.

So when we want to study mechanism of magnetic properties used as proxy of heavy metals, the weak magnetic samples were measured and analyzed, we should be carefully.

The mechanism may be mainly which the origin of the magnetic fraction in the anthropogenic particulate pollution is connected with the high-temperature technological processes during production and/or processing materials which have significant Fe contents.

4 Conclusion

Based on this study, heavy metals contamination of surface materials can be surveyed and monitored by environmental magnetism and chemical analysis, in Antaibao open pit coal mine, Shanxi Province, China. Magnetic properties can be used as proxy of heavy metals in the surface materials of AOPCM.

Referrence

1. Hao Zhi, Modern Miner. 7(1999) (in Chinese)
2. Hong Yu, Energ. Environm. Protect., 13, 2 (1995) (in Chinese)
3. S.Wang, Jin Y., & Wang H. Environm. Sci., (1987) (in Chinese)
4. R.Thompson & F.Oldfield, Environmental magnetism (London, George Allen & Unwin 1986).
5. J. King, S. Banerjee, J.Marvin and O.Ozdemir, Earth Planet. Sci. Letter 59 (1982)
6. National Environmental Monitoring Center of China (NEMCC), Background values of elements in China soils. (Beijing, Chinese Environmental Science Publish House. 1990)
7. Wang Cong, Liu Shiliang, Qinghe Zhao, et al., Ecotoxicology and Environmental Safety,82(2012)
8. P.R. Beckwith, J.B.Ellis, D.M.Revitt and F.Oldfield, Physics Earth Planetary Int.42 (1986)

Study of the destructive activity of microorganisms isolated from soil contaminated by pesticides

Azhar Malik[1,2,*], *Gulzhamal* Abdieva[1,2], *Perizat* Ualieva[1,2], and *Nuraly* Akimbekov[1,2]

[1]Al-Farabi Kazakh National University, Faculty of biology and biotechnology, Biotechnology Department, 71 al-Farabi Ave., Almaty, Republic of Kazakhstan
[2]Scientific Research Institute of Biology and Biotechnology Problems, 71 al-Farabi Ave., Almaty, Republic of Kazakhstan

Abstract. Chemical pollution of natural and agricultural habitats is a growing global problem and a serious threat to the sustainability and health of people. Among various chemical ecotoxicants of anthropogenic origin, organochlorine pesticides are among the most stable and dangerous for the environment and humans. Pesticides are chemical plant protection products, the intensive and not always justified the use of which has led to the fact that recently they are also considered among the most dangerous pollutants. One of the serious environmental problems is the pollution of natural objects with organic pesticides with high toxicity and persistence. Toxic substances can enter the environment and pose a threat to all living organisms, including the microbial population of the soil. Microorganisms isolated from ecosystems exposed to long-term treatment with pesticides can quickly decompose these compounds. The problem related to the decomposition of pesticide residues in the soil is an important environmental issue. Microorganisms are able to most effectively decompose xenobiotic substances foreign to the biosphere. The ability of microorganisms to decompose pesticides is associated with biochemical reactions and a high level of their adaptation. In this study, the microbial diversity of the soil in the territories adjacent to the pesticide burial sites was studied.

1 Introduction

At present, there is an acute environmental problem related to the effects of many years of use in agriculture of pesticides, chemical agents for controlling pests of agricultural plants. Pesticides, which are composed of insecticides, herbicides, and fungicides, are a group of potentially toxic substances that can disrupt the microbial structure and function in habitats.

As it is known, at present, their use is practically prohibited, but due to their high toxicity with respect to soil bio-objects and water bodies, another problem has arisen, the leveling of which is connected with the need to create new environmentally hazardous objects - storages (warehouses) for huge amounts of unused pesticides.

Considering the above, special storage facilities in a number of regions were built for pesticides available in warehouses. In particular, they are in the Talgar district, Almaty region.

Although, all the studies conducted earlier in this direction was devoted to study the target activity of pesticides in the process of their application in agriculture. Nowadays the attention of researchers is also related to the problems associated with the influence of pesticide storage facilities on the environment, on the physiological activity of microorganisms and other living objects of water and soil, which can serve as markers of change, based on the presence of pesticides and their decomposition products in environmental objects. Pesticides can have a direct toxic or beneficial effect on microbial communities, similar to those reported for higher organisms. Some microbes, in particular, bacteria, can use pesticides as a source of nutrients that promote their growth and survival, while sensitive species can be damaged or destroyed by pesticides [1].

It is possible to assess the intensity and nature of the effects of toxic substances by their metabolic activity, such as pesticides, on metabolic processes in the cells of soil microorganisms directly, i.e. metabolic potential of soil microbial communities in pesticide burial areas.

It is equally important that the study of the microbial communities of soils contaminated with pesticides is necessary both for the assessment of biological risk and for the selection of promising agents for bioremediation activities in contaminated areas. The method of cleaning soils contaminated with pesticides using microorganisms–biodestructors is undoubtedly effective and cost- effective [2].

In connection with the above purpose, further research was the study of the microbial diversity of the soil from the pesticides burial site, the screening of effective destructive microorganisms and the study of the destructive potential of crops [3].

* Corresponding author: azhar.malikkyzy@gmail.com

We have conducted studies on the microbial diversity of soil samples in Almaty region (Fig.1), adjacent to the burial places of pesticides (test samples) and background areas (control samples).

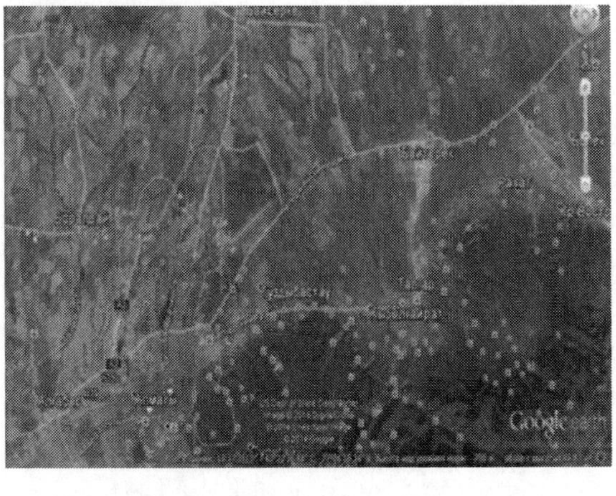

Fig.1. Soil sampling points in Almaty region

2 Materials and methods

Determination of the destructive activity of microorganisms

The search for destructors was carried out among cultures from the dominant populations of microorganisms. Indication of the destructive strains in the composition of the total soil microbiota was carried out on solid medium M9 with the addition of a pesticide as a carbon source 0.01%, 2,3,5-triphenyl tetrazolium chloride (TTCH) as an indicator of dehydrogenase activity of bacteria [4]. The destructive activity of the cultures was evaluated by the activity of growth and preservation of cell viability in the presence of organochlorine compounds [5].

The use of a wide range of pesticides suggests different mechanisms of action of these substances on prokaryotic and eukaryotic cells of microorganisms, on heterotrophic microorganisms, and the spectrum of these mechanisms is very wide. Therefore, the strains of isolated pure cultures showed different growth activities in the medium with the organochlorine preparation [6].

Sequencing

Extraction and PCR amplification of genomic DNA The genomic DNA sample was extracted by SDS, and then by agarose gel electrophoresis, which determines the purity and the concentration of DNA. An appropriate amount of sample is taken in a centrifuge tube and the sample is diluted to 1 ng/ml of sterile water. Using low genomic DNA as a template, PCR was used to provide amplification based on the selection of selection methods using specific barcode primers, master mix for high-performance Phusion® PCR with GC buffer from New England Biolab and high-precision enzymes. Efficiency and accuracy.

Primer related area: Primers 16S V4-region (515F and 806R): identification of bacterial diversity; Primers 18S V4-region (528F and 706R): identification of eukaryotic microbial diversity; Primers for the ITS1 region (ITS5-1737F and ITS2-2043R): identification of fungal diversity; In addition, the amplified region also includes 16S V3-V4 / 16S V4-V5; Archaea 16S V4; 18S V9 and ITS2.

Mixing and cleaning PCR products

The PCR product was determined by electrophoresis using a 2% agarose gel.

Samples were mixed according to the concentration of the PCR product. After thorough mixing, the PCR product was purified by electrophoresis using 1 × TAE concentration of 2% agarose gel. Group. The product cleaning kit uses the Thermo Scientific GeneJET glue regeneration kit to restore the product.

Library creation and sequencing

The library was constructed using the Rxns Kit 48 library of Thermofisher Ion Plus library fragments. The created library was quantified by Qubit and tested by the library, and then sequenced using Life Ion S5 TM from Thermofisher or Ion S5 TM XL.

3 Results and discussion

The screening of active microorganisms-destructors of persistent organic pollutants and their decomposition products was performed on all strains of pure cultures isolated from the soil adjacent to the pesticides burial sites. Strains K3, AK5 isolated from the burial site of pesticides have high destructive activity against pesticides. The results are presented in Fig. 2.

As can be seen from Figure 2, the number of cells of the studied strains after 24 hours was 6,4 x 10^4 for K3, while for AK5 – 1,3 x 10^5 CFU/g. On the fifth day (120 hours) the growth of the cultures K3 decreased to 5,6 x 10^4, while the number of cells of AK5 reached 2,5 x 10^5 CFU/g.

As can be seen, from Figure 3, the strain of cultures of microorganisms-destructors on the medium with the addition of pesticide form surface colonies of 1-2.5 mm in size. When growing crops on M9 medium with pesticides, the colonies had a larger size – 1.8-3 mm. The colonies were round, with a smooth edge and a smooth surface, matt red. The structure of the colonies is homogeneous. Almost all strains formed S-type colonies, sometimes mucous (O) colonies after surface sowing. Strains K3, AK5 were observed round and smooth, shiny colonies. On M9 medium with pesticides, all strains formed different zones of enlightenment around the colonies.

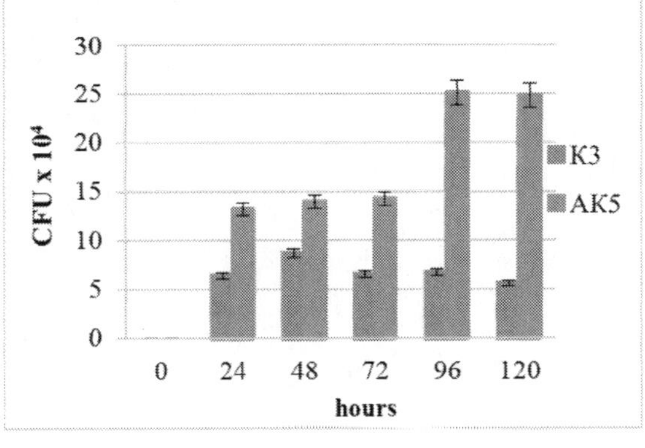

Fig.2. Growth of the culture medium with the addition of pesticides as the sole carbon source.

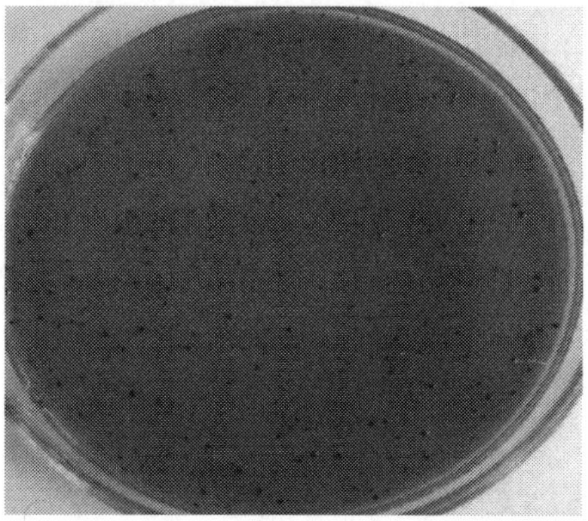

Fig. 3. The growth of cultures of microorganisms-destructors on the medium with the addition of pesticide

It should be noted that on a solid medium M9 all colonies were only smooth, shiny S-forms of red color. Cultures on M9 medium with addition of pesticide, as the only carbon source, change the initial white-matte color to red, which is due to the formation of reduced triphenyl formazane (TFF) in the medium. The scientific literature provides numerous examples of the transformation of various pesticides under the action of microorganisms in certain conditions and certain soils [7]. Organochlorine preparations under the action of microflora are subjected to deep decomposition with splitting of aromatic rings.

A study of the chemical contaminant destructor strains revealed that the K3 and AK5 bacterial strains are capable of destroying the preparation, which was indicated by the coloration of the colonies and the environment around them to become red, indicating the formation of reduced TFF.

Since under aerobic conditions the first stage of biodegradation of xenobiotics is oxidative metabolism reactions catalyzed by various oxidoreductases, the main of which are dehydrogenases, the identification of these enzymes in microorganisms indicates the destructive potential of the culture. Molecular genetic identification of microorganisms was performed by Sanger sequencing. A 16S rRNA gene fragment of about 700 bp was amplified by PCR. As a result of molecular genetic identification, selected strains of microorganisms were identified: strain K3 was assigned to the species Bacillus aryabhattai, strain AK5 to the species Bacillus subtillis.

It is known that soil fertility and self-purification directly depend on the activity of microbiological processes, however, as a result of high soil intoxication, the autochthonous microflora is inhibited [8]. Therefore, the development of complex technologies aimed at restoring the basic functions of the soil and increasing their fertility is of considerable scientific interest, both for theoretical and for applied microbiology [9-10]. Currently, biological remediation methods are considered as priorities for solving the problem of cleaning contaminated soils.

To date, research in the field of molecular ecology of microorganisms is united under the common name – metagenomic studies. The object of the study of metagenomics is a metagen, i.e. genetic material obtained directly from the environment without prior cultivation of microorganisms. [11,12,13].

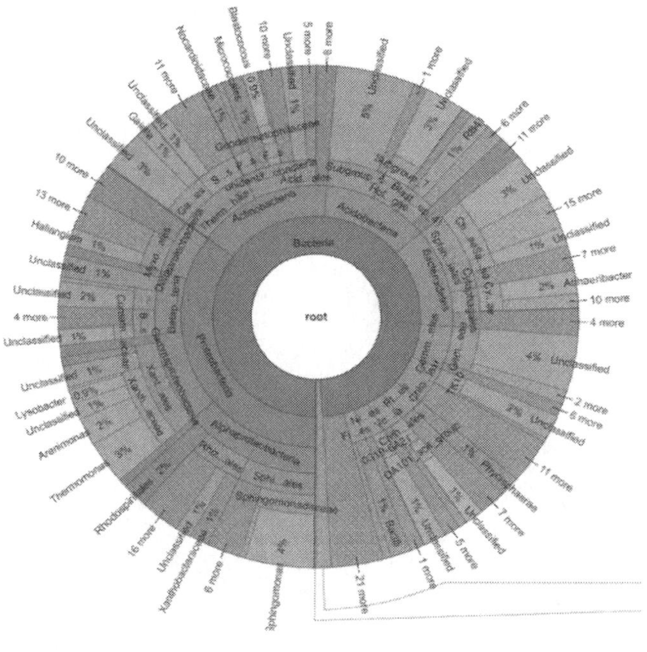

Fig. 4. The Krona graph of the relative abundance of annotated taxa in the Kyzylkairat.

Figure 4 presents the results of the study of the total microflora of Kyzylkairat soil samples using metagenomic analysis of Illumina high-throughput sequencing. As a result of metagenomic analysis, the following physiological groups of microorganisms dominate in soil samples contaminated with pesticides: Proteobacteria 36%, Actinobacteria 16%, Acidobacteria 13%, Bacteroidetes 12%.

One of the urgent tasks of modern biotechnology is the creation of biologics based on destructive strains isolated from native microflora to solve a complex of tasks related to the rehabilitation of soils contaminated with xenobiotics.

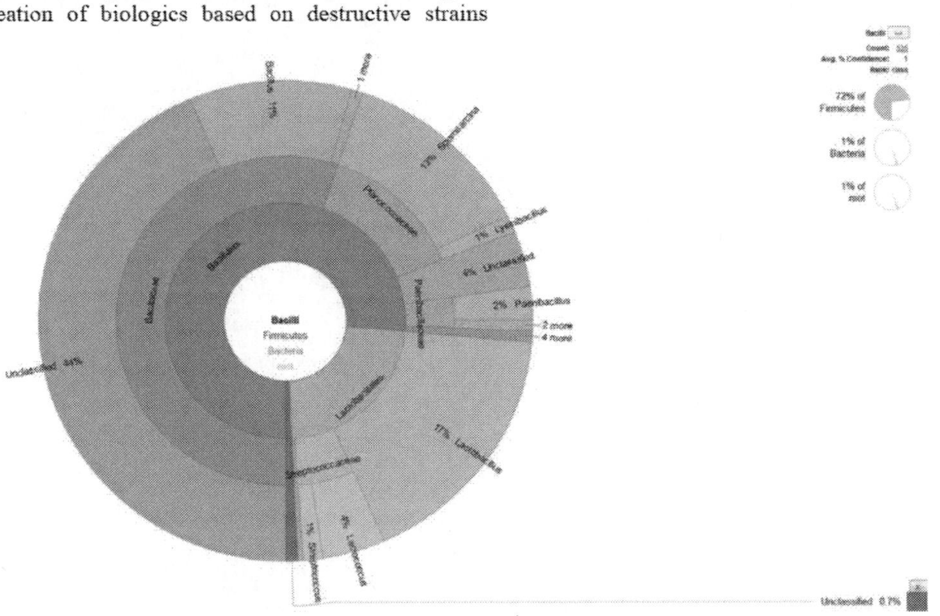

Fig. 5. Krona taxonomy web visualization illustrating the share of bacteria of the genus Bacillus

In the study, as a result of studying the destructive activity of strains isolated from contaminated soil samples from Amangeldi №1 and Kyzylkairat, 2 strains were selected that have higher destructive activity against persistent organic pollutants. Additionally, molecular genetic identification results demonstrated that selected cultures of microorganisms are assigned to the genus Bacillus.

Figure 5 shows the visualization of the crown web taxonomy using the example of Bacillus. The composition of the microbial diversity of soil samples is dominated by spore-forming gram-positive bacteria of the Bacillaceae family of 72%. In the microbiota of Kyzylkairat soil samples, the share of bacteria of the genus Bacillus was 14%. These indicators indicate that the destructor dominates in the microflora of the polluted soil - organochlorines of organochlorine pesticides of the genus Bacillus.

4 Conclusions

Thus, the conducted studies indicate the need for continuous monitoring of the state of microbial diversity in the territory of adjacent to the pesticides burial sites, as well as studying the possibility of using microorganisms as indicators of environmental pollution. The search and selection of promising cultures of microorganisms capable of active degradation of persistent organic pollutants is the basis for the development of bioremediation measures to clean the soil from residual pesticides. The present study, on the one hand will provide reliable information not only on the impact of pesticides on the environmental situation of the investigated region, but also have predictive value, as they allow to judge about possible changes of the microbial potential in the presence of pesticides. Of equal important is the fact that the study of the microbial communities of soils contaminated with pesticides is necessary both for the assessment of biological risk and for the selection of promising agents for remediation in contaminated areas.

References

1. T. Doolotkeldieva, M. Konurbaeva, and S. Bobusheva, Environ Sci Pollut Res Int. **25**, 31848 (2018)
2. T.P. Baczynski, D. Pleissner, T. Grotenhuis. Chemosphere, **78**, 22 (2010)
3. R. Yusiasih, M. Efendi, M. Pitoi, H.R. Hariyadi. Earth *and Environmental Science* **160**, 012023 (2018).
4. E. V. Vasnetsova, O. Yu. Ksenofontova, D. A. Tikhonova, E. A. Filimonova, K. V. Savina, *Izv.*

Sarat. un-t. New ser. Ser. Chemistry. Biology. **16**, 3 (2016)

5. E.N. Kulikova-Khlebnikova, Yu.V. Robertus, A.V. Kivatskaya, R.V. Lyubimov. Bull. of the Altai State Agr. Univ. 106, 59-63 (2013)
6. A.V. Kolupaev. Abs. of thesis. work. **44**, 18-30 (2010)
7. Ye.V. Vasnetsova, O. Yu. Ksenofontova, D. A. Tikhonova, Ye. A. Filimonova, K. V. Savina. *New sir Ser. Chemistry.* **16**, 3 (2016)
8. M. Geetha, M.H. Fulekar, Afr J Environ Sci Technol. 2, (2008)
9. M. Hayatsu, M. Hirano, S. Tokuda, Appl Environ Microbiol, 66, 1737 (2000)
10. F. Kafilzadeh, M. Ebrahimnezhad, Y. Tahery, Iran. Osong Public Health Res Perspect, **6**, 39 (2015)
11. K. Yu, T. Zhang, PLoS ONE, **7** (2012)
12. M. Wexler, A. W. Johnston, Methods Mol Biol **668**, 77 (2010)
13. L. P. Wackett, Envir. Micro. Wiley Online Library **14**, 3066 (2012)

AUTHOR INDEX